P9-AFZ-854

Computerized Engine Control and Diagnostics

Computerized Engine Control and Diagnostics

TOM WEATHERS, JR.
CLAUD C. HUNTER

PRENTICE HALL, Englewood Cliffs, N.J. 07632

Library of Congress Cataloging-in-Publication Data

Weathers, Tom.
Computerized engine control and diagnostics / Tom Weathers, Jr., Claud C. Hunter.
p. cm.
Includes index.
ISBN 0-13-162660-4
1. Automobiles—Motors—Control systems. 2. Automobiles—Electronics equipment. I. Hunter, Claud C. II. Title.
TL214.C64W43 1989
629.2′5—dc19 88-37434
CIP

Editorial/production supervision and
interior design: **Eileen M. O'Sullivan**
Cover design: **Wanda Lubelska**
Manufacturing buyer: **David Dickey**

A Division of Simon & Schuster
Englewood Cliffs, New Jersey 07632

Printed in the United States of America

10 9 8 7 6 5 4 3 2 1

ISBN 0-13-162660-4

PRENTICE-HALL INTERNATIONAL (UK) LIMITED, *London*
PRENTICE-HALL OF AUSTRALIA PTY. LIMITED, *Sydney*
PRENTICE-HALL CANADA INC., *Toronto*
PRENTICE-HALL HISPANOAMERICANA, S.A., *Mexico*
PRENTICE-HALL OF INDIA PRIVATE LIMITED, *New Delhi*
PRENTICE-HALL OF JAPAN, INC., *Tokyo*
SIMON & SCHUSTER ASIA PTE. LTD., *Singapore*
EDITORA PRENTICE-HALL DO BRASIL, LTDA., *Rio de Janeiro*

Contents

Preface

As we have noted in the preface to several related books, a revolution has taken place in the automotive industry. Virtually every automobile produced in the world today has at least one on-board computer. Functions that were once handled by mechanical devices, or even by the driver, are now managed by an electronic brain.

Most people are not aware of the revolution. Yet, its effects have been significant, and will be even more significant in the years to come.

For the average driver, the revolution probably will not seem real until the manufacturers automate some function that was always in the driver's domain. It might be a fail-safe braking function that completely takes over the vehicle under emergency conditions. Or, it might be a government-mandated speed control that won't allow you to exceed safe speeds during dangerous traffic or weather conditions.

However, for the purposes of this book, the most significant effect has already taken place. It has to do with our ability to understand automobiles. As indicated in Chapter 1, the mental tools used to understand traditional systems are no longer enough for systems managed by digital computers. To grasp these systems, you need to become a digital thinker. You must be able to break the world into little pieces, to think like a computer.

So, the basic purpose of this book is to help you understand automotive computer-controlled systems. To do this, we cover some background theory; we explain how computers work and we show some examples of how they are used to control engine operation, instrumentation functions and automated braking systems (ABS). Because of the fast pace of developments in this field, there is a lot that we don't cover.

However, if you understand what is included in this book, you will have a headstart on understanding new developments.

Although some service-related chapters are included at the end, the book does not tell you how to repair computer-based systems. This information is contained in factory repair manuals.

Also, be aware that even though many manufacturers have generously signed permissions allowing us to use art from their manuals, they do not endorse (directly or indirectly) anything contained in this book. You need to consult their materials for their approach to service. In some cases, to avoid any possibility of confusion, we have agreed not to include any materials from certain manufacturers or from groups within those manufacturers.

Tom Weathers, Jr.
Claud C. Hunter

Computerized Engine Control and Diagnostics

1

Digital Versus Analog Systems

INTRODUCTION

Automotive computers represent a fundamentally different way of looking at the world in comparison to the systems they replaced. That is why they can perform tasks not possible with the previous systems, and why many mechanics have trouble diagnosing problems with computer-controlled systems (even though computer systems are probably more trouble-free than mechanical systems).

Therefore, it is appropriate to start our book on automotive computers with a bit of ''philosophizing'' that examines this basic issue. It will help later in understanding how and why computers are used in automobiles. It will also give you some idea why learning about this subject is so important.

TWO WAYS OF LOOKING AT THE WORLD

Although some people might not agree, you could say that there are just two basic ways of looking at the world. We can call one way the analog view and the other the digital view. The analog view tends to see things as parts of a unified whole. This is the world of sunrises and sunsets, rainbows and poetry, and precomputerized cars.

The digital view, on the other hand, deals with the pieces. It breaks the world into manageable units. All scientific theories represent the digital view. So do all numbers and words (when stripped of their emotional content). Automotive computers are digital, even though they are used in vehicles, which are analog devices.

ANALOG VERSUS DIGITAL CONTROLS IN A HYPOTHETICAL SYSTEM

To see the distinction on a practical level, let us compare the same function as it might be performed by an analog system and a digital system. The camshaft in an automotive valve train is an analog control device. High points on the camshaft push against rods and mechanical parts to open the valves leading into the combustion chamber. The shape of the cams and the gears that drive the camshaft control valve opening.

Now suppose the components in a typical valve train were replaced with a set of electrically operated solenoids. Let us say there is one solenoid for each valve. Energize the solenoid and the valve opens; de-energize the solenoid and the valve closes. Controlling the solenoids in this hypothetical (so far) system is a digital computer. Sensors located in the engine compartment provide information to a program running in the computer that "decides" when to open and close the valves.

What are the differences between these two systems? First of all, there is the matter of understanding. Almost anyone can comprehend the camshaft-based analog system. We can virtually see the flow of forces from the camshaft through the linkage to the valves. A novice might not understand right away that the valve spring rather than the cam closes the valve, but once that is pointed out, it is not difficult to picture.

Now look at the system controlled by the digital computer. Aside from the solenoids, which some people readily comprehend, all you have are some wires and sensors and the box containing the computer. You cannot "see" what is going on. The only way you can understand this system is to read a description in a book or have somebody tell you. That is why we have trouble diagnosing computer-controlled systems. We cannot see what is going on.

There is also a difference in the way the two systems represent information. In the analog system, the control program is represented as a physical shape. The camshaft is directly analogous to the function it performs. In the digital system, the program is a coded statement describing the function to be performed. Physically, this statement is nothing more than a set of invisible electronic patterns inside the computer.

Looking at it another way, you can compare the analog device to a special-purpose animal—a beaver, for instance. You can compare the digital device to a more general purpose creature—for example, a human being. Figure 1-1 shows the comparison. You can tell just by seeing the beaver that it probably chews trees (because it has big, strong teeth) and that it slaps and manipulates mud (because of its wide, flat tail). The beaver's function is apparent from its very shape. However, looking at the human, especially if it is not accompanied by any tools or artifacts, does not tell you much. When using a chain saw or a mortar trowel, a human performs some of the beaver's functions. But give the same human another set of tools (and the necessary skills) and he or she can perform another set of functions altogether. Like most digital systems, the human is very flexible.

This brings up another difference between digital and analog systems: the ability to respond to changes. Suppose, in our valve train

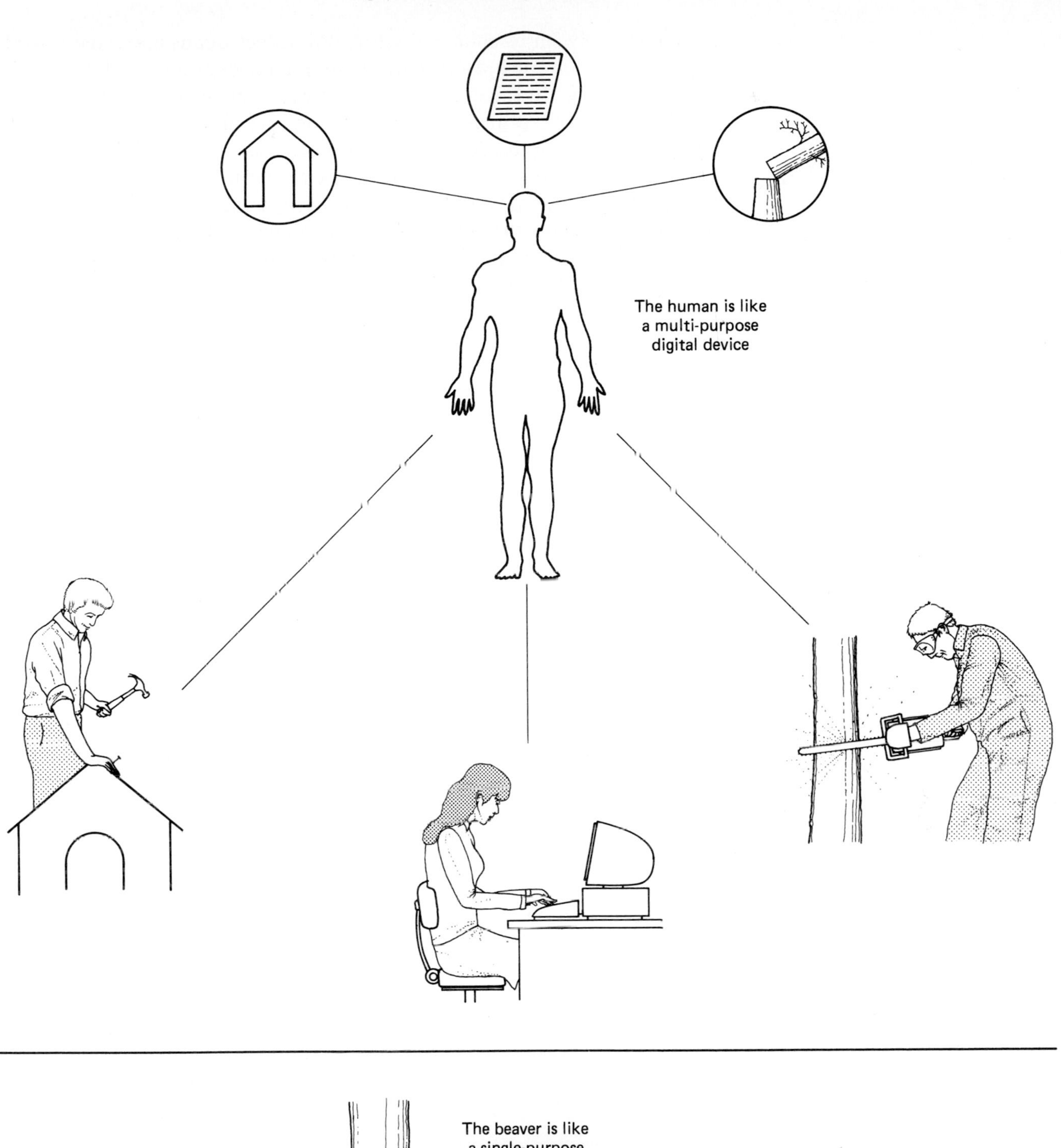

The beaver is like
a single-purpose
analog device

FIG. 1-1 General-purpose, digital human versus specialized, analog beaver.

system, we wish to change the time the valves opens and closes. In the analog version, this would mean replacing the camshaft with a different "grind" unit. In the digital system, it would require changing only the description coded into the program running in the computer. In fact, the program can be designed from the outset to handle change. Using information from various sensors, the program can adjust valve openings in response to changes in load, speed, temperature, and so on. We could even allow the vehicle operator to select various engine operating modes corresponding to hot-rod derived camshaft configurations: street grind, $\frac{3}{4}$ race, and full-race.

Of course, all this flexibility can cause some problems. The analog system, although inflexible, is a direct representation of its function. The digital system is only a statement that describes the world and the function to be performed. It can never be totally complete and, if poorly done, can be quite wrong. Computer people are always looking for better ways to describe the world. The errors they make along the way are called "bugs."

SUMMARY

What have we learned from all this philosophizing? From a practical point of view, probably the most important thing is that computer systems may be inherently more difficult to understand than the mechanical systems they replaced. In order to diagnose and repair computerized systems, you are going to have to spend more time studying and thinking. People who expect to deal successfully in a computerized world must become digital thinkers. The shade tree approach, as described in Chapter 16 of this book, will no longer work.

Although it may not be too clear yet, you have also learned why computers came to be used in cars. As digital, general-purpose devices, they are better able to handle varied, changing information. This too, is examined in later chapters.

2

Electricity, Magnetism, and Some Basic Devices

INTRODUCTION

In one way or another, all the chapters in this book deal with electricity. Therefore, to get started on the right foot, we review the separate but related concepts of electricity and magnetism. Then we look at some basic electrically operated devices, including solenoids, basic electric motors, stepper motors, induction coil pickup sensors, and Hall-effect devices.

ELECTRICITY

What is electricity?

We often use the word *electricity* as if it described a real, physical object. However, that is not exactly what the word means. How many times have you seen, tasted, or touched electricity? Never, although you may have seen and touched the effects of electricity.

When we talk about electricity, we are actually talking about movement, a flow of tiny particles deep within the heart of matter. The visible or detectable signs of electricity—sparks, shocks, readings on test instruments, and so on—are a result of this flow. When the particles are in motion, they have an effect on the material through which they flow as well as on the surrounding space.

Nature in balance

Why do the particles move in a wire, a computer circuit, electrical control device, or anywhere else? It has to do with the universal tendency of conflicting forces to try to stay in balance. The planets remain in their orbits because the centrifugal force of their movement is balanced by the force of gravity from the sun. Water stays in an artificial lake because the weight of the water is balanced by the strength of the dam.

All these are examples of forces in balance. However, what happens if something changes, if, for instance, the dam breaks? Then the forces are no longer in balance. In the case of the broken dam, water rushes downhill, affecting everything in its path, until the lake is emptied and a new balance is achieved.

It is this urge for balance that causes electrical flow. A battery, alternator, or some other device builds up the number of tiny electrical particles in one part of a circuit and reduces the number in another part of the circuit. If the circuit is complete—in other words, if there is an uninterrupted path between the out-of-balance parts—current flows. Excess particles flow to the part of the circuit containing fewer particles. The current continues to flow as long as the battery or alternator maintains the imbalance. Like the water flowing downhill from the broken dam, everything that lies in the path of the current is affected.

Atomic structure

The next question is: What are these particles? If you have been exposed to the subject previously, you have probably already guessed that the particles are electrons. Electrons, together with protons, neutrons, and other particles, make up atoms, which are the basic building blocks of nature. Atoms or pieces of atoms combine in various ways to create all the material in the universe.

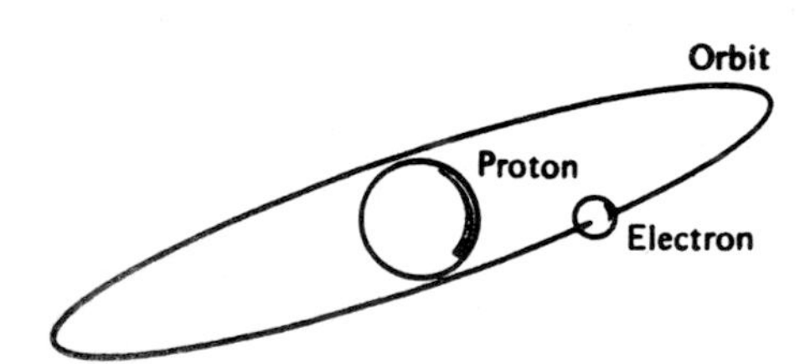

FIG. 2-1 Hydrogen, the simplest element. (From T. Weathers and C. Hunter, *Diesel Engines for Automobiles and Small Trucks*, Reston Publishing Company, Inc., Reston, Va., 1981.)

The simplest atomic combination is the element hydrogen, which is a very light gas at room temperature (Fig. 2-1). It has one proton at the center of the atom and one electron whirling around the proton, something like a planet whirling around the sun. Hydrogen was one of the first elements created at the beginning of the universe. Other elements, including helium, oxygen, copper, gold, and lead—more than 100 altogether—were formed as additional protons, neutrons, electrons, and other parts were added to the original hydrogen atoms (Fig. 2-2).

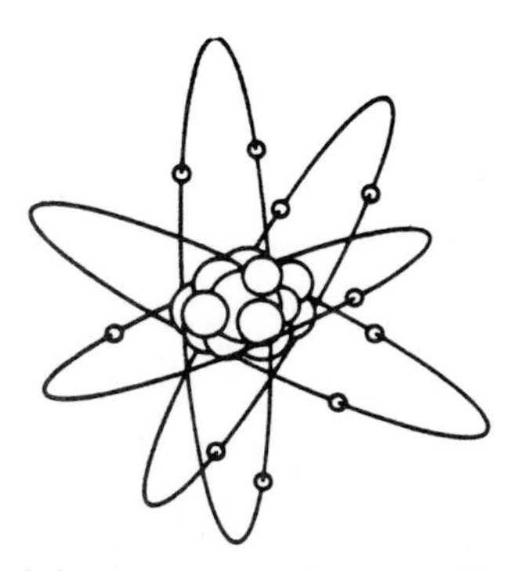

FIG. 2-2 A more complex atom. (From T. Weathers and C. Hunter, *Diesel Engines for Automobiles and Small Trucks*, Reston Publishing Company, Inc., Reston, Va., 1981.)

The atomic parts "stick" together to create atoms because nature strives for balance everywhere, from the very small to the very large. In atoms, the balance we are interested in is between protons and electrons. Protons are said to have a *positive charge,* like the south end of a magnet. Electrons are said to have a *negative charge,* like the north end of a magnet. For an atom to be balanced electrically, it must have an equal number of electrons and protons.

When an element is created, protons are packed together in the center, or nucleus, of the atoms. Electrons are attracted in equal numbers to orbits surrounding the nucleus. Because the electrons have the same charge and tend to repel one another, each orbit can contain only so many electrons, two in the first orbit, eight in the second, and so on.

The last or outermost orbit is the most important in our understanding of electricity. The number of electrons in this orbit determines whether an element will be a conductor, a nonconductor, or a semiconductor of electricity. It is in this orbit that electrical flow takes place.

Electrical flow

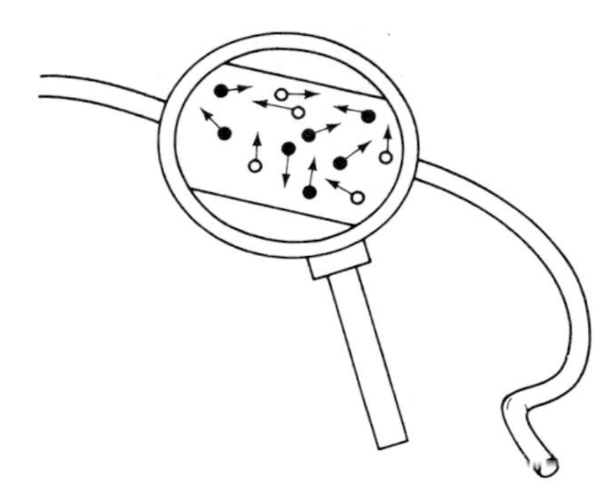

FIG. 2-3 Random electron movements (From T. Weathers and C. Hunter, *Diesel Engines for Automobiles and Small Trucks*, Reston Publishing Company, Inc., Reston, Va., 1981.)

To see how electrical flow works, we will take an imaginary trip into the atomic structure of a copper wire. Copper, a conductor of electricity, has 29 electrons. That gives enough electrons to fill all but the last orbit, which has only one electron. Because these single electrons are so far from the nucleus and because they are not bound in a fixed pattern with other electrons, they tend to drift from atom to atom. From our imaginary vantage point, the outer orbit electrons might resemble a swarm of fireflies, wandering here and there in response to random influences (Fig. 2-3).

As long as the wire is not hooked into a live electrical circuit, the pattern of electron movement remains random. An atom in one place loses an electron. Another electron, from somewhere else, takes its place, attracted by the extra positive charge exerted by the atom that has lost its electron. Since the movements are random, the total of all the electrical charges on all the atoms remains balanced.

However, what happens if the ends of the wire are attached to the positive and negative terminals of a battery? The negative pole contains an excess of electrons and the positive pole has fewer electrons than it needs to be balanced. The excess electrons at the negative pole repel the electrons in the wire, causing them to flow toward the other end. At the same time, the electrons at the other end of the wire are attracted to the empty orbits at the battery's positive pole (Fig. 2-4).

It is like a game of tag played near the speed of light, where nobody (or no electron) ever catches anybody. The chemical activity inside the battery creates a constant imbalance between the two poles. As long as the battery remains active and the circuit is complete, the electrons will chase each other from empty orbit to empty orbit, through the battery and through the wire. If the battery weakens, the current slows down. If the wire is cut, the flow stops altogether. However, the battery still pulls electrons away from the end of the wire connected to the positive pole and builds up electrons at the end connected to the negative pole.

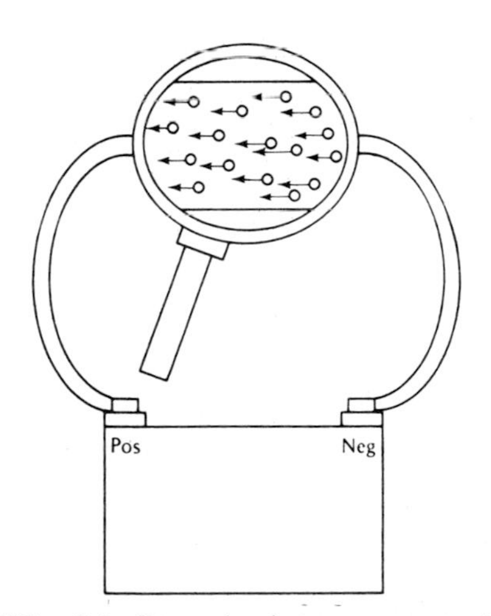

FIG. 2-4 Directed electron movement. (From T. Weathers and C. Hunter, *Diesel Engines for Automobiles and Small Trucks*, Reston Publishing Company, Inc., Reston, Va., 1981.)

Electrical flow takes place wherever there is an overabundance of electrons in one place, an underabundance in another place, and an uninterrupted path between the two. The potential for electrical flow exists whenever there is an imbalance but no complete path for current to flow. For instance, even though no current flows when the ignition switch in a car is off, the potential for current flow still exists because there is an electrical imbalance on either side of the switch.

Current flow can be temporary or sustained. The sudden discharge of electricity represented by a lightning bolt is obviously temporary. An electrical imbalance is created between a cloud and the ground. When the potential becomes great enough, the excess electrons are able to create a path through the atoms in the air.

Most sustained current flow is the result of some manufactured device: a battery, alternator, generator, and so on. They convert either chemical or mechanical energy into electrical potential. That is, they create regions of electrical imbalance, which, when connected to a complete circuit, will cause electrical flow.

Conductors, nonconductors, and semiconductors

The number of electrons in the outer orbit of an atom determines how readily substances containing the atom will allow electrical flow.

Conductors. Substances containing atoms with one to three outer orbit electrons are usually considered to be good conductors. The electrons in their outer orbits are loosely held and easily put into motion. Copper, gold, iron, and silver are examples.

Nonconductors. Materials containing atoms with five or more outer electrons are poor conductors under normal circumstances. Their electrons are tightly held in a pattern with other electrons and are not easily dislodged. Nonconducting materials, such as rubber, glass, and many plastics, are used as insulators to channel and direct electrical flow—making sure that it goes only where it is supposed to go.

Semiconductors. Semiconductors are very stable elements. Their atoms, containing four outer electrons each, join together in rigid patterns that are not easily disturbed and that do not readily support electron flow. However, this holds true only as long as the element remains in its pure state. If small amounts of impurities are added, the semiconductor becomes a conductor. Depending on the kind of impurity added, current flow will depend primarily on the presence of positively charged ''holes'' (actually vacant spaces in outer orbits) or on free electrons floating through the semiconductor. Semiconductors are discussed more completely in Chapter 3.

Effects of electron flow

Electron flow alone does not have much practical application. The effects of electron movement are what we put to everyday use. For instance, electron flow jostles all the electrons in a circuit. These bouncing electrons cause atoms to vibrate, which results in heat. All current flow involves some degree of heating. If uncontrolled, electrical heating can be dangerous. A wire can melt its insulation and cause a fire. However, by using the right kinds of materials in properly designed circuits, the heat can be put to practical use.

A related effect of the atomic disturbance is light. Every time an electron bounces back and forth between the orbits in an atom, light

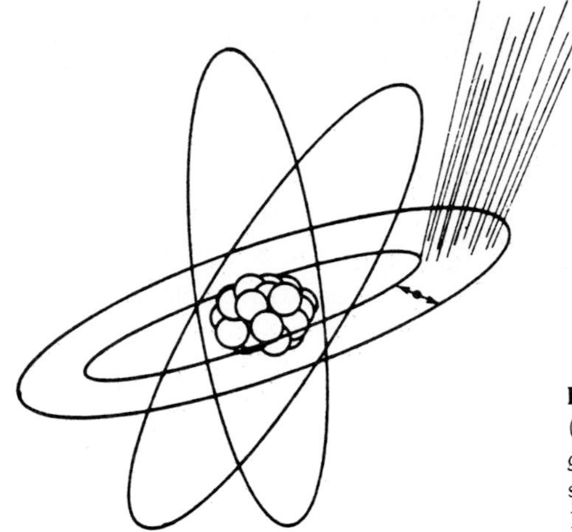

FIG. 2-5 Light emission from an atom. (From T. Weathers and C. Hunter, *Diesel Engines for Automobiles and Small Trucks*, Reston Publishing Company, Inc., Reston, Va., 1981.)

flashes from the atom (Fig. 2-5). As Thomas Edison proved, by using the right kinds of materials in the right circumstances, practical electric lights are possible.

Besides having an effect on the atoms that make up a conductor, electron flow also causes a change in the surrounding space. When current flows through a wire, revolving rings of magnetic force are created around the wire. These lines of magnetic force, as noted in the second half of this chapter, have many applications, such as in electric control devices.

Electrical measurement

We have now seen what electrical flow is, why and how it works, and some of its effects. The last item to be considered before going on to magnetism is electrical measurement.

Volts. Voltage is a measure of electrical pressure. It is an indication of the pressure exerted on each electron in a circuit by a source of electrical potential. As the following example shows, it is not a measure of the total electrical force felt by all the electrons in a circuit.

To see what this means, imagine you have two 12-volt batteries, a large one that is used in a full-size pickup truck, and a small one that is used in a compact automobile. Both have the potential for exerting 12 volts of force of pressure on each electron in a circuit. In other words, the electrical imbalance existing at the positive and negative poles of both batteries will have the same disruptive effect on a given electron in a given atom. However, the larger battery has the potential for disturbing more electrons and more atoms.

Amperes. Amperage is a measure of total electrical flow. It is an indication of the number of electrons flowing past a given point in a circuit. Amperage (or the number of electrons in motion) depends on electrical pressure (voltage) and on electrical resistance (ohms, discussed next).

Ohms. An ohm is a unit of electrical resistance. It is used to express the varying ability of different materials to support current flow. Some factors affecting resistance are (1) the size of the conductor—just as a larger water pipe will let more water flow than a small pipe, a large conductor will let more electrons flow; (2) the length of the conductor—the longer the conductor, the greater the resistance; and (3) the nature of the conductor—as we have already seen, some materials support current flow better than others.

Ohm's law

As you might imagine, ohms, amperes, and volts are related quantities. If one quantity changes, at least one of the others must also change. For instance, in a given circuit:

1. If voltage changes (goes up or down) and ohms of resistance stay the same, the amperage must also change (go up or down).
2. If the ohms go up (as a result of a frayed or otherwise damaged wire) and the voltage at the source does not change, the amperage must go down. Not as many electrons can flow.
3. Conversely, if the ohms of resistance drop, the amperage must increase.

These relations are expressed in a statement called Ohm's law. That law and associated formulas are given in Fig. 2-6.

Ohm's Law:

An electrical pressure of 1 volt is required for 1 ampere of current to flow past 1 OHM of resistance.

Stated as a formula:

$$E = I \times R$$

or

$$I = \frac{R}{E}$$

or

$$R = \frac{E}{I}$$

E = Voltage (or Electromotive Force)

R = Resistance

I = OHMS (or Impedance)

FIG. 2-6 Ohm's Law. (From T. Weathers and C. Hunter, *Diesel Engines for Automobiles and Small Trucks*, Reston Publishing Company, Inc., Reston, Va., 1981.)

MAGNETISM

What is magnetism?

Magnetism is a mysterious force operating equally well through air, empty space, or solid matter. No one knows exactly how it works. Useful theories have been devised that help predict how magnetism behaves and how it can be put to practical application. We will examine some of these theories because they are needed to help understand automotive electrical operation. Like electricity, magnetism cannot be felt, seen, touched. It is more a description of an effect than of a thing.

What is a magnet?

A magnet itself is easier to describe than magnetism. Magnets are objects with the power to attract or repel other magnets and to attract iron and certain materials made from iron. Magnets are directional; that is, they have distinct ends, or poles. If a straight or bar magnet is allowed to hang freely in a horizontal position, one end (called the *north pole)* will always point in a northerly direction and the other end (called the *south pole)* will always point toward the south. No matter how a magnet is cut, shaped, or altered, it (or its pieces) will have two magnetically different poles.

Theory of magnetic operation

Scientists believe that magnetic properties originate at the atomic level. What follows is a slightly altered version of the explanation most often given.

Every atom in the universe is thought to spin on its axis, like the earth spins on its axis, or a toy spins about its handle. The spinning action (together with electron movement) causes the atoms to be surrounded by lines of magnetic force. No one knows exactly what these lines of force are, or if they even exist in the usual physical sense. However, for our purposes, we can visualize them as being the paths taken by imaginary particles flying around and through atoms (Fig. 2-7). The particles travel out one end of an atom's axis (its north pole), circle around the atom, reenter at its south pole, then go through the atom to start the trip again.

Atomic Axis

N

S

Lines of Force (Imaginary Particles) Passing Through Atom

FIG. 2-7 Magnetic force lines passing through a simple atom. (From T. Weathers and C. Hunter, *Diesel Engines for Automobiles and Small Trucks*, Reston Publishing Company, Inc., Reston, Va., 1981.)

In most materials, the atomic poles point in different directions. The lines of force surrounding adjacent atoms do not line up; the imaginary particles flying around neighboring atoms bump into one another as often as they travel along parallel paths.

However, something happens in magnetic materials (Fig. 2-8). A substantial number of the atomic poles point in the same direction. This means that a substantial number of the imaginary particles travel in the same direction. Instead of expending their imaginary energy in headlong collisions, the nonexistent particles fly along parallel paths. They join forces to fly from atom to atom, building up enough momentum to actually escape, for awhile, the confines of the magnetic material.

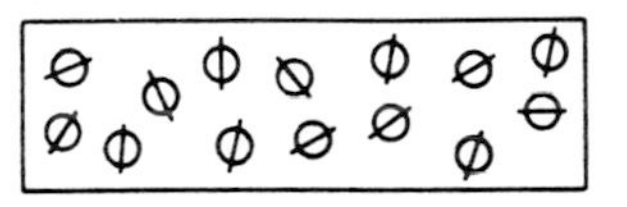

Random Atomic Alignment

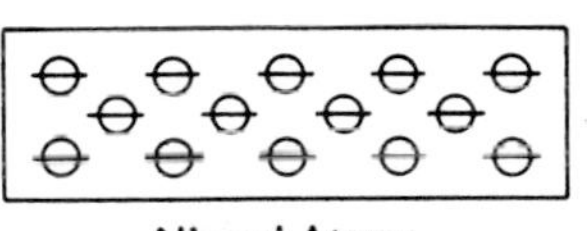

Aligned Atoms (Magnet)

FIG. 2-8 Aligned versus unaligned atoms. (From T. Weathers and C. Hunter, *Diesel Engines for Automobiles and Small Trucks*, Reston Publishing Company, Inc., Reston, Va., 1981.)

The action duplicates on a large scale what takes place on a small scale within atoms. Imaginary particles fly along parallel paths within a magnet, building up enough power and energy to leave the magnet at its north pole. Then the particles circle back to reenter the magnet at its south pole and start the trip again. The paths taken by the particles represent lines of magnetic force, which extend in three dimensions on all sides of the magnet (Fig. 2-9).

The flight of the imaginary particles helps to explain how magnetic materials attract and repel each other (Fig. 2-10). When the opposite poles of two magnets are placed together, the imaginary particles travel in the same direction, just as they do within the atomic structure of the two magnets. Consequently, the particles join forces, pulling the

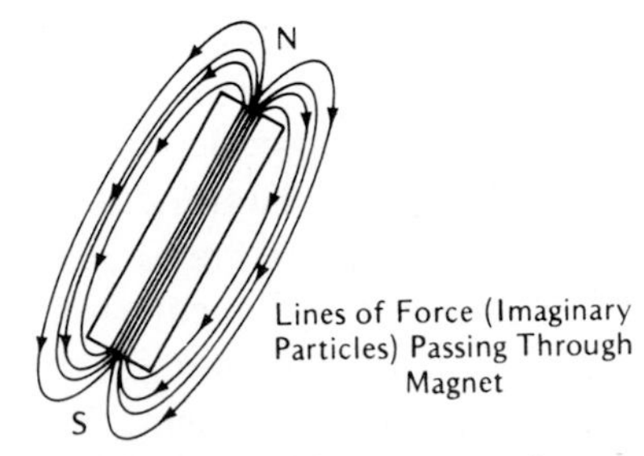

FIG. 2-9 Lines of force passing through a bar magnet. (From T. Weathers and C. Hunter, *Diesel Engines for Automobiles and Small Trucks*, Reston Publishing Company, Inc., Reston, Va., 1981.)

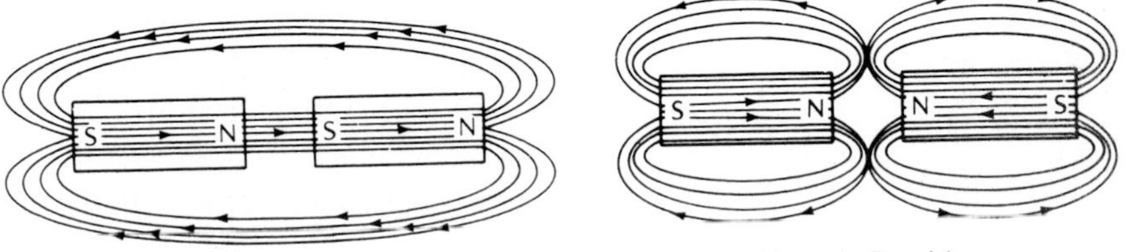

FIG. 2-10 Attraction and repulsion between two magnets. (From T. Weathers and C. Hunter, *Diesel Engines for Automobiles and Small Trucks*, Reston Publishing Company, Inc., Reston, Va., 1981.)

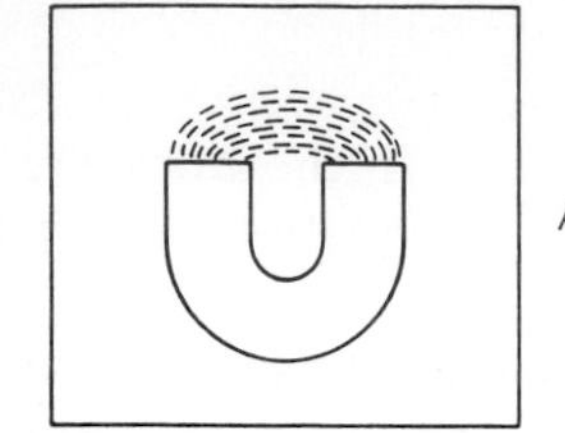

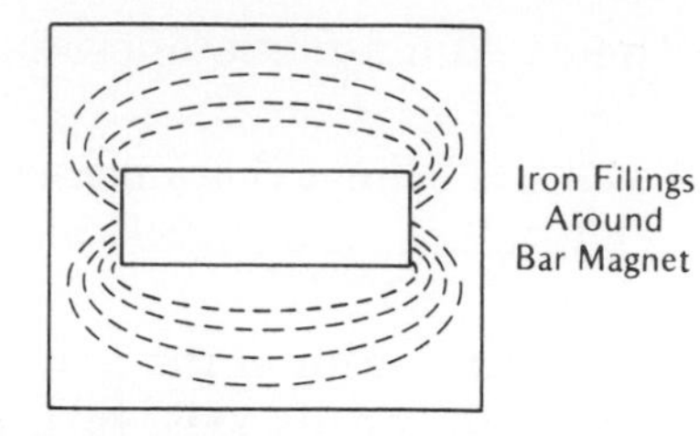

FIG. 2-11 Iron filings around horseshoe and bar magnets. (From T. Weathers and C. Hunter, *Diesel Engines for Automobiles and Small Trucks,* Reston Publishing Company, Inc., Reston, Va., 1981.)

two magnets together. In effect, this creates a single magnet with double the power of the individual magnets.

However, if the like poles of two magnets are put in close proximity, the particles move in opposite directions. As a result, they bump together to push the magnets apart.

Magnets are attracted to iron objects when the lines of force from the magnet penetrate the atomic structure of the iron material. The iron atoms tend to line up with the lines of force, causing the imaginary particles to fly along with same paths. As the particles join forces, they tend to pull the two objects together.

These magnetic flight paths (whatever they represent) can actually be seen by sprinkling iron filings on a piece of paper that has been placed over a magnet. Tapping the paper lightly loosens the filings so that they can follow their natural attraction to the lines of force. The pattern of filings represents the shape of that part of the three-dimensional magnetic field sliced through by the paper. Figure 2-11 pictures the magnetic fields surrounding horseshoe and bar magnets.

Kinds of magnets

There are three basic kinds of magnets: permanent, temporary, and electromagnetic.

Permanent magnets. Permanent magnets are often made from the mineral magnetite, which is a naturally occurring magnet. Its atoms are aligned sufficiently to produce a coherent magnetic field. Commonly called *lodestone,* magnetite was discovered hundreds of years ago. Early mariners used it as a crude compass to help chart their course when traveling out of sight of land.

Temporary magnets. Temporary magnets are created by placing certain iron or iron-based materials in the presence of strong magnetic fields. The lines of force from the magnetic field "pull" the iron atoms into alignment. Depending on the strength of the magnetic field and on the nature of the material, the atoms may remain in alignment after the magnetic field is removed. However, as the name implies, these kinds of magnets may not last very long. Any sort of shock can disturb the fragile atomic alignment; this causes the atoms to become jumbled again and no longer capable of producing a coherent magnetic field.

Electromagnets. Electromagnets depend on a peculiar property of electricity: whenever current flows through a conductor, the atoms in the conductor line up sufficiently to produce a coherent magnetic field.

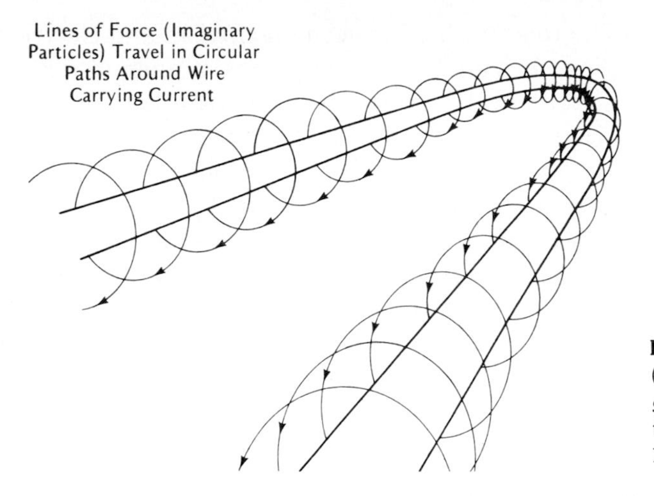
Lines of Force (Imaginary
Particles) Travel in Circular
Paths Around Wire
Carrying Current

FIG. 2-12 Force lines going around wire. (From T. Weathers and C. Hunter, *Diesel Engines for Automobiles and Small Trucks*, Reston Publishing Company, Inc., Reston, Va., 1981.)

If the conductor is a wire, the lines of force form concentric rings around the wire (Fig. 2-12). In other words, the imaginary particles fly in circular paths around the wire. The direction in which the nonexistent particles fly, clockwise or counter-clockwise, depends on the direction in which the current is flowing.

Electromagnets are created when current-carrying wires are formed into a coil (Fig. 2-13). This is how it works: current flows in the same direction in adjacent loops of the coil. Consequently, the imaginary particles fly in the same direction in adjoining loops. As we have seen before, whenever these particles fly in the same direction, they join forces. So the lines of force merge to circle all the loops, traveling around the outside of the coil, going back through the inside of the coil, then going around the outside again.

An electromagnet is like a permanent magnet in many ways. It has a north pole (where the particles come out of the coil) and a south pole (where the particles return). An electromagnet can attract or repel other magnets. It can also attract iron.

However, there are still some important differences between the two. First, electromagnets can be turned on or off by turning the current on or off. Second, the polarity of the electromagnet (which end is north and which is south) can be reversed by reversing the current flow. Finally, the strength of the magnet can be varied by varying the current flowing through the coil or by increasing the number of loops in the coil.

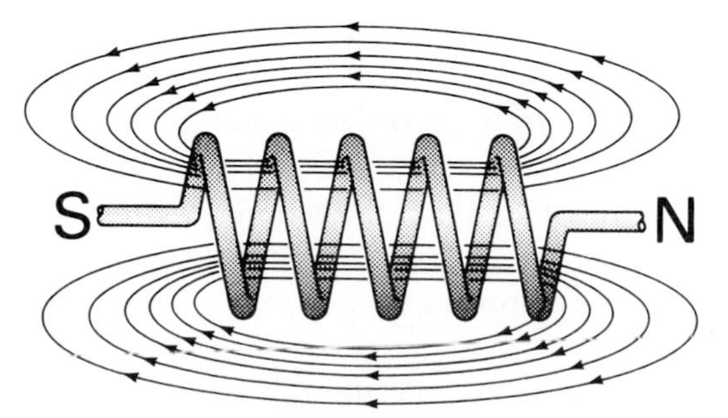
S
N
When Wire is Coiled,
Force Lines Merge to
Travel Around and Through
Entire Coil

FIG. 2-13 Wire formed in an electromagnetic coil. (From T. Weathers and C. Hunter, *Diesel Engines for Automobiles and Small Trucks*, Reston Publishing Company, Inc., Reston, Va., 1981.)

INDUCED CURRENT FLOW

The relationship between magnetism and electricity is strange and complex. As noted earlier, electrons moving along a wire will create a magnetic field. However, it is also possible to do just the opposite, to create electrical movement in a wire by using a magnetic field.

To see how this works, imagine a horseshoe magnetic being passed over a wire. The magnet's lines of force, as they travel between the poles, cut across and wrap around the wire. The lines penetrate the very atomic structure of the wire. As a result, the electrical balance of the atoms is upset and electrons are put into motion along the wire (as long as the wire is attached to a complete circuit).

The direction of the electron flow is determined by the motion of the lines of force across the wire. If (relatively speaking) the lines of force cut down across the conductor, the current will flow in one particular direction. But if the lines of force move up across the conductor, the current will flow in the opposite direction.

Note: It should be remembered that it does not make any difference if the conductor moves across the lines of force or if the lines of force move across the conductor. All that matters is that there is relative motion between the two. When there is, current flow is induced in the conductor.

BASIC DEVICES

There are several basic electrical devices so widely used in control systems that they must be introduced now so the concepts will be available to you through the rest of the book.

OUTPUT

The first three devices—solenoids, basic electric motors, and stepper motors—are used by control computers to perform physical tasks. These output devices are like the muscles in animals or humans, responding to commands from an electronic brain. Additional applications for these and other output devices are described in Chapter 11.

Solenoids

A typical *solenoid* consists of a coil (or coils) of wire wrapped around a hollow tube. When electrical current flows through the coil, a strong magnetic field is generated. The field, as it flows around and through the hollow tube, causes the solenoid to act like a magnet, with the ability to attract iron-based objects. However, unlike permanent magnets, a solenoid can be turned on or off very rapidly. A solenoid is often used to draw an object inside its hollow core. Such objects, referred to as *core rods,* are attached by a linkage system to the mechanical device operated by the solenoid (Fig. 2-14).

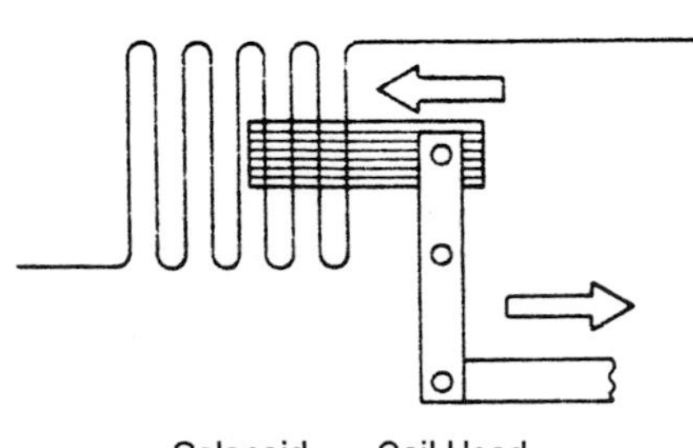

Solenoid . . . Coil Used to Actuate Mechanical Linkage

FIG. 2-14 Solenoid. (From T. Weathers and C. Hunter, *Diesel Engines for Automobiles and Small Trucks,* Reston Publishing Company, Inc., Reston, Va., 1981.)

Electric motors

Electric motors contain two sets of magnets, the field and the rotor (or armature). The field may be an electromagnet or a permanent magnet. The rotor is always an electromagnet.

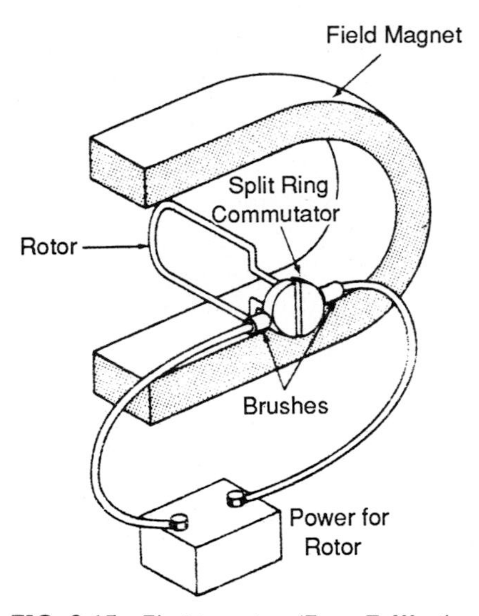

FIG. 2-15 Electric motor. (From T. Weathers and C. Hunter, *Diesel Engines for Automobiles and Small Trucks,* Reston Publishing Company, Inc., Reston, Va., 1981.)

Figure 2-15 illustrates a very simple electric motor. The field is a horseshoe magnet and the rotor is a single loop of wire placed between the ends of the horseshoe magnet. The rotor is connected by a switch (called a split-ring commutator) to the power source.

In terms of the imaginary magnetic particles described earlier, this is how the motor works: particles travel from the north to the south pole of the field magnet. At the same time, particles circle the rotor loop, going around one way as the current flows down one side of the loop, and revolving the other way as the current flows back.

The end view of the motor shown in Fig. 2-16 shows how the particles interact to turn the rotor. The top part of the rotor is pulled to the left because the particles from the field and the rotor move in the same direction on the left side of the loop and in the opposite direction on the right side of the loop. The particles on the left try to join up, which pulls the top of the loop to the left. At the same time the particles on the right push apart, which pushes the top to the right.

A similar action takes place at the bottom of the loop. However, since the current (and hence the particles) move in the opposite direction, the pushing/pulling is to the right.

These forces move the rotor halfway around, as shown in Fig. 2-17. Momentum then takes the rotor on a little farther. However, unless something happens, the forces acting on the rotor will cancel out at this point. They will work in opposite directions and the rotor will stop.

That is where the split-ring commutator comes into play. As shown in Fig. 2-18, it reverses the current flow through the loop. The two halves of the commutator swap connections with the power source. What was once the negative connection to the rotor becomes the positive lead, and vice versa. As a result, the particles or lines of force circle the loop in opposite directions. Therefore, the forces acting on the rotor are reversed, causing it to continue on around. This cycle is repeated with every revolution of the rotor.

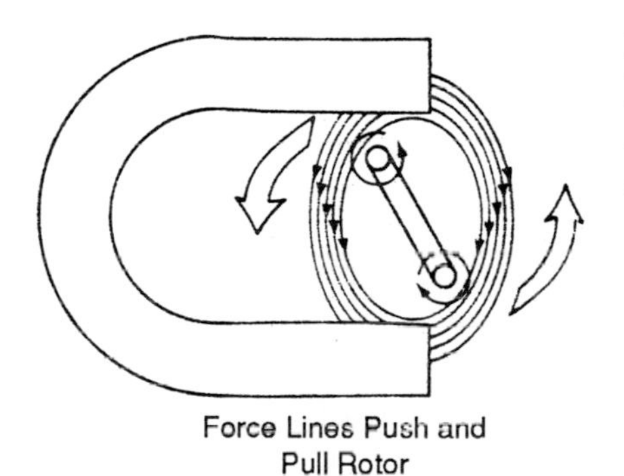

FIG. 2-16 End view of a motor showing the reaction between field and rotor force lines. (From T. Weathers and C. Hunter, *Diesel Engines for Automobiles and Small Trucks,* Reston Publishing Company, Inc., Reston, Va., 1981.)

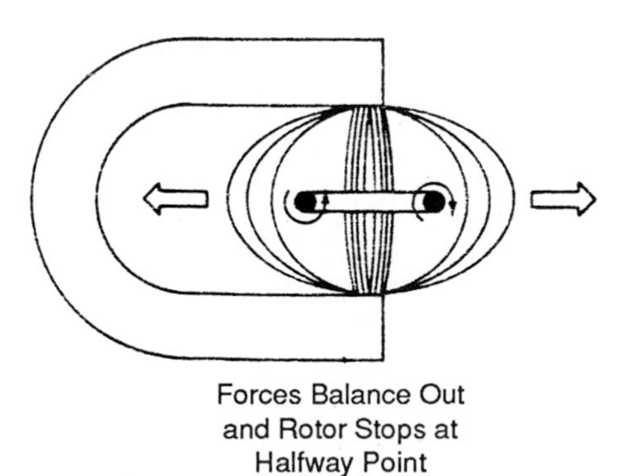

FIG. 2-17 Rotor after turning halfway around. (From T. Weathers and C. Hunter, *Diesel Engines for Automobiles and Small Trucks,* Reston Publishing Company, Inc., Reston, Va., 1981.)

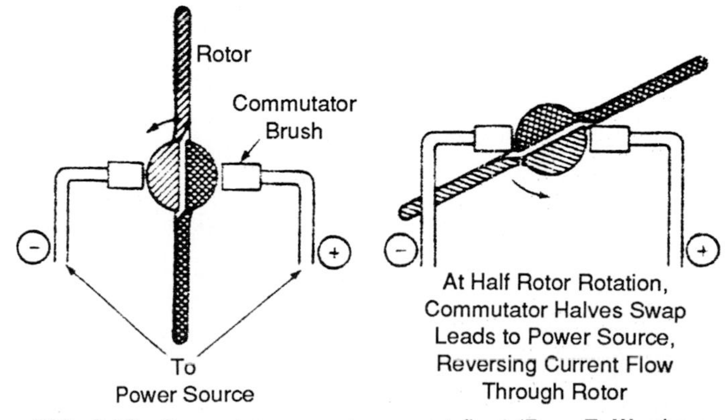

FIG. 2-18 Commutator reversing current flow. (From T. Weathers and C. Hunter, *Diesel Engines for Automobiles and Small Trucks,* Reston Publishing Company, Inc., Reston, Va., 1981.)

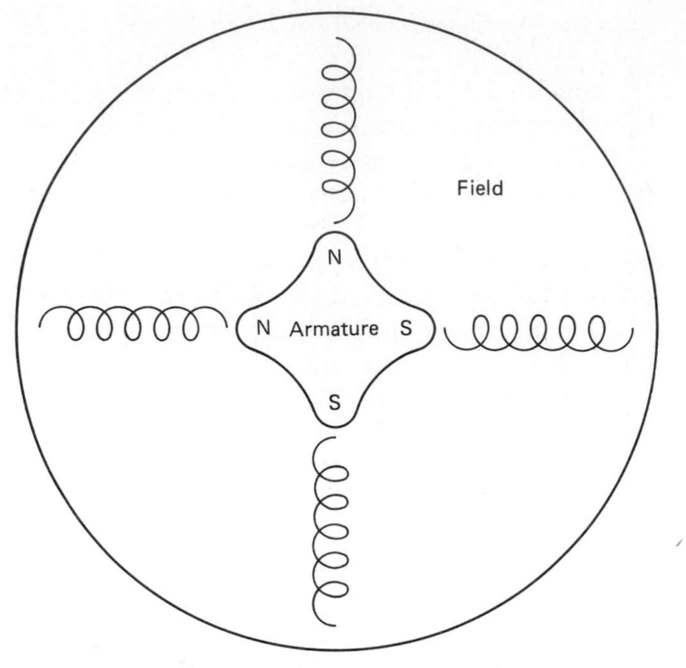

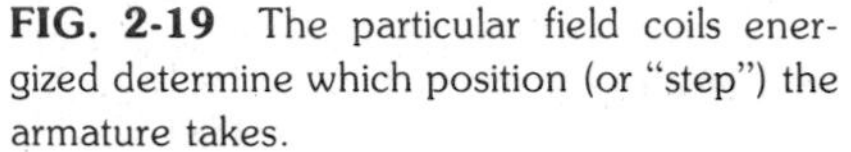
FIG. 2-19 The particular field coils energized determine which position (or "step") the armature takes.

Stepper motors

A special type of electric motor is the *stepper motor.* Unlike the basic motor just described it does not produce continuous circular motion. Instead, a stepper motor produces movement in precise increments or steps.

Figure 2-19 is a simplified representation of a stepper motor. The armature is a star-shaped permanent magnet. The points are the poles of the magnet: north and south, north and south. The field consists of four electromagnetic coils.

When the field coils are energized, the lines of force from the coils pass through the armature. As a result, the poles of the armature try to line up with the lines of force from the coils. Energize one set of coils and the armature will move to one position. Energize another set of coils and the armature will move to another direction. By selectively energizing the coils, the armature is caused to move in precise steps, or increments. So long as a given pattern of coils remains energized, magnetic attraction ensures that the armature stays in one position.

INPUT

Input devices are like the sense organs in animals. They convert changes in physical conditions (such as temperature, air pressure, and movement) into corresponding fluctuations in electrical voltage or current. These fluctuations are detected by the computer, which records or makes decisions based on what it "senses." Often a command to an output device is issued as a result of an input signal.

Various types of input sensors, or transducers as they are sometimes called, are used in automotive systems. The next paragraphs examine two important devices, induction coil pickup sensors and Hall-effect sensors. Additional details and specific applications of these devices appear in remaining chapters throughout the book. Other input devices are specifically examined in Chapter 10.

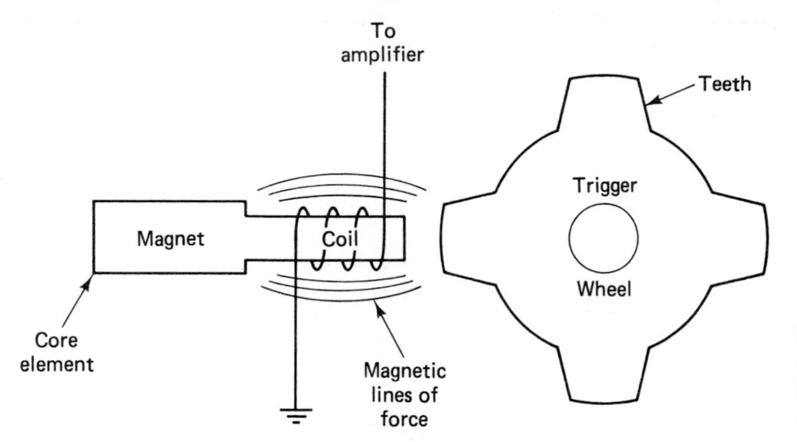

FIG. 2-20 As trigger wheel teeth bypass core element, magnetic lines of force balloon in and out across coil. This induces voltage in coil.

Induction coil pickup sensors

Induction coil pickup sensors are used to detect the speed or position of rotating components, particularly distributor driveshafts and crankshafts. They use the principles of induced current flow described earlier in this chapter.

The main components of this sensor are a pickup assembly and a trigger wheel, as shown in Fig. 2-20.

- The pickup assembly consists of a core element enclosed in a coil wrapping at one end and attached to a permanent magnet at the other. Two leads are attached to the coil.
- The trigger wheel is attached to the rotating shaft. When used in distributors, the wheel usually has one tooth per cylinder.

As the trigger wheel rotates, its teeth pass by the pickup coil. When the teeth and the core piece line up, the lines of force from the permanent magnet are concentrated. When the teeth and the core piece move apart, the lines of force diminish. As a result, lines of force balloon in and out across the pickup coil. This causes a pulsating dc current to be transmitted through the two leads attached to the coil.

In distributor applications, the current from the two leads goes to an electronic device called the *amplifier module*. Although the current is not very strong, it is sufficient to operate the base of a transistor. The transistor acts like a solenoid switch to control current flow through the ignition coil. See Chapter 3 for an explanation of semiconductors. See Chapter 7 for additional information on ignition systems.

Hall-effect sensors

When employed in automotive applications, Hall-effect sensors are most often used to detect the speed or position of rotating components, such as camshafts, crankshafts, distributor drive shafts, and the like.

The Hall-effect takes place on a thin plate, typically made from a semiconductor. Looking at the simplified representation shown in Fig. 2-21, the plate has four electrical leads. Located near the plate is a

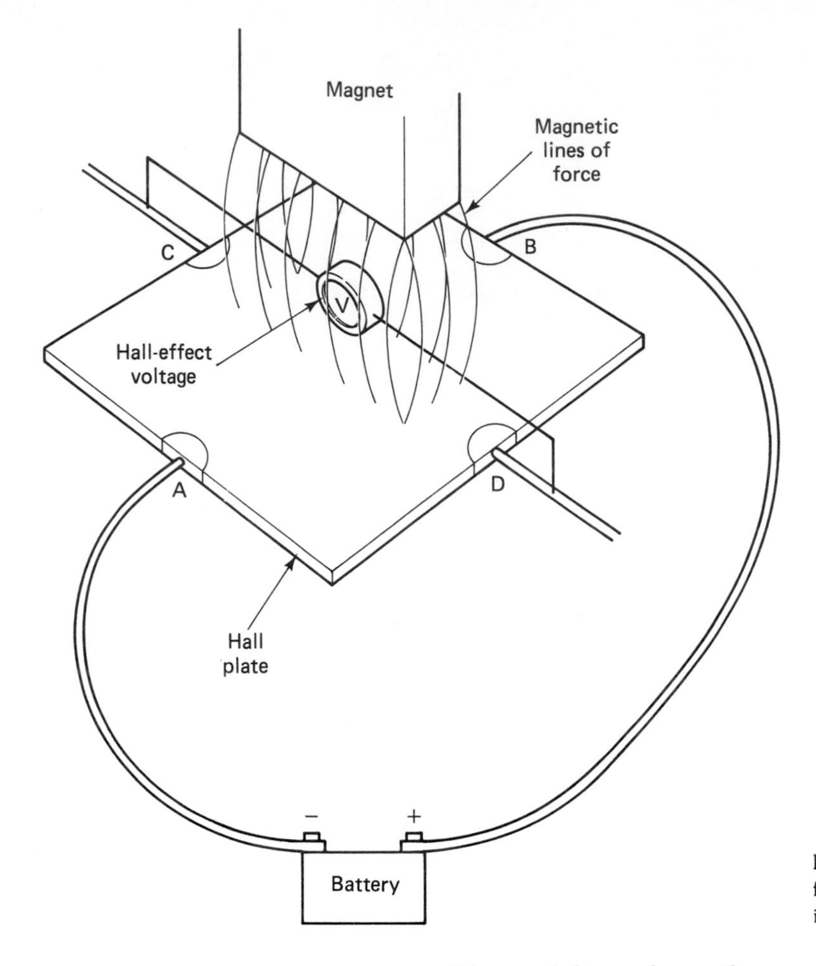

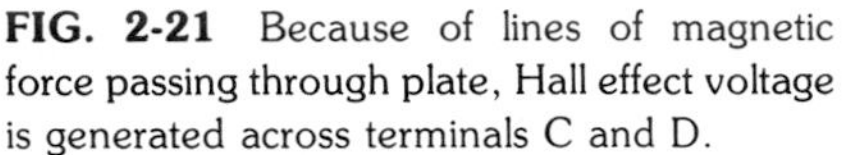
FIG. 2-21 Because of lines of magnetic force passing through plate, Hall effect voltage is generated across terminals C and D.

magnet. Lines of force from the magnet pass through the plate in a vertical direction.

Two of the leads attached to the plate are connected to a source of electrical potential, in this case a battery. Because of the effect of the lines of magnetic force on the electrons moving across the plate between the first two leads, a slight potential exists between the second two leads. This potential is the Hall-effect. Other factors being equal, the strength of the potential depends on the strength and orientation of the magnetic field. Therefore, changes in the magnetic field can be used to produce signals from the sensing unit.

The typical Hall-effect unit found in automobiles includes a small permanent magnet, the Hall-effect sensor itself, and one or more flat, ferrous vanes attached to the rotating component. An example is shown in Fig. 2-22. As the component rotates, the vanes pass through a gap between the magnet and the sensor. The vane, which has a high degree of magnetic permeability, diverts the magnetic flux lines from the permanent magnet. This causes a change in the lines of force passing through the thin plate of the sensor. As just described, this results in a change in electric potential in the sensing element. The change in potential becomes the signal produced by the unit.

In most cases, the Hall-effect sensor is fabricated on a single integrated chip (IC) containing the Hall-effect plate, an hysteresis signal conditioner, and a transistor. The signal produced on the plate is di-

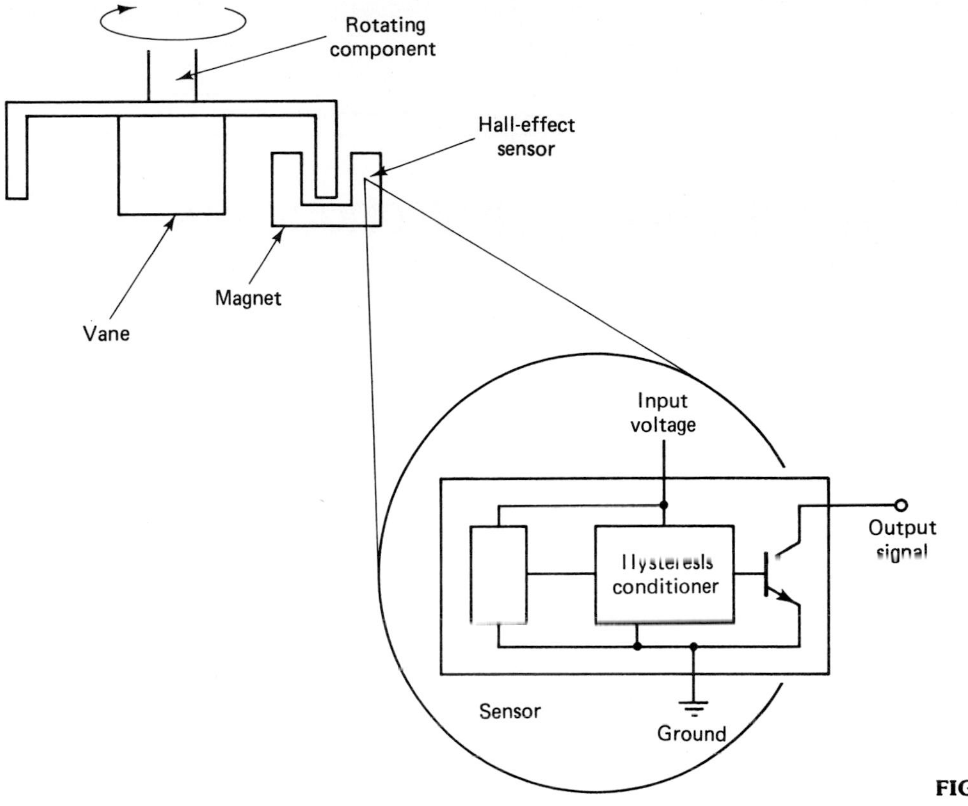

FIG. 2-22 Hall-effect sensor.

rected to the hysteresis unit which produces a digital, two-state output, i.e., either on or off. This output is directed to the base of a transistor, turning it on or off, thus controlling the flow of current through the emitter-collector circuit. The square-shaped voltage level shifts in that circuit is the final output of the Hall-effect sensor. (See Chapter 3 for details on transistors and semiconductors.)

In appearance and operation, a Hall-effect unit is somewhat similar to the induction pickup sensor just described. They are both used in similar applications. The main difference is accuracy at higher speeds. Hall-effect units are said to suffer less magnetic distortion at higher speeds than induction coil pickup sensors.

3

Transistors and Semiconductors

INTRODUCTION

In 1956 the Nobel prize for physics was awarded to three American researchers: J. Bardeen, W. H. Brittain, and J. Shockley. Before 1956, few people outside the scientific community appreciated the importance of what had been done. Since then, transistors and other semiconductors have been used in everything from portable radios to Apollo moon rockets. Semi-conducting devices have also found their way into automotive applications. All alternating charging systems use them, as well as all on-board computers.

In order to understand automotive systems based on semiconductors, it is necessary to appreciate some of the theory involved. This chapter introduces you to the subject.

ATOMIC STRUCTURE

Semiconductors, as a category of matter, fall somewhere between conductors and nonconductors. The differences are primarily due to the number and arrangement of electrons and to the way atoms are joined. The structure of a conductor promotes electrical flow, whereas the bonding between atoms of nonconductors inhibits current movement. Semiconductors may be either conductors or nonconductors, depending on certain circumstances and on the presence of carefully introduced impurities. This "either-or" property of semiconductors makes them valuable as one-way current valves, electrical switches, and so on. To see how semiconductors work, it will be necessary to look again at atomic structure, paying particular attention this time to the behavior of outer-shell electrons.

Reviewing basic concepts: all matter is composed of molecules, which are composed of various kinds of atoms (Fig. 3-1). Atoms, in

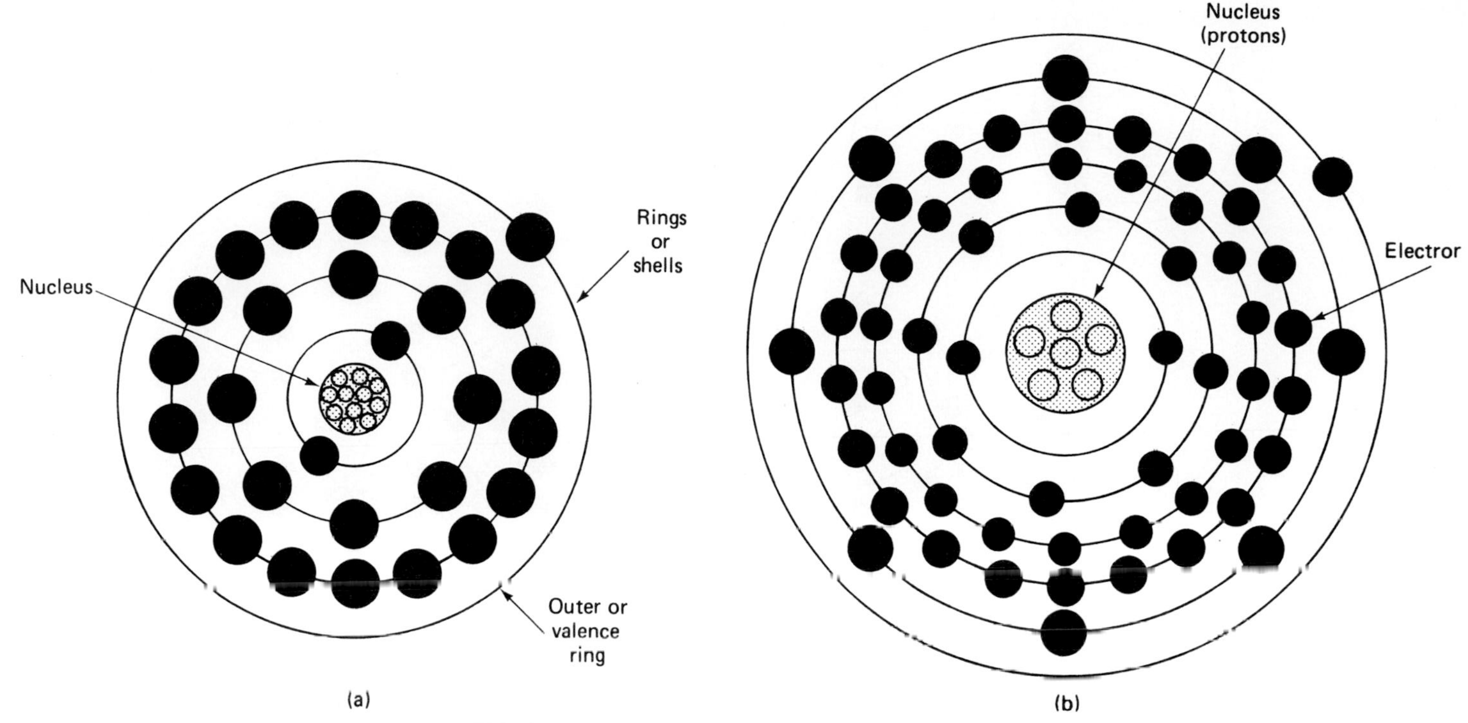

FIG. 3-1 (a) Atomic nomenclature. (b) Uranium atom.

turn, are composed of electrons, protons, and neutrons. Positively charged protons and no-charge neutrons occupy the nucleus of atoms. Negatively charged electrons move in circular orbits around the nucleus, much the same way that planets orbit the sun in our solar system. Current flow is the movement of these electrons from one atom to another under the influence of an EMF (electromotive force).

Electrons occupy different orbits or shells around the nucleus. Following a definite pattern, some electrons orbit close to the nucleus and some orbit farther away. The number of electrons in any given orbit depends on the position of the orbit (first, second, third, etc.).

The electrons in the inner orbits are tightly bound to the nucleus. As a rule they do not enter into reactions with the electrons from other atoms. However, this is not the case with electrons in the outermost or valence orbits. The number of electrons in the valence orbit determines the electrical nature of a substance, whether it will be a conductor of electricity, a nonconductor, or a semiconductor.

CONDUCTORS, NONCONDUCTORS, AND SEMICONDUCTORS

An "ideal" valence orbit contains eight electrons. No valence orbit will contain more than eight, but many will have less. When that happens, an atom will try to "lend" or "borrow" electrons to achieve the satisfied state.

Atoms with one to three valence electrons tend to lend electrons [Fig. 3-2(a)]. They are conductors of electricity. Their valence electrons are rather loosely held and with a sufficient application of outside energy can be put into motion. For instance, the single valence electron of a copper atom can be easily made to drift to the valence orbit of the

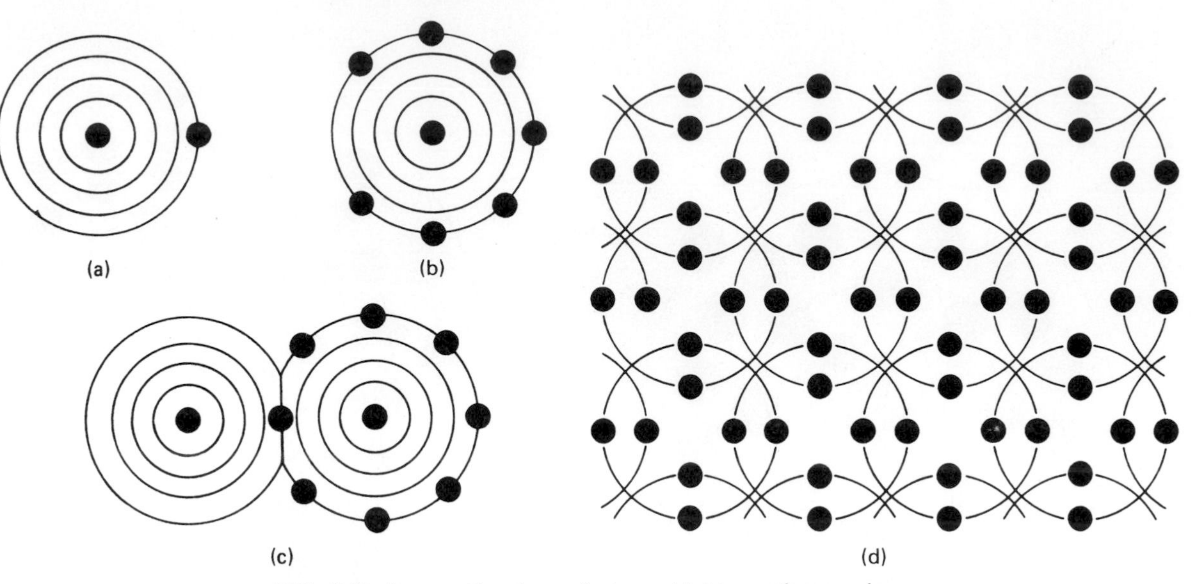

FIG. 3-2 Atoms with valence electrons. (a) Atom with one valence electron. (b) Atom with seven valence electrons. (c) "Shared" valence electrons to achieve "balanced" eight-electron valence orbit. (d) Crystal lattice work of semiconductors atoms. Neighboring atoms share four valence electrons to achieve balanced valence orbits.

next copper atom. Then its valence electron will be repelled to the next atom, and so on. The result is electron movement or current flow.

Atoms with five to seven valence electrons tend to borrow electrons from other atoms [Fig. 3-2(b)]. They are nonconductors. Their valence electrons are more tightly held to the nucleus and cannot be put into motion easily. The valence orbits of two or more atoms combine to achieve the satisfied valence state. Atoms with one to three valence electrons join atoms with five to seven valence electrons (by the borrowing and lending process), so the total valence electrons of the combined atoms equals eight [Fig. 3-2(c)].

Atoms with four valence electrons combine with one another to achieve the ideal condition. For instance, the four valence electrons of a germanium atom can combine with the four valence electrons of a neighboring germanium atom. The result is a complex, lattice-work crystal of satisfied valence orbits, each having eight electrons [Fig. 3-2(d)].

Because of this structure, atoms with four valence electrons are called semiconductors. In their pure state, considerable voltage or high temperatures are required to break loose electrons from the satisfied valence combinations. However, something dramatic happens to the conductivity of semiconductors when impurities are added. Depending on the kind of impurity, a semiconductor can be made into a carrier of negative charges or positively charged "holes." The process is called *doping* and the result is a conducting semiconductor. The remainder of this chapter will examine the formation and behavior of semiconductors.

NEGATIVE, N-TYPE SEMICONDUCTORS

Negative, or N-type semiconductors are created by adding atoms with five valence electrons to the parent material (whose atoms contain four valence electrons). Four of the added electrons combine with four elec-

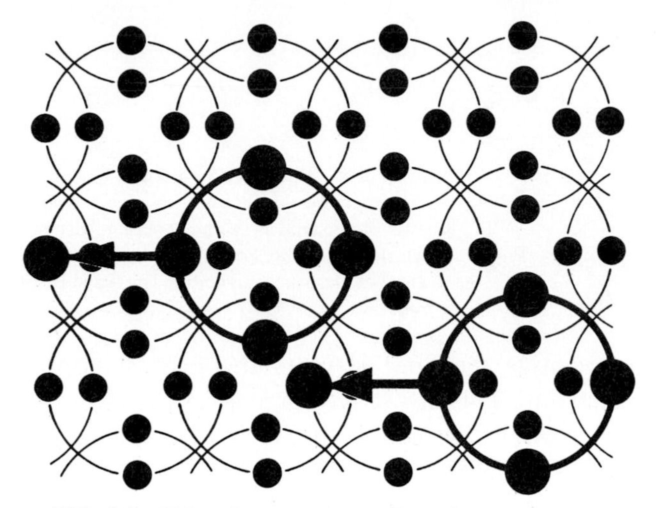

FIG. 3-3 N-doped semiconductor. Extra electrons support current flow in the presence of EMF

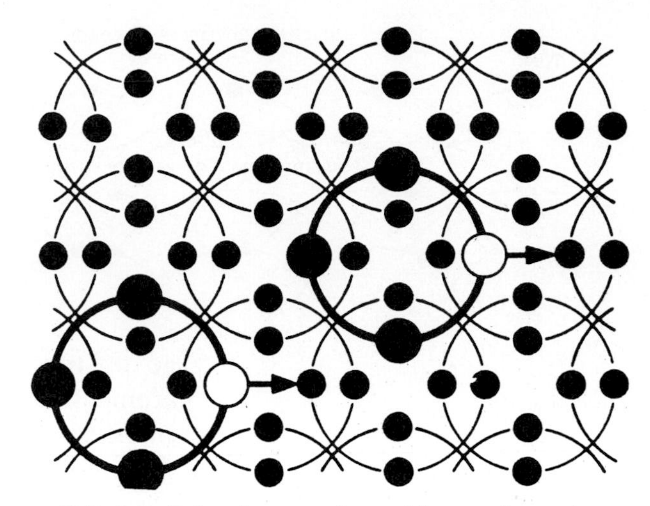

FIG. 3-4 P-doped semiconductor. Missing valence electron creates a "positive hole" that will support current flow.

trons of a parent atom to form a stable, combined valence orbit. However, the fifth added electron has nowhere to go since no valence orbit can contain more than eight electrons. This electron drifts through the lattice structure of the combined atoms. Under the influence of an EMF, it will support current flow (Fig. 3-3).

N-type semiconductors are created by adding materials such as phosphorus, antimony, and arsenic to the parent semiconductor. These additives are sometimes called N-type doping agents. Usually, the doping agent is combined with the parent semiconductor at a ratio of 1 atom of doping agent for every 10 million atoms of the parent material.

POSITIVE, P-TYPE SEMICONDUCTORS

Positive, or P-type semiconductors are created by adding atoms with three electrons in their valence orbits (aluminum, indium, boron). These atoms, as before, enter the lattice structure of the parent semiconductor. However, this time, electrons are missing from the valence orbits of the combined atoms. Some of the valence orbits will only have seven electrons instead of eight. The empty spaces in P-type semiconductors are considered to be positively charged holes (Fig. 3-4) because the "unsatisfied" valence orbits will have a tendency to attract free electrons into the holes, the same as if an actual positively charged particle were present.

CURRENT THEORY: ELECTRON VERSUS HOLE MOVEMENT

In order to understand semiconductor operation, it is necessary to add to the previously described theory of current flow. Going back to the example of a copper wire attached to the positive and negative terminals of a battery, we have said that the absence of electrons at the positive terminal tends to pull electrons out of the wire and that the excess of electrons at the negative terminal pushes electrons back in to replace those pulled out. The pulling action is due to magnetic attraction between unlike charges, and the pushing action is due to magnetic repulsion between like charges.

If we include the valence theory introduced in this chapter, we can

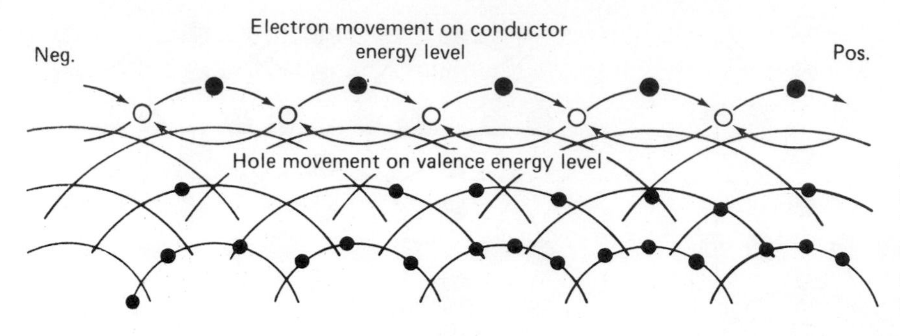

FIG. 3-5 Whenever an electron goes up to the conductor level, it leaves a hole behind. This hole attracts a neighboring electron, which leaves a hole behind that attracts another electron, and so on. The result is current flow.

go on to say that the positive terminal attracts the nearest copper atom's single-valence electrons. This attractive force gives the electrons sufficient energy to move from the so-called valence energy level up to the conductor level (Fig. 3-5). At this energy level, the electrons become free, able to drift under the influence of an EMF toward the battery's positive terminal.

However, something else happens when an electron moves up the conductor level. It leaves behind an empty space or hole in the valence orbit. Since this hole will attempt to capture an electron, it is considered to be positively charged. It will attract negatively charged electrons.

When a shifting electron creates a positive hole, the hole has a tendency to fill itself by attracting another electron from a neighbor atom. Then, when that electron is excited out of its valence energy level another hole is created. That hole attracts another electron, which creates another hole, which attracts another electron, which creates another hole, and so on. Consequently, as electrons move from the negative to the positive side of the circuit along the conductor level, positive holes will move along the valence level from positive to negative. This means that current flow can actually be described in two ways: as the movement of negative electrons and positive holes.

In the early days of automotive engineering, current flow was usually said to flow from positive to negative (although it was not explained in terms of holes). Now, most experts say that current moves from negative to positive. To explain diode and transistor operation, both negative and positive flow concepts are needed.

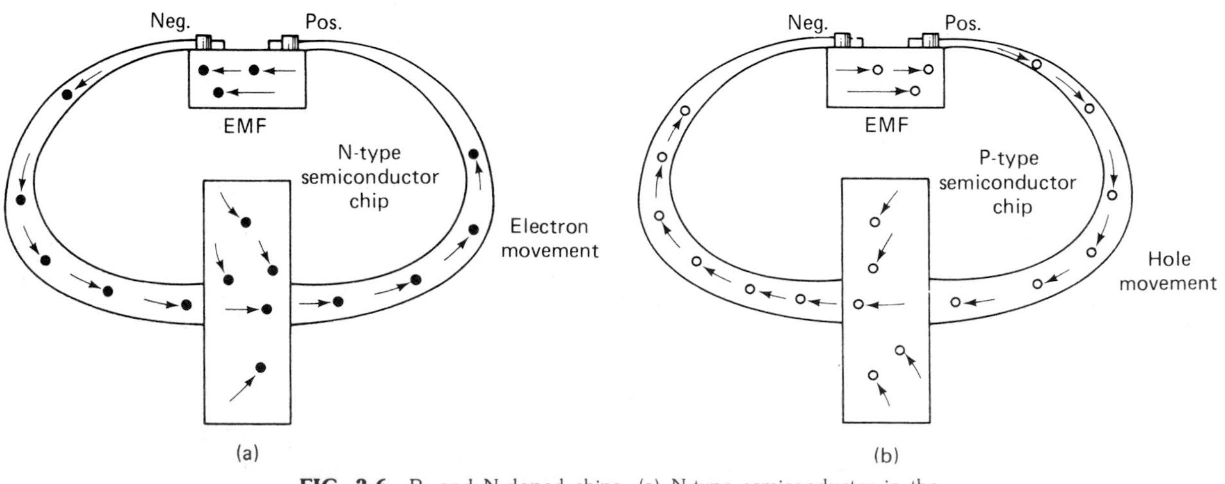

FIG. 3-6 P- and N-doped chips. (a) N-type semiconductor in the presence of an EMF. (b) P-type semiconductor in the presence of an EMF.

CURRENT FLOW IN N- AND P-TYPE SEMICONDUCTORS

Both N- and P-type semiconductors support current flow (Fig. 3-6). In N-type semiconductors, current flow depends primarily on the movement of free electrons contributed by the doping agent. When a source of EMF is connected to an N-type conductor, the negative side of the circuit pushes electrons through the semiconductor and the positive side attracts the free electrons. [Fig. 3-6(a)].

In P-type semiconductors, current flow depends primarily on the empty holes formed by the doping agent. When P-type material is attached to an EMF, the negative side of the circuit attracts the positive charged holes and the positive side repels the holes [Fig. 3-6(b)].

DIODES

In most automotive applications, semiconductors are not used singly. Diodes and transistors may be packaged in individual units or they may be simultaneously formed by the hundreds and thousands on fingernail-size IC chips. Because they are easier to visualize, most of the following examples show individually packaged semiconductor components. ICs, such as those used in on-board computers, are discussed at the end of the chapter.

N- and P-type semiconductors joined in two layers are called *diodes*. Diodes are used in on-board computers. To change or rectify ac to dc current and, with special types of diodes, to help control the voltage output of alternators.

There are two basic ways a diode may be introduced into a circuit. The N half of the diode may be connected to the negative side of the circuit and the P half, to the positive side [Fig. 3-7(a)]. Or, the N half may be connected to positive and the P half, to negative [Fig. 3-7(c)]. Depending on how the diode is connected, it will either allow current to flow or it will act as a barrier to electron movement.

One key to understanding diode (and, later, transistor) operation is to examine the behavior of positive and negative charges at the junction between P- and N-type layers.

When a diode is not attached to a circuit [Fig. 3-7(b)], the positive holes from the P side and the negative charges from the N side are drawn toward the junction. Some charges cross over to combine with

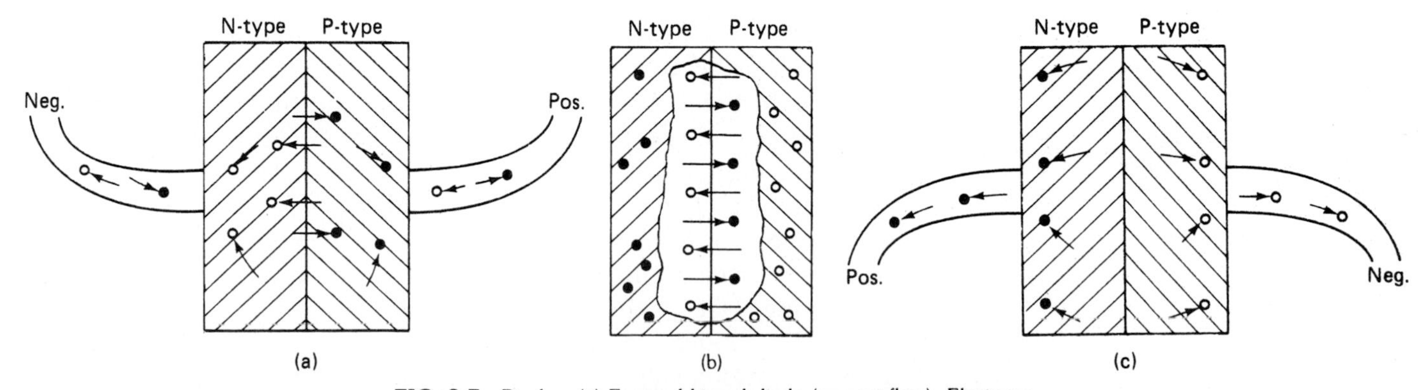

FIG. 3-7 Diodes. (a) Forward-biased diode (current flow). Electrons and holes cross P/N junction. (b) Unconnected diode. Internal EMF due to crossover at junction. (c) Reverse-biased diode (no current flow). Electrons and holes pulled from junction. There is no crossover.

opposite charges from the other side. However, when the charges cross over, the diode halves are no longer electrically balanced. In other words, when an electron from the N side goes over to the P side, it leaves a positive charge behind on the N side. The same kind of thing happens when a hole goes from P to N. Consequently, each half of the diode builds up a network of internal charges opposite to the charges at the PN junction. The attraction (or internal EMF) between the opposite charges tends to limit further diffusion of charges across the junction.

When the diode is attached to an external EMF source, the situation changes. If the diode is connected P to positive and N to negative, there will be current flow. The negative pole will push electrons across the barrier as the positive pole pushes holes across. The diode is said to be forward-biased.

However, if the diode is attached P side to the negative pole and N side to positive, there will be no current flow. The negative pole will attract the positive holes away from the junction and the positive pole will attract the electrons away. As a result, no charges will cross over the junction. The diode is then said to be reverse-biased. These features give the diode the ability to act as an ac to dc rectifier or as a one-way current valve.

SEMICONDUCTOR BREAKDOWN/ZENER DIODES

Most charge movement or current flow in a diode is the result of the impurities added to the parent semiconductor. It is called extrinsic flow because it is external to or apart from the basic atomic structure of the pure semiconductor. It depends on the extra electrons or holes added by the impurity.

Current flow that depends on the electrons provided by the parent semiconductor itself is called *intrinsic flow* because it is intrinsic or basic to the parent material. Intrinsic flow in most semiconductors is limited to the few electrons that can slip along the atomic cracks and flaws within the crystal structure of the parent material. In other words, these few electrons find pathways through the otherwise satisfied crystal network.

Intrinsic flow in semiconductors increases as the temperature goes up (as opposed to metal conductivity, which increases as the temperature goes down). In most cases, the intrinsic current flow in semiconductors does not become significant until the material nears its melting point—when the atoms are vibrating so much that the valence bonds are about to break apart. However, certain carefully prepared, heavily doped semiconductors can be made to conduct intrinsic current at lower temperatures.

Note: The temperature of semiconductors is raised by increasing the voltage impressed against it. The greater the voltage, the greater the force given to electrons moving between atoms and the more the atoms will vibrate. Heat is the result of atomic vibration.

Zener diodes are constructed from semiconductors which will allow intrinsic current flow above certain voltage/heat levels. Below these levels, a Zener diode behaves in a normal manner, allowing only extrinsic, forward-biased current to pass. However, when sufficient, reverse-biased voltage is applied, the diode will "break down" and allow intrinsic current to pass in the opposite direction. Zener diodes are particularly useful in voltage-regulating devices and as protective shunt switches for other circuit components, usually transistors.

TRANSISTORS

Transistors are three-layer semiconductor chips. The two principal combinations are NPN and PNP. In effect, a transistor is made up of two diodes, each sharing a center layer. No matter how the transistor is connected into a circuit, one of the diodes will be reversed-biased and the other forward-biased.

The three layers in a transistor have certain designations. The outside layer of the forward-biased diode (the layer whose polarity is the same as the polarity of the circuit side to which it is attached) is called the *emitter* (Fig. 3-8). The outside layer of the reversed bias diode is called the *collector*. The shared center layer is called the *base*. Each layer—emitter, collector, and base—has its own electrical lead for connecting to different parts of a circuit.

Common materials used for the emitter and the collector are germanium, N doped with phosphorus and P doped with boron. The base section, also commonly made from silicon is usually only lightly doped,

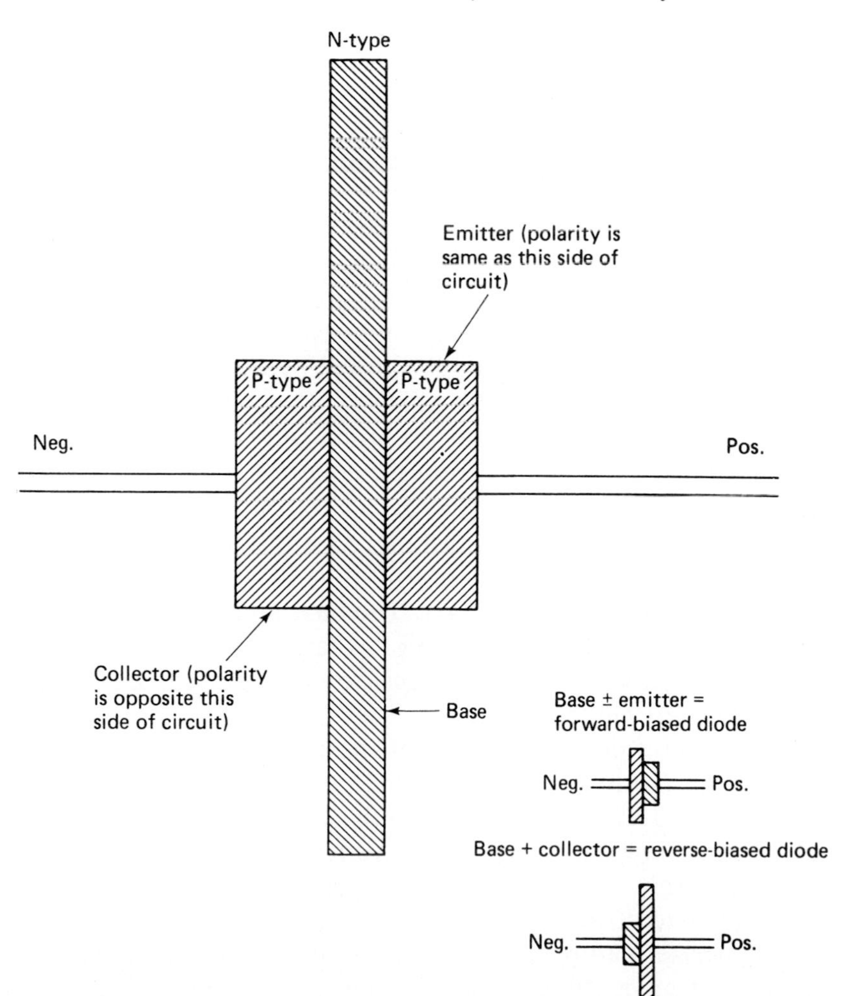

FIG. 3-8 Transistor components.

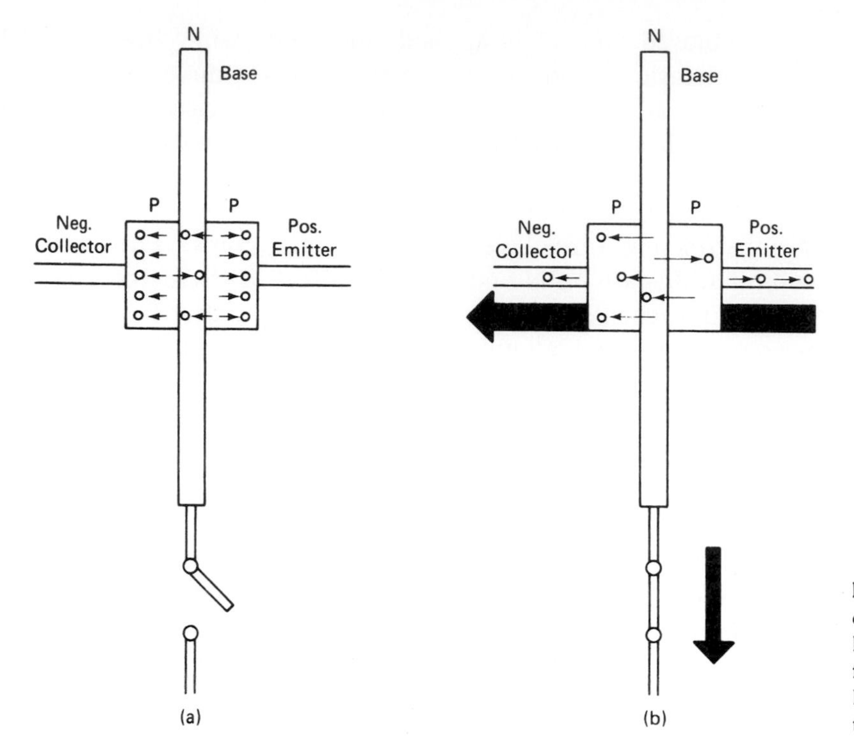

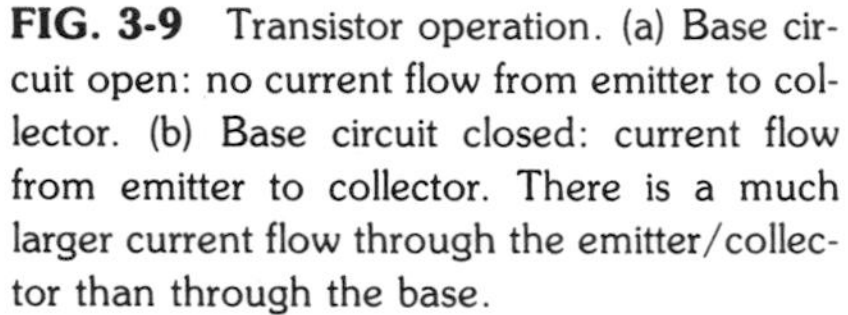
FIG. 3-9 Transistor operation. (a) Base circuit open: no current flow from emitter to collector. (b) Base circuit closed: current flow from emitter to collector. There is a much larger current flow through the emitter/collector than through the base.

just enough to give a certain minimal number of free charges. The base is very thin compared to the other layers. And its lead is usually attached to a surrounding ring somewhat removed from the emitter/base/collector junctions.

The base section provides a key to transistor operation. To see how it works, examine Fig. 3-9. Figure 3-9(a) diagrams a PNP transistor with the lead from its base layer connected to an open circuit leg (which removes the base section from a source of charges). The positive holes in the P-type collector are pushed by the positive charges in the attached circuit to the junction between the collector and the base. On the other side of the transistor, magnetic attraction draws the positive holes in the collector away from the collector base junction. Many (of the already limited number) of free electrons in the base section are drawn to the emitter/base junction. As a result of this counterbalancing arrangement of charges, few positive or negative charges can pass completely across the base layer. The base layer does not have enough electrons to support anything more than a minor current "leakage" through the layers. And the transistor assumes the character of the reverse-biased diode pair.

Current will flow only when the base circuit is completed [Fig. 3-9(b)]. Then it provides sufficient electrons to support hole movement from the emitter across the base to the collector. However, the amount of current flowing through the layers is different. Because the base layer is so thin and because the base circuit is attached to a ring relatively removed from the emitter-base junction, the holes speeding out of the emitter can pass more easily into the collector. Only a limited number of holes will go from the emitter through the base.

The base circuit acts as a control for the emitter-collector circuits. When the base circuit is open, no current passes; and when it is closed, current flows. And because few charges are allowed to pass through the base, a limited amount of current flow in the base can be used to control much heavier current flows in the emitter circuit.

Using this feature, transistors can serve two main purposes. They can act as amplifiers, converting low-voltage control signals (at the base) into higher voltage and current patterns passing through the emitter-collector circuit of the transistor. The amplification is used in radios to increase weak signals produced by radio waves. Transistors can also act as selectable switches (off–on switches). Digital computers, such as those that control many operations in automobiles, depend upon these features.

Development of integrated circuits

One of the most significant aspects in the history of semiconductors has been the continuing reduction in the size of components. From the beginning, semiconductors have always had the potential for being very small because most of the action takes place in the microscopic regions near oppositely doped elements. In general, any part of a semiconductor not associated with this activity is wasted space.

The problem has been finding ways to take advantage of this potential for miniaturization. Conventional fabrication techniques do not work because mechanical assembly operation, no matter how small the assembly tools used, would still dwarf the components in even a moderately miniaturized circuit.

So, the first semiconductors, although much smaller than the components they replaced, still had a lot of wasted space. An example of these so-called discrete components is a point contact–type diode (not to be confused with old-style automotive ignition points). It is manufactured by dipping wires into molten blobs of N- and P-doped semiconductor material and then fusing the blobs together to create an NP junction (Fig. 3-10).

The real breakthrough in miniaturization occurred when scientists adapted two unrelated technologies to the production of semiconductor devices. One technique came from the manufacture of mirror backings. Uniform but very thin backing can be obtained by exposing the smooth glass surface to a cloud of vaporized metal. The metal is deposited on the glass in a manner not possible with any sort of manual application technique. Using a highly modified version of this process, electronics engineers can create microscopically thin layers of doped semiconductor materials. They can also form conductive metal layers to replace conventional wires as well as nonconductive layers to replace ordinary electrical insulation.

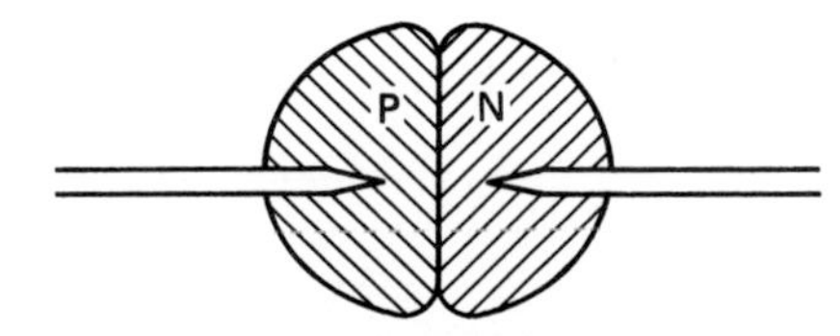

FIG. 3-10 N- and P-doped semiconductor material, fused together.

Of course, producing uniform layers of semiconductors, conductors, and nonconductors does not by itself result in practical electronic devices. Some means is necessary to create organized regions in the different layers of the different materials and to do it on a very small scale.

The other technique comes from the application of certain photographic principles. Photographers and photoengravers have long possessed techniques for selectively removing small amounts of materials from light-sensitive materials. By applying similar principles, scientists are able photographically to etch tiny circuit patterns in vapor-deposited layers. The process may involve drawing a large-scale picture of the circuit (perhaps room size) and then taking a photograph of the drawing. The photograph is reduced until it is the size of the actual circuit. The reduction is used as a light mask for selectively exposing

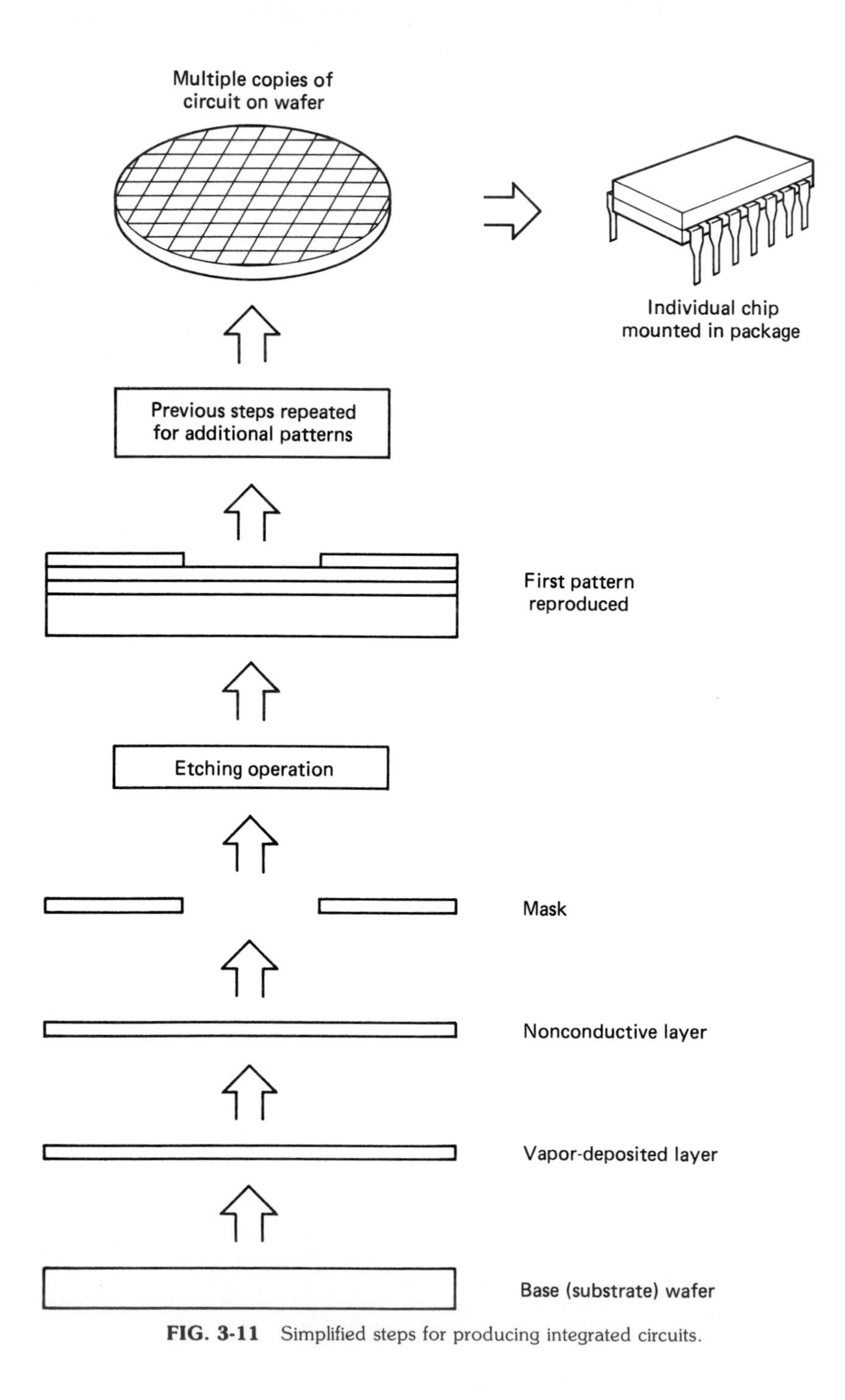

FIG. 3-11 Simplified steps for producing integrated circuits.

the portion of the vapor-deposited material to be etched away. Masks can also be used for selective vapor deposition.

Following is a series of simplified steps for producing integrated circuits (Fig. 3-11):

1. First a rod of doped semiconductor material is "grown." In many cases, the rod is about 3 inches in diameter.
2. A thin, wafer-like section is sliced off the rod. This wafer is used to create 100 to 1000 or more identical circuits, all at the same time.
3. The wafer, or substrate, is exposed to a cloud of oppositely doped semiconductor material. The result is a two-layer semiconductor.
4. The vapor-deposited layer is exposed to an oxygen-rich atmosphere. As the oxygen reacts with the semiconductor, a third, nonconductive layer of material is formed.
5. A mask, corresponding to the first part of the circuit to be formed, is placed over the nonconductive layer.
6. Reacting to light shining through the clear parts of the mask, the exposed portions of the nonconductive layer are chemically altered.
7. The nonconductive layer is subjected to an etching operation. The chemically altered portions are etched away; the other portions of the layer are unaffected. As a result, the pattern of the mask is reproduced on the wafer. Depending on how many times the basic pattern is repeated on the mask, 100 to 1000 or more copies of the circuit are reproduced simultaneously.
8. The structure may now be exposed to another vapor-depositing step to create "wells" of oppositely doped semiconductor material in the parts exposed by the first etching operation.
9. The masking, etching, and depositing steps are repeated until the desired circuit pattern has been created.
10. One of the last steps is to join the semiconductor regions electrically. This is done by etching out connecting patterns, and then vapor-depositing a grid of conductive metal. The resulting pattern serves the same purpose as wires in a mechanically produced electrical device.
11. After the last operation, the wafer is broken along scored lines to separate each individual chip.
12. Finally, the chip is placed on a nonconductive supporting package. The conductive ends of the chip circuit are joined to connector pins protruding from the package.

SOLID STATE DEFINED

The preceding process for producing chips is often called *planar technology,* and the results are usually referred to as *solid-state electronics.* The term planar is employed because all the components are created in single planes. Solid state refers to the difference between semiconductor devices and the vacuum tubes used previously. Quite simply, semicon-

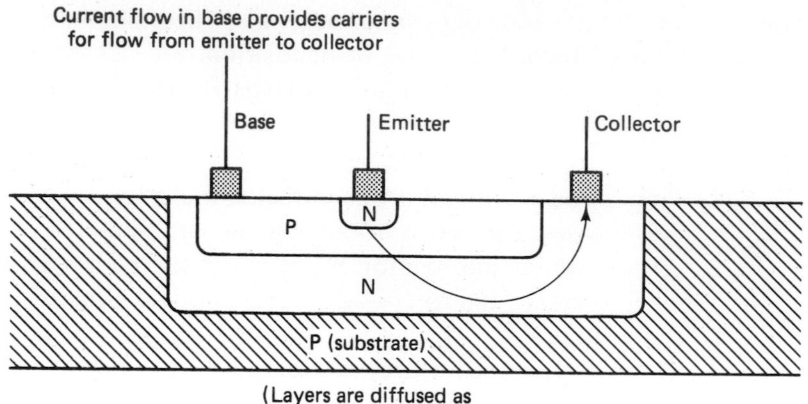
Current flow in base provides carriers
for flow from emitter to collector
Base
Emitter
Collector
P
N
N
P (substrate)
(Layers are diffused as
"wells" into P substrate)

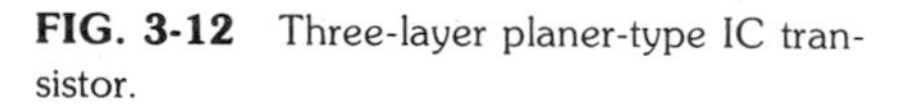
FIG. 3-12 Three-layer planer-type IC transistor.

ductors are solid and compact, whereas vacuum tubes are hollow and relatively bulky. The number of components formed on a single circuit determines the degree of integration. The acronym LSI means *large-scale integration* and describes circuits containing many elements. Very large scale integrated (VLSI) circuits may contain hundreds of possible circuits.

Most semiconductor components used in computers are contained on ICs. A single IC chip may contain part of the computer's memory and another, the control processor unit (CPU). In some cases an entire working computer might be fabricated on a single chip.

Bipolar versus mosfet devices

Within the broad category of planar, solid-state technology, there are two main groups of semiconductor devices: bipolar and field-effect. (The latter may be referred to as MOSFET, describing the metal oxide silicon commonly used to create field-effect semiconductors.) The first mass-produced semiconductors were of the bipolar variety. They depended upon the action of both holes and electrons at the junction between N and P layers. Earlier in this chapter we saw how individually packaged bipolar transistors work. Planar-type transistors found in ICs work the same way, but they look different.

Notice the three-layer (NPN) planer-type IC transistor shown in Figure 3-12. The two outside (N) layers are the emitter and collector. The center (P) layer is the base. As in individually packaged units, when no current flows in the circuit attached to the base layer, no carriers are available to support the passage of electrons from the emitter to the collector. Therefore, current is blocked. However, when current does flow in the base circuit, carriers become available for electron movement between the emitter and collector. Note Fig. 3-12.

Figure 3-13 shows a PMOS (positive metal oxide silicon) field-effect IC transistor. It also has three main elements, called the *source, drain,* and *gate.* As in a bipolar transistor, the flow of current from the source to the drain is controlled by the activity in another circuit—the gate circuit. When no current is present at the gate, current cannot flow from the P-doped source through the intervening N-doped region to the P-doped drain. However, when current is flowing through the gate,

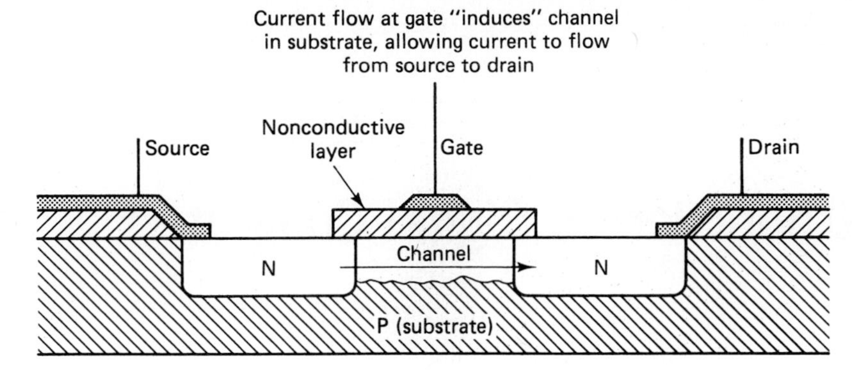

FIG. 3-13 A PMOS field effect IC transistor.

the resulting electromagnetic field "induces" a conductive channel through the N-doped region. As a result, current can flow from the source to the drain (Fig. 3-13).

MOS chips are relatively recent developments in planar technology. They are made possible by improvements in diffusion and etching processes. Compared to bipolar products, MOS chips offer several advantages and disadvantages.

On the positive side, MOS transistors generally use less power than bipolar transistors. This means less heat is produced, since fewer electrons are jostling through the circuits, disturbing other atomic elements. As a result, more components can be packed into smaller spaces. MOS circuits also require fewer masking steps. 10 or fewer for MOS, versus 10 or more for bipolar devices.

On the negative side, MOS components are generally slower than bipolar devices. Speed in this case refers to the time required to switch a transistor on or off. Transistor speed, which is one of the limiting factors in computer operation, is usually stated in terms of nanoseconds (ns). A single nanosecond is one billionth of a second. Automotive on-board computers use a number of MOS-type chips.

4

Introduction to Digital Computers

GENERAL

This chapter introduces you to the general theory and operation of the digital computers used in modern automobiles. Strictly speaking, it is not necessary to understand these computers in order to service the systems they control. Defective automotive computers are never repaired at the local level. They are treated as "black boxes," simply to be replaced should a malfunction occur. However, understanding something about digital computers will help you acquire a comfortable feeling for current and future automotive systems. Without some minimal understanding of what is going on, you will remain totally in the dark, working strictly on a "monkey-see, monkey-do" basis.

DATA REPRESENTATION

What are computers? They are often described as information processors. However, if you think about it, such a statement does not make any sense. How can a machine perform a physical action on something as insubstantial as information? The answer is that computers process the physical conditions that stand for information. Even though a machine cannot add 2 and 2 to get 4, it can combine two sets of physical conditions that represent 2 and 2 and come up with a result that represents 4.

Generally, there are two ways to represent information, analog and digital. Automobiles use both types of representations. The following paragraphs discuss the differences.

ANALOG DATA REPRESENTATIONS

Analog devices represent information as measurable shapes, movements, or conditions. For instance, the contour of a camshaft repre-

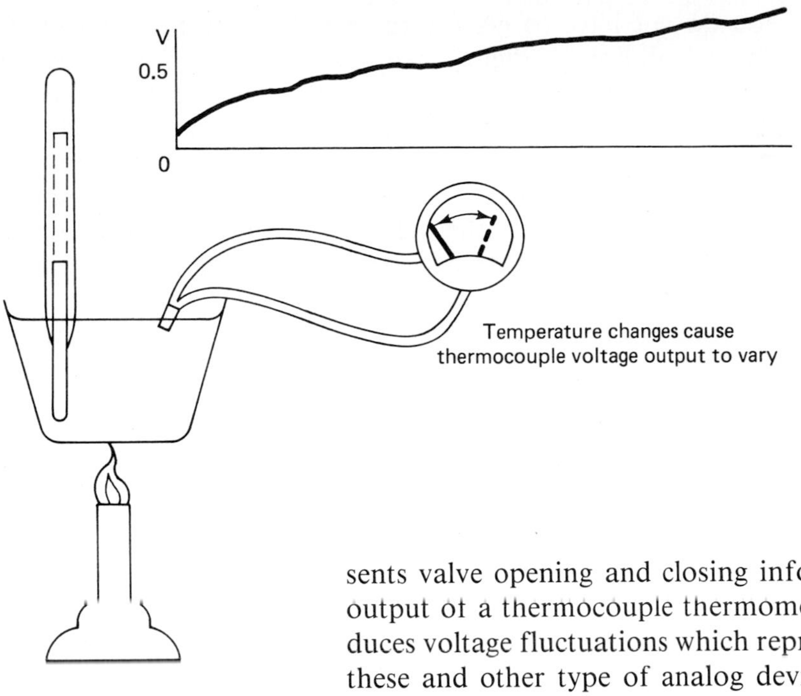

FIG. 4-1 Voltage fluctuations equal information.

sents valve opening and closing information. Another example is the output of a thermocouple thermometer. As shown in Fig. 4-1, it produces voltage fluctuations which represent temperature information. In these and other type of analog devices, the representation is directly analogous to the information—hence the term analog.

Most information coming into and leaving automotive computers is an analog representation because the real world is an analog place. However, once it gets inside the computer, analog data is changed to digital form. Figure 4-2 depicts the relationship between analog and digital data representations in an automotive computer.

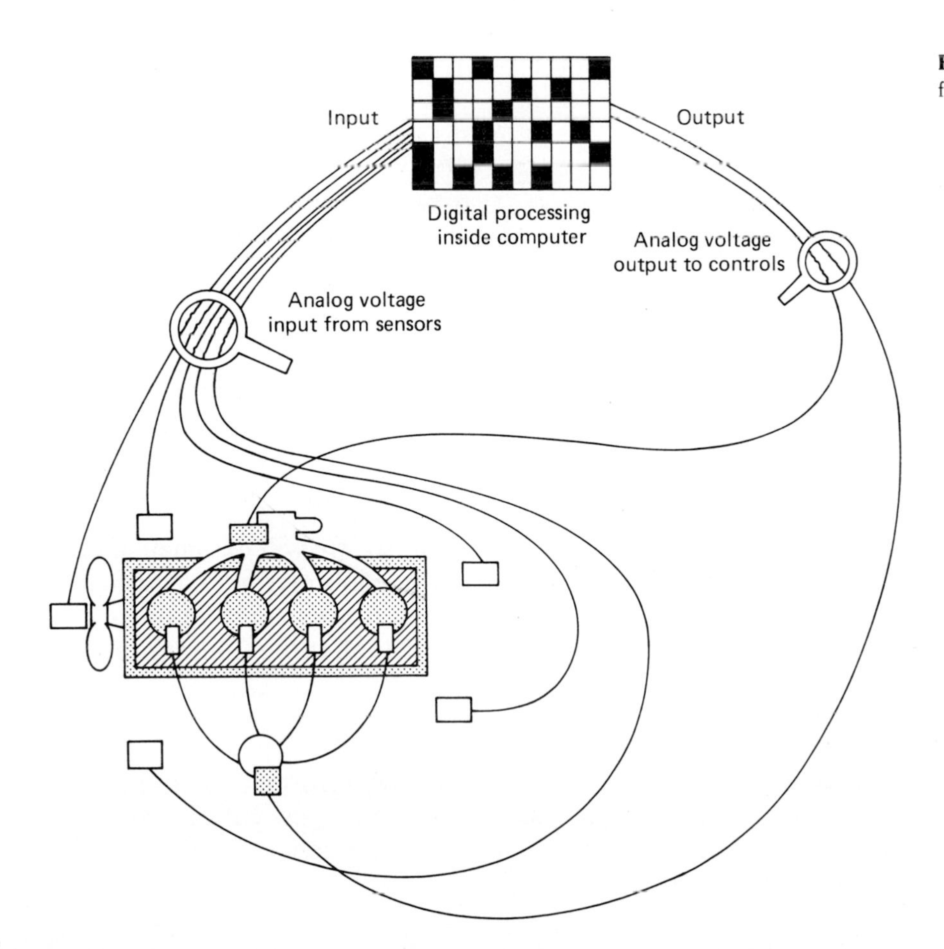

FIG. 4-2 Analog and digital information flow.

DIGITAL DATA REPRESENTATIONS

You do not have to go any further than the tips of your fingers to find an example of digital data representation. Patterns of raised and lowered fingers have been used since time immemorial to represent digital data. In fact, the word *digital* came from the word *digit,* which means, among other things, finger or toe. Children use fingers as number processors. Adults use fingers to express attitudes and to prescribe courses of actions for other adults. In both cases, real-world information has been reduced to an abstract set of codes—in this case, finger patterns.

So, on a simplified level, we can say that human fingers comprised the first digital computer. Carrying the example further, we can compare modern, digital computers to collections of hands and fingers packed into very small spaces. Such a truly "digital" system might even work, aside from some problems with care, feeding, and worker morale. Before going on, let us see how the design and operation of this imaginary system might be handled.

DESIGNING A DIGITAL COMPUTER FROM DIGITS

The basic building block of our digitized system is naturally enough, a digit, or finger. Certain collections of fingers are used to perform the following various computer functions.

Data representation

As noted at the beginning of this chapter, we need to represent information in some physical manner. That is easy in our manual system. We will use up and down finger patterns to represent numbers and letters, the fingers on two hands standing for one character. For example, all fingers down equals 0, one up equals 1, two up equals 2, and so on.

Data storage

Finger patterns are used to store two kinds of information, data to be worked on and the directions for doing the work. The directions are called a *program;* the data are simply called the *data.* The place where the information is stored is called *memory.* In our system, memory can be visualized as a field of waving fingers, something like stalks in a field of grain. So that information does not get lost, address locations are given to each section of this memory field.

Data processing

We now need a place to process the data, so, we will designate another field of fingers as the processor. Addition, subtraction, counting, and other data manipulation occur as the tiny fingers curl and uncurl. Because this field is centrally located, we will call it the *central processor.*

Data transfer

An interesting point arises. How do we get information to and from memory and the central processor? We cannot just cut off the fingers and pass them back and forth. In computer terms, that would be a very destructive read operation. However, it is not really the fingers themselves we are interested in. It is the information represented by the pattern of raised and lowered fingers. The solution becomes apparent. We will use a connecting road. Like the memory and central processor regions, this road is made up of waving fingers. If we want to get information from memory, the appropriate patterns are passed down the road in braille or by touch. The process might look like wind rippling across a field of grain (or like electrons rippling along the valence orbits of a conductor). We will call this transfer component the *data/address bus*.

Data synchronization

Another factor needs to be considered. We cannot have fingers waving and wagging at random, one group starting up here and another group starting up later somewhere else. Things must happen in unison. To get the necessary rhythm, we will designate one or two of the biggest sets of fingers and hands as the beat keeper. At regular repeating intervals, they will pound the ground, producing a thump, thump sound. All operations will be done in time with this signal, which we will call the *clock function*.

Data input/output

Finally, our computer must be able to communicate with the outside world. For that we will add a person. The person uses touch to pass information to and from the address/data bus and uses voice to communicate with the world beyond the computer. We will call this person the I/O interface.

Figure 4-3 pictures the components in the digited digital computer.

OPERATING THE DIGITED, DIGITAL COMPUTER

Now that we have designed our living computer, let us see how it works. Suppose, that for some reason, you cannot remember how to add 2 and 3. However, strangely enough, you remember how to operate the computer.

First, in order to get started, we need to put the numbers somewhere. You tell the I/O interface to store the number 2 at memory address A and the number 3 at address B. The I/O interface passes this information to the wagging fingers in the data/address bus; there is a ripple of movement, then suddenly, you see that memory address A has two fingers sticking up and address B has three fingers sticking up.

Now, you need a program or some directions for performing the operation. You come up with these steps:

1. Move the pattern (not the fingers) stored at address A to the central processor.
2. Move the pattern stored at address B to the central processor.
3. Combine the two patterns.
4. Move the result (the answer) to memory address C.

This is your computer program. Since it is also information, you store the steps at addresses D through G.

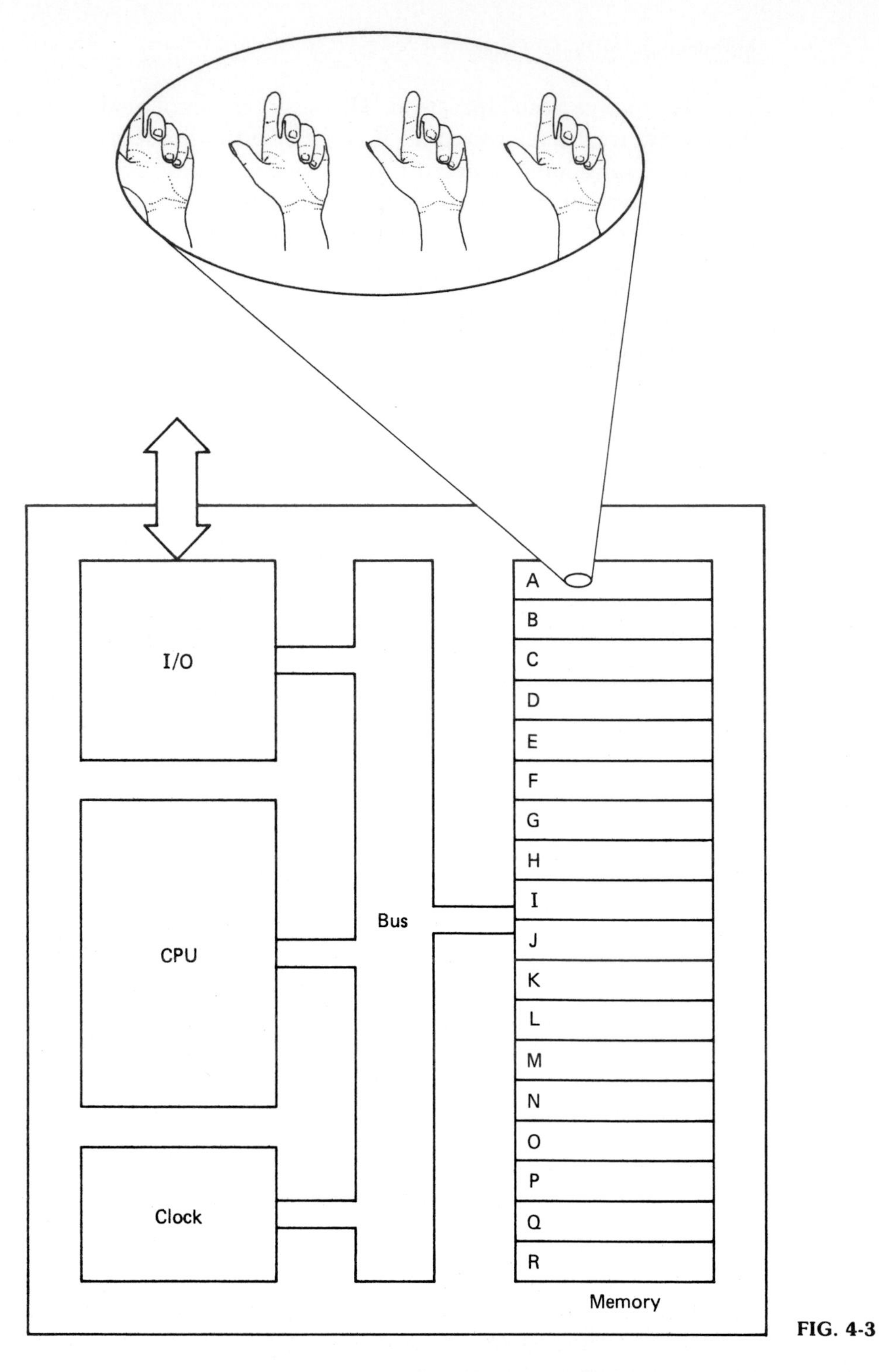

FIG. 4-3

Finally, you tell the I/O interface to tell the central processor to run the program. There is a flurry of activity along the address/data bus as the processor requests the contents of addresses, performs the actions indicated, and passes the result back into memory. After the last beat of the clock, when everything is still, you ask the I/O interface, "What is the answer?"

This disembodied spectre passes a silent request down the address/data bus. Five fingers appear at address C, there is a returning wave down the bus, and the I/O interface solemnly announces, "The answer is five." You wonder, "Why didn't I think of that?"

Now that the computer is complete and has been demonstrated to work, there is one last thing to do. The ADD function we created is so powerful and so useful that we will store it directly in the central processor. We will also create a number of other basic operations, such as SUBTRACT, COMPARE, MOVE, and store them directly in the central processor as well. These instructions will become the building blocks of other, more complex programs. We will call this collection the *instruction set* of the central processor.

LOOKING AT REAL COMPUTERS

The elements of the imaginary, digited digital computer are similar, in purpose at least, to the main components of an actual electronic computer.

DATA REPRESENTATIONS

In the imaginary computer, patterns of raised and lowered fingers represent information. In a real computer, information is represented by electrical patterns. Looking at electrical flow through a wire, a square-shaped voltage pulse serves the same purpose as an upraised finger.

The patterns in a digital computer are binary in nature. Things occur in twos. A switch is on or off; voltage is present or not present; current is flowing or not flowing. Binary is ideal for machines. Figure 4-4 compares binary and decimal representations.

The basic element in a binary circuit is a 1 or 0, or, in computer terms, a *bit*. In microcomputers, 8 bits are called a *byte*. Using standard coding schemes, the pattern of bits in a byte is used to represent a single alphabetic character, number, or special control character. The most

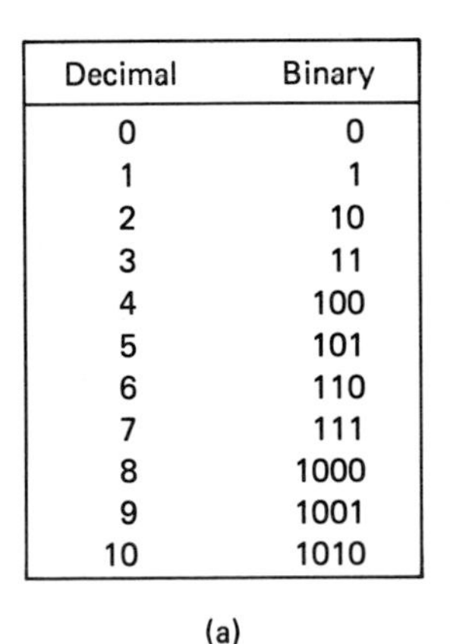

Decimal	Binary
0	0
1	1
2	10
3	11
4	100
5	101
6	110
7	111
8	1000
9	1001
10	1010

(a)

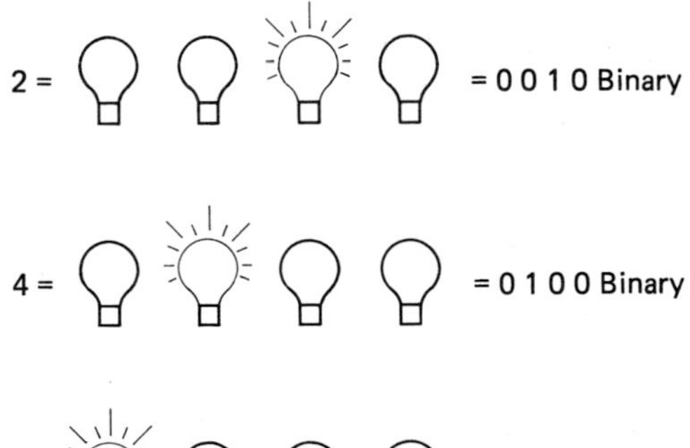

(b)

FIG. 4-4 (a) Comparing decimal and binary numbers. (b) Each "off" bulb equals 0, each "on" bulb equals 1. Binary numbers, or "bits," can be represented by any two-state (on-off) condition.

common microcomputer coding standard is ASCII, which stands for American Standard Code for Information Interchange.

There are three common ways to physically represent binary information: (1) as patterns of voltage pulses moving along a conductor, (2) as patterns of tiny on–off switches in electronic circuits, and (3) as patterns of polarity in magnetic media.

We do not have to worry (yet) about magnetic media in automotive systems. Typically, only general-purpose microcomputers use magnetic storage media, such as diskettes, fixed disks, and tapes. However, we do need to be concerned about voltage and circuit patterns. The next sections discuss these topics.

VOLTAGE PATTERNS

Compared to analog voltage representation, digital voltage patterns are square-shaped and regular. In other words, the transition from one voltage level to another is very abrupt. Analog signals, on the other hand, tend to be irregular because the real-world data which the analog signal directly represents also tends to be irregular. You can look at an analog pattern and get some idea of the condition being represented. A digital signal is an abstract code, with voltage pulses standing for binary 1s or 0s, which, in turn, stand for ASCII characters or symbols. Figure 4-5 shows digital voltage patterns. Figure 4-1 at the beginning of the chapter shows analog patterns.

SERIAL VERSUS PARALLEL

There are two ways to pass digital voltage patterns along a conductor, in a serial or parallel manner. Using serial transmission, the voltage pulses in a byte follow one another along a single wire (or conductive path). Using parallel transmission, the bits proceed at the same time along multiple, parallel wires, one wire for each bit.

Parallel transmission is faster; serial transmission uses fewer wires and is less expensive. Most communication between the components within a computer is parallel. Data and address buses transmit data in a parallel fashion.

Most (although certainly not all) communication between com-

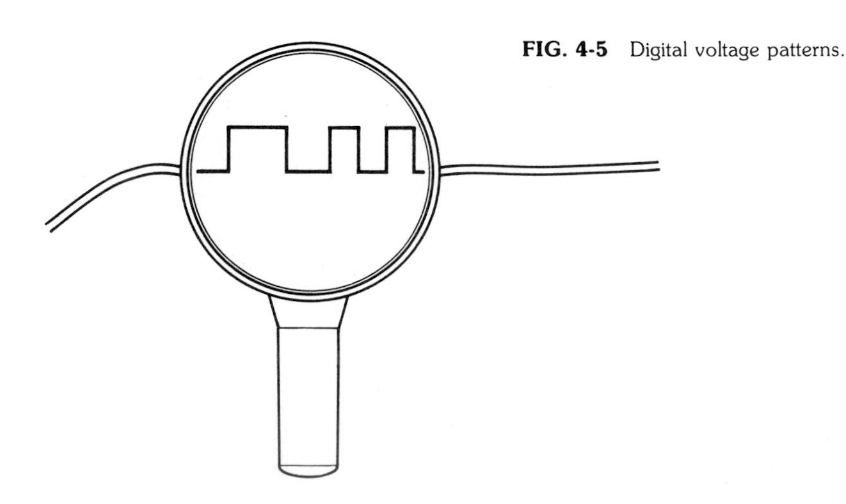

FIG. 4-5 Digital voltage patterns.

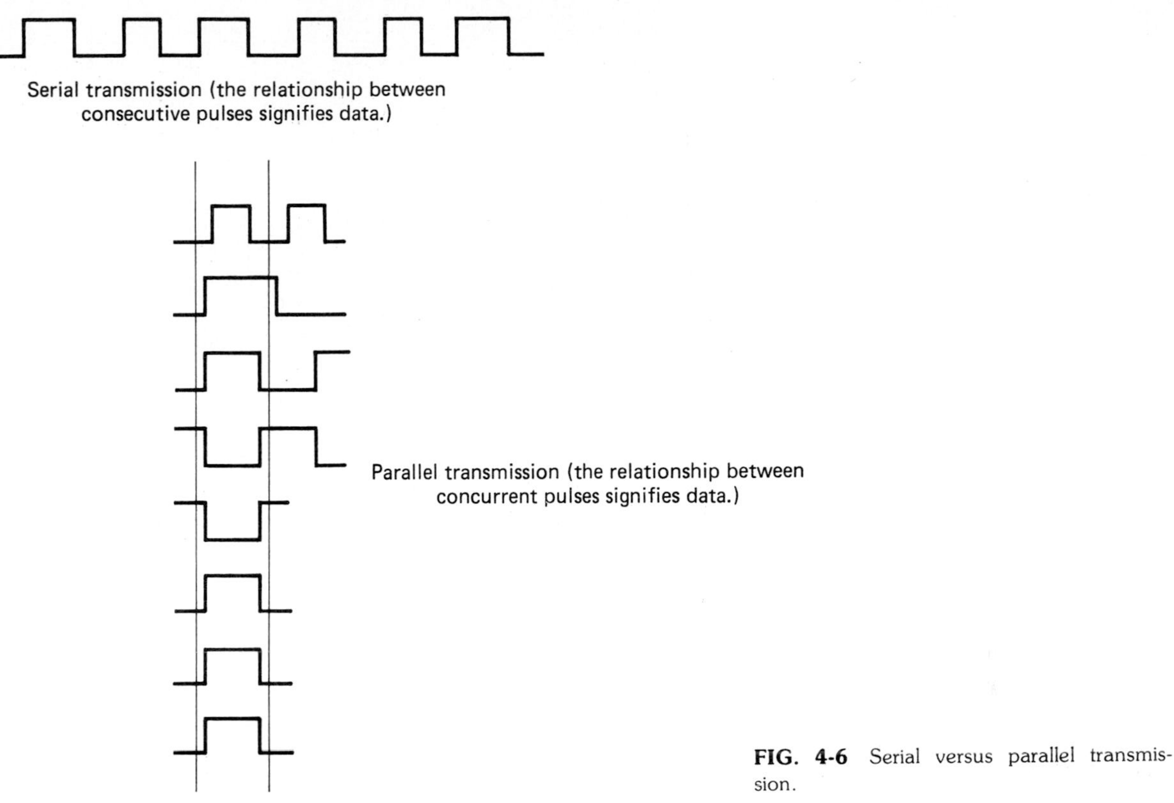
Serial transmission (the relationship between consecutive pulses signifies data.)
Parallel transmission (the relationship between concurrent pulses signifies data.)

FIG. 4-6 Serial versus parallel transmission.

puters and devices in the outside world is serial. For example, automotive engine-control computers often have a serial port, which they use for communication with a separate diagnostic computer located apart from the vehicle. Figure 4-6 compares serial versus parallel transmission.

ELECTRONIC CIRCUITS

ORGANIZATION OF COMPONENTS

You cannot understand a computer in total; nobody can. However, you can understand how the components are organized. Everything begins with gates. They are the basic building blocks of computers. Gates are combined into functional groups. These groups are incorporated into the circuitry of chips, sometimes one function per chip, sometimes many. Chips are assembled onto boards, again by functional groups. Depending on the complexity of the computer, there may be one board or several. The boards are then packaged into an enclosure. Although the actual computer stops here, the description of the computer may be extended to include input and output devices.

GATES

As indicated earlier; the fundamental component for manipulating the voltage pulses that stand for binary 1s and 0s is the gate. The pattern of pulses coming into a gate determine the pattern of pulses leaving the gate. The following figures and paragraphs describe various kinds of basic gates. The *truth tables* that accompany each gate show all possible outputs from the gate for all the possible inputs.

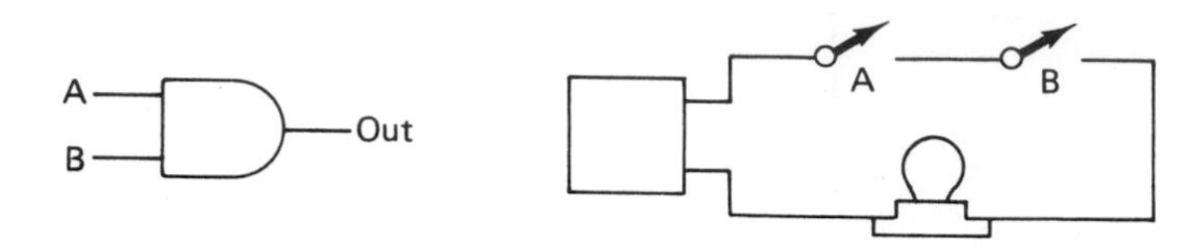

AND gate, like a series circuit

Truth table

A	B	Out
0	0	0
0	1	0
1	0	0
1	1	1

FIG. 4-7 AND gate.

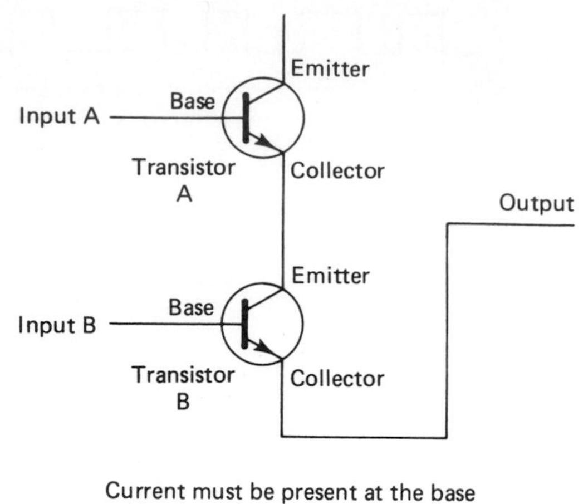

Current must be present at the base of both transistors A and B before any current flow is present at the output of the AND gate

FIG. 4-8 AND gate circuit.

AND gate: The AND gate has two inputs and one output. As noted in the truth table, the output is high (a positive voltage pulse) only if both inputs are high. An AND gate is something like a circuit with two switches wires in series. Figure 4-7 shows the symbol for an AND gate. It also shows a series circuit used to obtain the AND junction. Figure 4-8 shows how two transistors are used to create an AND gate.

OR gate: A high signal at *either* input produces a high output in an OR gate. As shown in Fig. 4-9, the OR gate is somewhat like a parallel circuit.

NOT gate: A NOT gate, as shown in Fig. 4-10, simply reverses what comes in. High results in low output; low results in high. The NOT gate is called an *inverter.*

NAND and NOR gate: Using a NOT gate, the functions of AND and OR gates can be reversed. The results, as shown in Fig. 4-11 are called NAND and NOR gates.

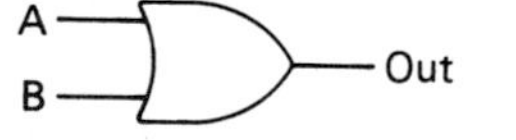

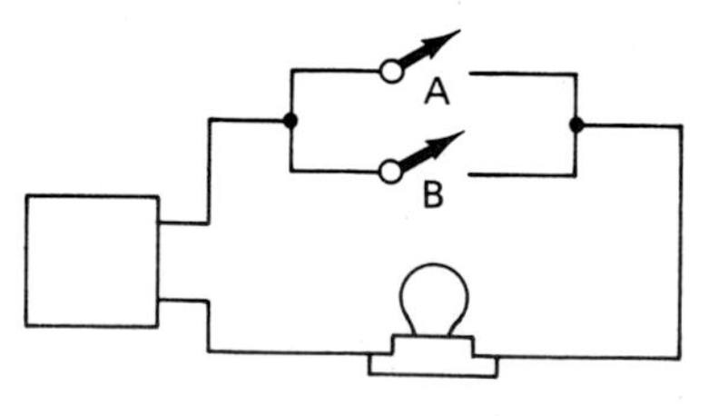

OR gate, like a parallel circuit

Truth table

A	B	Out
0	0	0
0	1	1
1	0	1
1	1	1

FIG. 4-9 OR gate.

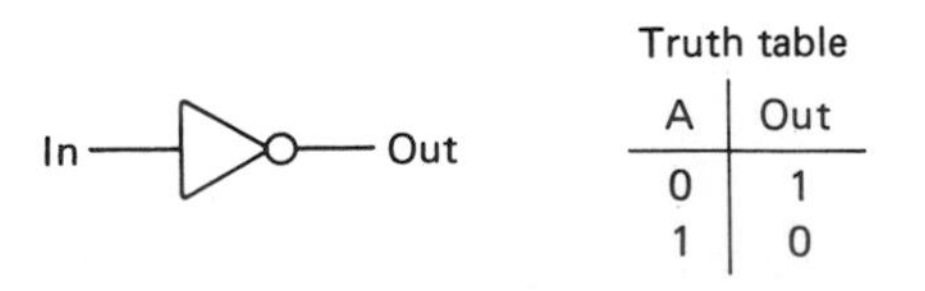

Truth table

A	Out
0	1
1	0

FIG. 4-10 NOT gate.

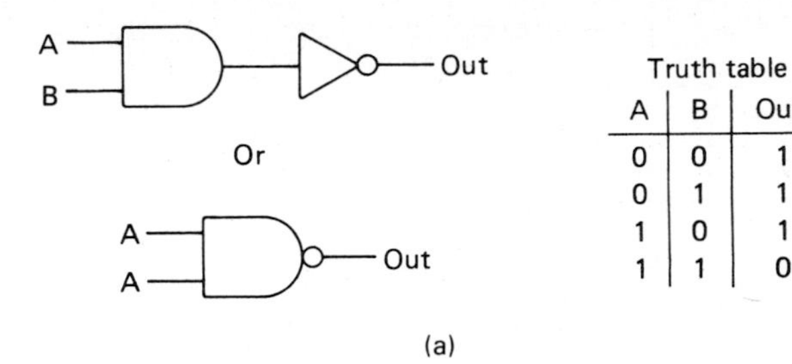

Truth table

A	B	Out
0	0	1
0	1	1
1	0	1
1	1	0

(a)

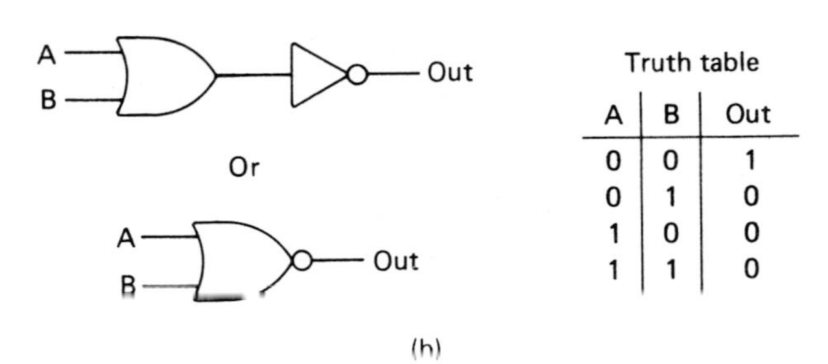

Truth table

A	B	Out
0	0	1
0	1	0
1	0	0
1	1	0

(b)

FIG. 4-11 (a) NAND gate, (b) NOR gate.

EXCLUSIVE-OR gate: Shown in Fig. 4-12, this combination of AND and OR gates produces a true output only when both inputs are different.

GATE FUNCTION GROUPS

The gates just described are combined to perform the next level of processing. Here are some of the resulting combinations.

Full adder circuit

The *full adder circuit* is a combination of AND and EITHER-OR gates, as shown in Fig. 4-13, and can be used to add binary numbers. Individual bits (voltage pulses) enter at A and B. Carry bits from a previous addition enter at the carry-in point. The carry-in bit is combined with the bits at A and B to produce a sum and a carry-out bit.

A
B
Out

Symbol for exclusive — OR gate

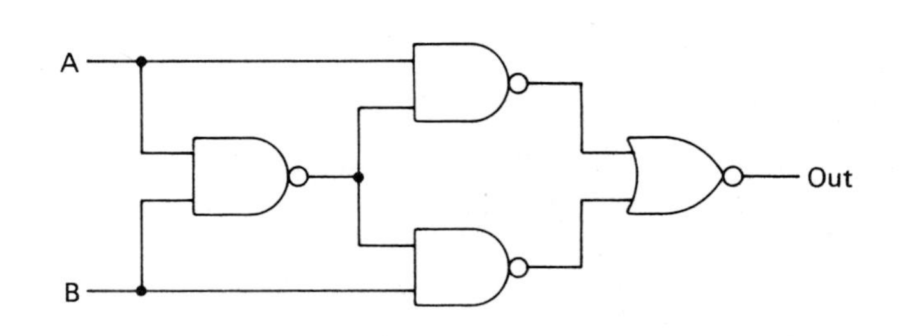

Truth table

A	B	Out
0	0	0
0	1	1
1	0	1
1	1	0

FIG. 4-12 EXCLUSIVE-OR combination.

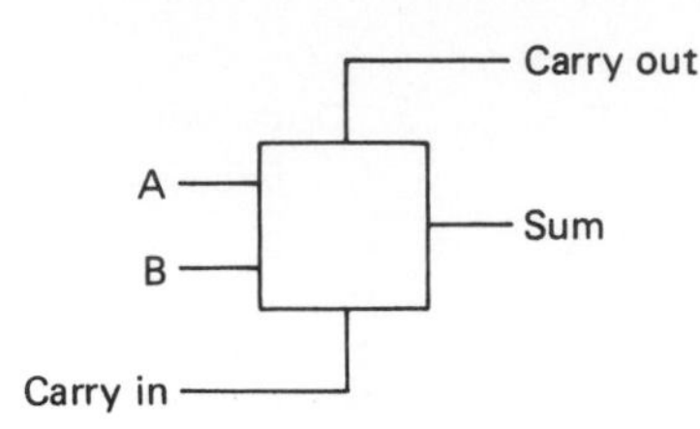

Symbol for full adder

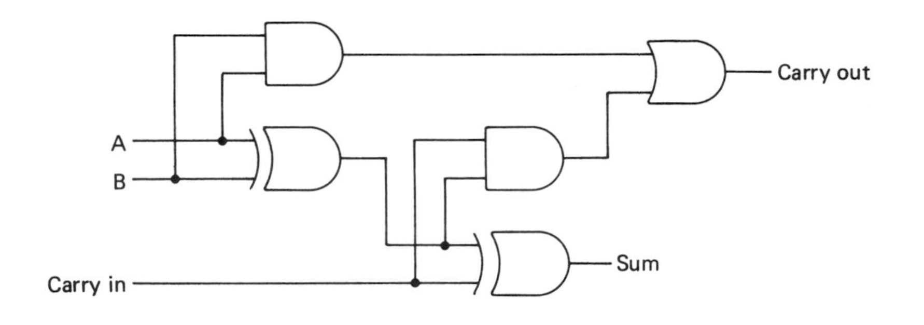

Truth table

A	B	Carry in	Carry out	Sum
0	0	0	0	0
0	1	0	0	1
1	0	0	0	1
1	1	0	1	0
0	0	1	0	1
0	1	1	1	0
1	0	1	1	0
1	1	1	1	1

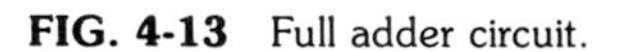

FIG. 4-13 Full adder circuit.

Decoder

Created from combinations of AND gates, *decoders* are used to provide certain output for a given combination of inputs (Fig. 4-14). For instance, suppose you want to turn on a switch when a given temperature is reached. Representing temperature information as binary patterns, a decoder could be set to produce a true (positive) output on receipt of the correct bit pattern.

Multiplexer

A decoder has to examine all its inputs before making a decision about the output. A *multiplexer* is able to examine one of many inputs. In the illustration shown in Fig. 4-15, the bit pattern at DCBA determines which input line is examined. Of course, a program has to provide the bit pattern that tells the multiplexer which line to examine.

For the most part, the components just examined are used to process data. They are typically known as combinational logic devices, since they operate based on a combination of inputs at one time. Flip-flops employ what is known as sequential logic. The output from the flip-flop is determined by the sequence of inputs. Stated differently, a given input affects the output produced by the next input. This feature make flip-flops ideal for storing bit patterns.

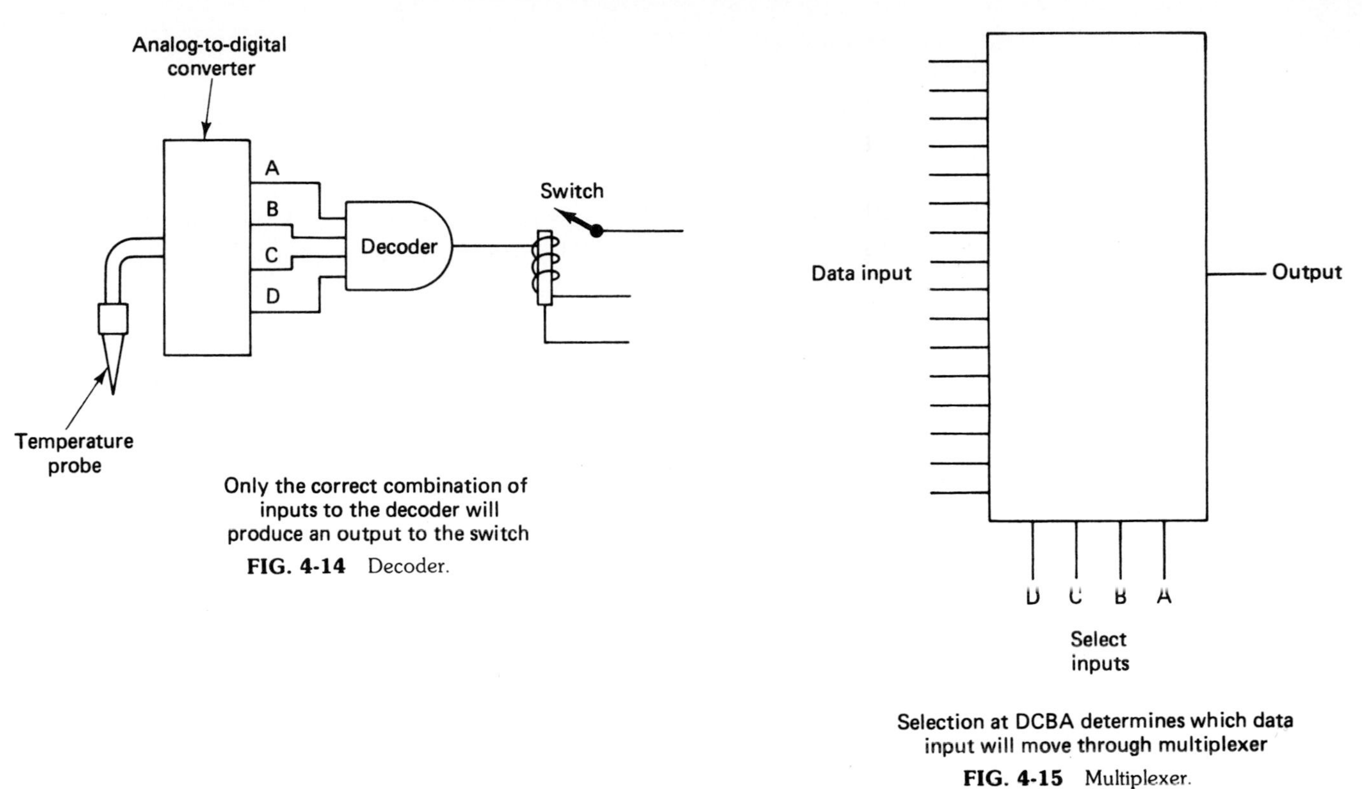

Only the correct combination of inputs to the decoder will produce an output to the switch

FIG. 4-14 Decoder.

Selection at DCBA determines which data input will move through multiplexer

FIG. 4-15 Multiplexer.

Basic RS flip-flop

Various kinds of flip-flops are used. Figure 4-16 pictures a typical example. It is called a basic *RS* flip-flop. As noted on the accompanying truth table, its output changes every time a certain pattern of inputs appear.

Clocked flip-flop

Other kinds of flip-flops respond to one or more inputs plus a clock signal (Fig. 4-17). As noted in the example of the digital computer formed from digits, a clock signal is a pattern of pulses that time the action of many units throughout the computer. Most operations require a clock signal so they will occur in the proper order.

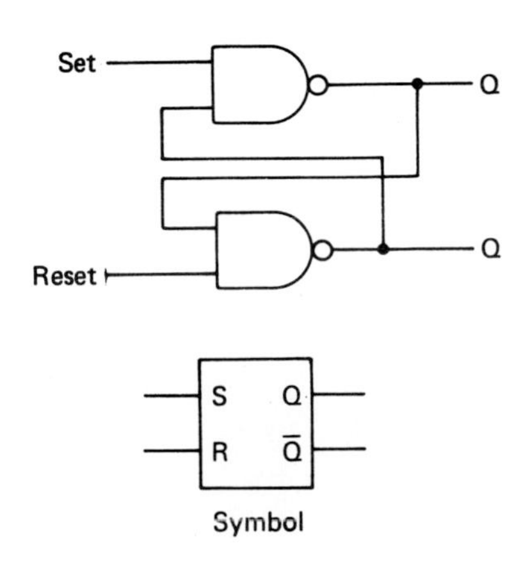

Truth table

S	R	Q	Q̄
0	0	Disallowed	
0	1	1	0
1	0	0	1
1	1	No change	

FIG. 4-16 *RS* flip-flop.

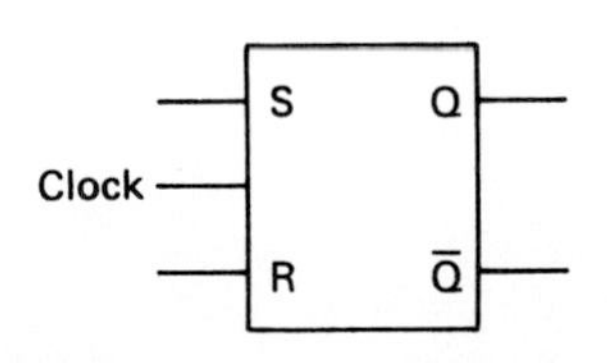

FIG. 4-17 Clocked *RS* flip-flop.

Registers

Flip-flops acting in a bucket-brigade manner are combined to form registers. In the example shown in Fig. 4-18, bits are transferred from one flip-flop to the next every time a clock pulse occurs. As noted in the illustration, it takes four clock beats to load 4 bits into the register. Registers are used in CPU's to hold data temporarily. As we will note in a moment, every CPU has certain basic registers.

Accumulators

Registers used to store the results of logic or arithmetic operations are called *accumulators*. Figure 4-19 shows a register/accumulator used in an ADDER circuit. Note that the circuit has two input registers A and B, an ADD gate combination in the middle, and an accumulator register C at the other end. Binary numbers coming from memory or other registers enter the input registers one at a time. As the numbers are pushed through the input registers, they are combined in the ADD device. Upon leaving the ADD device, the results are stored in the accumulator register. These results can now become input to other components.

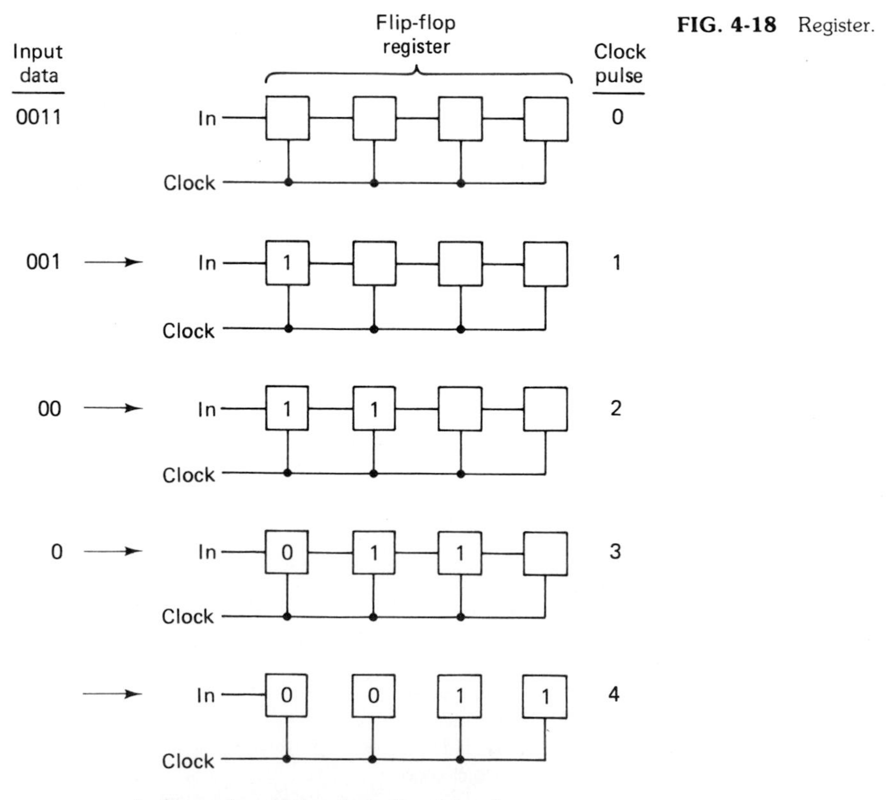

FIG. 4-18 Register.

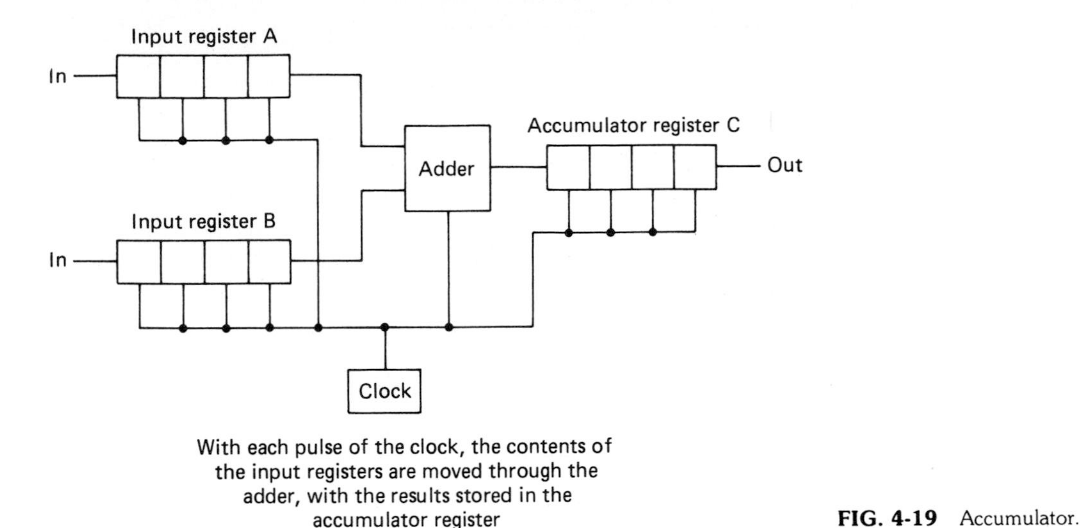
Input register A
In
Accumulator register C
Adder
Out
Input register B
In
Clock
With each pulse of the clock, the contents of the input registers are moved through the adder, with the results stored in the accumulator register

FIG. 4-19 Accumulator.

CHIPS (FROM GATE COMBINATIONS)

These and other gate combinations are further combined to form the circuitry of chips. Some fundamental chip-level functions include the following.

Memory

Memory chips contain electrical patterns that represent information. As noted in Fig. 4-20 and explained in the following paragraphs, most computers, including those found in automobiles, have several kinds of memory chips:

1. ROM (read-only memory) consists of a fixed pattern of 1s and 0s that can only be read. Once created, no new data can be written to a ROM chip. ROM is used to store programs (instructions) as well as permanent data that applies to a wide variety of vehicles. The same ROM memory chips may be used for an entire product line. ROM data are included in the design of the chip and are written during the chip-masking operation described in Chapter 3.

2. PROM (programmable, read-only memory) consists of a pattern of fusible links. Using a special PROM ''burning'' machine, the links are selectively melted to give the desired pattern of 1s and 0s. Like ROM, once created PROM can only be read; no new data can be written. Therefore, like ROM, PROM is used to store permanent data. However, PROM memory chips are likely to contain calibration data (idle speeds, distributor settings, etc.) for a line of vehicle types because data for a variety of different vehicles can be ''loaded'' into PROM chips faster and easier than into ROM chips. The drawback is that PROM chips cost more, making ROM better suited for mass production. PROM data are written after the chip has been fabricated but before it is used.

3. RAM (random access memory) is fabricated from tiny flip-flop elements formed into the chip. Each flip-flop will output a 1 or 0, depending on the previous input. So long as current continues to flow

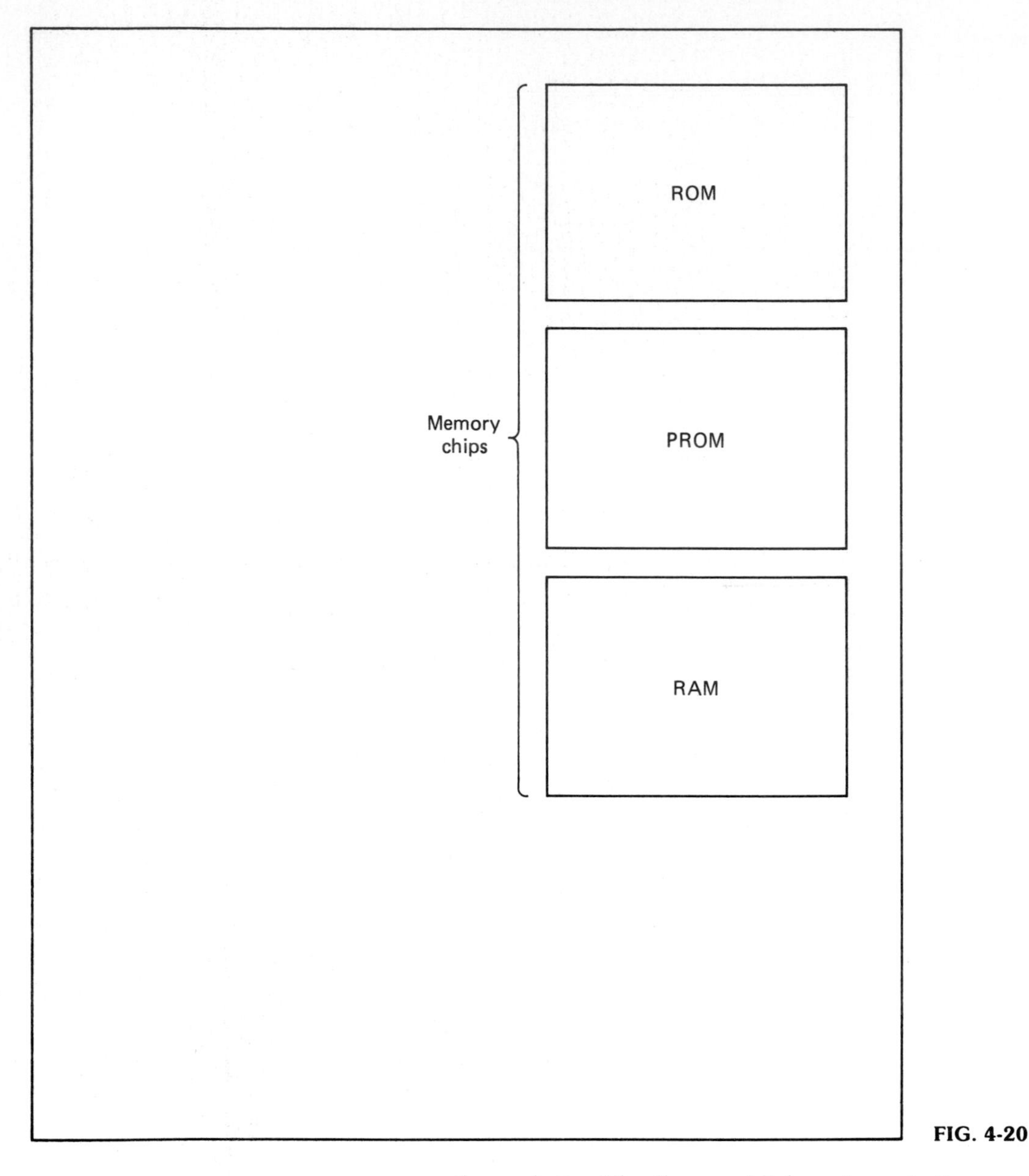

FIG. 4-20

through the flip-flop and it is not reset, the same value will be output. RAM memory however, is volatile. Turn off the power and the data patterns are destroyed. That is the disadvantage of RAM. The advantage, compared to ROM and PROM, is that data can both be read from and written to RAM. This makes RAM an ideal place to store ongoing information about the operation of the engine. RAM data can be read or written at any time.

RAM is also used by some on-board automotive computers to store trouble codes that result from periodic tests performed by the computer. Upon receiving a request from the vehicle operator or from a service technician, the computer will display these trouble codes on a dash-mounted display. The technician can then refer to a diagnostic manual to see what the trouble codes mean. Of course, if there is no battery backup to preserve the data in RAM, the information will be lost when the engine is turned off.

4. Erasable PROM is similar to PROM except that the contents can be erased and new data can be written back into the chip. The flip-flop element used to store data is controlled by an electron charge "buried" within an insulating layer on the chip. Under normal circumstances, the charge and the associated flip-flop do not change. However, with an application of sufficient energy from an outside source, the charge can be removed and the flip-flop reset. This energy can come from ultraviolet light photons or an electric field. In the former case, the PROM is known as an UVEPROM (ultraviolet erasable PROM). In the latter case, it is known as an EEPROM (electrically erasable PROM). Neither UVEPROMs nor EEPROMs are substitutes for RAM. Special equipment is required to erase and rewrite a UVE-PROM. An EEPROM can be rewritten by an on-board automotive computer; however, the write time is much slower than with regular RAM.

5. NVRAM (Nonvolatile random access memory) combines RAM and EEPROM (just above) on the same chip. During normal operation, data are written to and read from the RAM portion of the chip. However, if a power failure is detected, the data are written from the RAM into the EEPROM section of the chip. When the power is restored, data are written from the EEPROM back into the RAM. NVRAM is good for making sure critical RAM information is not lost. However, it should be noted that the EEPROM portion of the chip does not last forever. After a certain number of read/write operations, it wears out.

Processing unit (CPU/MPU)

These are the chips that manage, or oversee, the operation of most other chips. Programs are brought from memory to the CPU, one step at a time. While temporarily residing as a bit pattern in the Instruction Register, the instruction causes actions to take place in the ALU (arithmetic logic unit), other CPU registers, and components throughout the computer.

The processor is variously called the MPU (microprocessor unit) or the CPU. Typically, a CPU is the processing heart of a general-purpose microcomputer. It manages, or oversees, the operation of the rest of the computer, including, in some cases, several special-purpose MPUs.

An MPU is typically used for specific types of processing operations. For instance, in a personal computer (PC), special-purpose MPUs might be used to manage the operation of the disk drive and the display. Also, the processor in most process-control devices is referred to as the MPU.

Depending on the application, the same processor chip may be called a CPU in one device and an MPU in another. As special-purpose, control units, the processors in automotive computers are often called MPUs. However, as on-board computers perform more and more functions, the processor may become known as the CPU.

As noted in Fig. 4-21, a typical CPU has several main components:

- *Registers,* including the: (1) accumulator, used to store data temporarily before and after processing, (2) data counter, used to store the address of the next data element to be processed, (3) program counter, used to store the address of the next data element to be processed, and (4) instructions register, used to store the instruction currently contained in the CPU. *Note:* It may be helpful to view the CPU (and the entire computer) as a general-purpose machine whose current function is defined by the instruction currently located in the instruction register.
- *Control unit,* comprising circuitry that implements the instruction located in the instruction register. The total of the instruc-

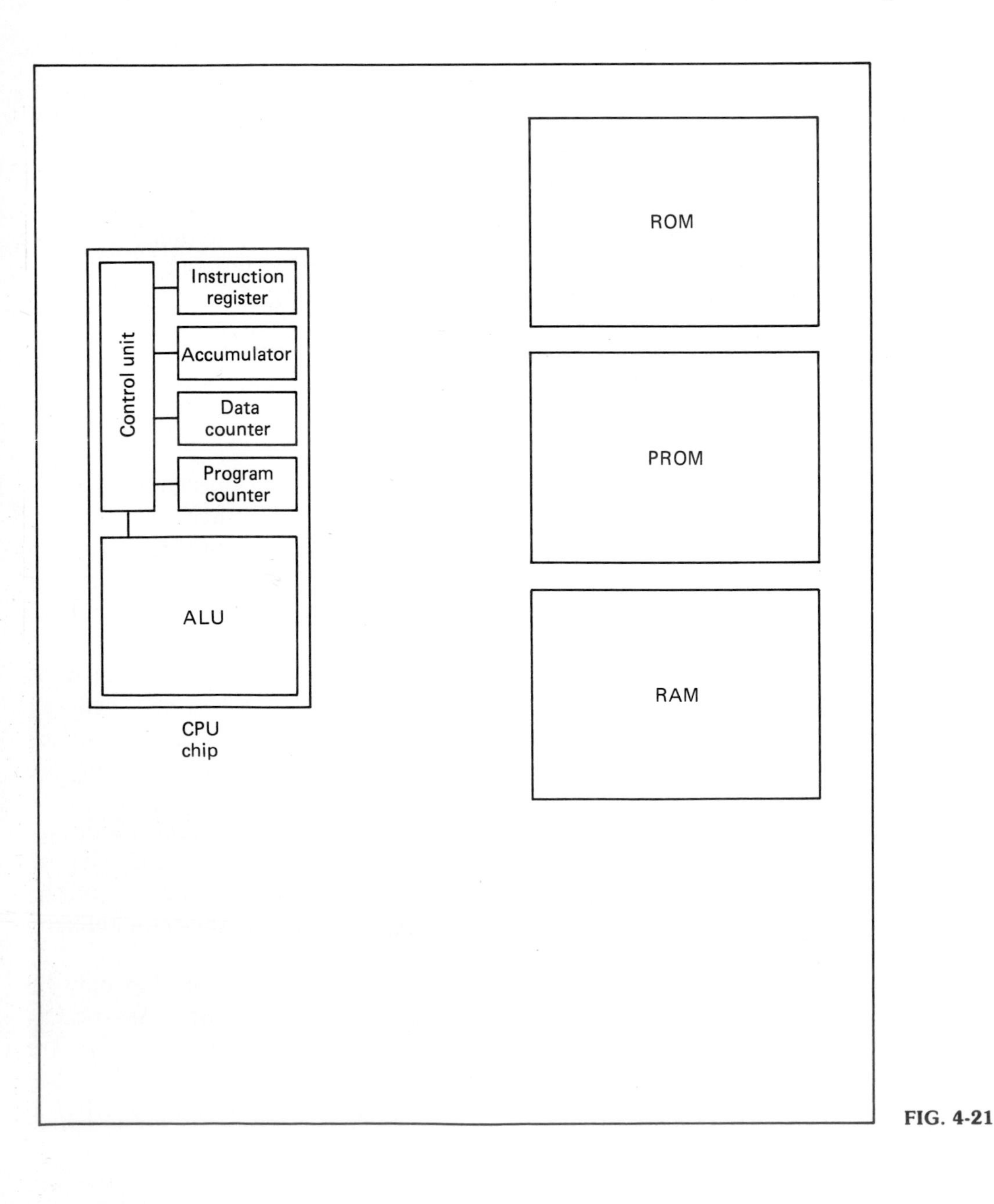

FIG. 4-21

tions that the control unit can implement is known as the instruction set of the CPU. This is where software becomes hardware (in human terms, where mind becomes body).

- *ALU (arithmetic logic unit),* upon receiving instructions from the control unit, performs the arithmetic and logic operations specified by the instruction in the Instruction Register. ADD, SUBTRACT, COMPARE, and the like, are performed by the ALU.

UART (universal asynchronous receiver/transmitter)

The UART chip sits (figuratively speaking) at the interface between the computer and the real world. It converts parallel data coming from the computer into a serial data flow. Serial data coming into the computer are converted into parallel data. These types of serial processing devices operate under a variety of names and perform various additional and/or specialized functions.

Clock

A computer may have two *clocks,* the process-timing clock mentioned earlier and a so-called real-time clock. The process clock function is implemented by one or more chips. The clock itself is a crystal that electrically vibrates when subjected to current at certain voltage levels. As a result of this vibration, the chip produces a very regular series of voltage pulses. Various circuits divide the pulses to get the desired timing rate. It should be noted that the timing pulses cannot exceed certain values; otherwise flip-flops and other devices are not given enough time between cycles to operate.

A real-time clock counts seconds, minutes, hours, and days, just as a wrist watch or household clock. The output of the real-time clock is used to start and stop devices or to perform other functions that have a relationship to real-world times.

A/D and D/A converter

The converter function (whether implemented on one or several chips) is essential to the operation of automotive process control computers. An A/D (analog-to-digital) converter changes analog data from sensors into digital signals. Typically, an input signal is a variable voltage pattern representing a real world event. The A/D converter assigns a numeric value to different voltage levels. These numeric values are passed on to the CPU when called for by a program instruction.

A D/A (digital-to-analog) converter performs just the opposite task as an A/D converter. It changes digital data back to an analog representation. Since many control outputs are binary in nature (an output device is turned on or off), there is less need for D/A conversion.

Bus architecture

In many microcomputers, information is passed between chips over bus lines. These are electrically connective paths, corresponding to wires printed on the boards to which the chips are attached. A bus may include several major lines:

- *Data bus.* The data bus is used to pass data between the CPU and memory chips. The data bus may also be used when I/O inputs are treated as sections of memory (although data from the outside may first have to pass through a UART and an A/D converter).
- *Address bus.* When an address is placed on the address bus—by the CPU, for example—a collection of components known as the *chip-select circuitry* accesses the indicated section of memory for either a read or write operation. If a read is indicated, the contents of this memory address are placed on the data bus. If a write has been specified for RAM, the data presently on the data bus are written to the address indicated. As noted earlier, I/O ports may be assigned a memory address. If so, they are accessed in much the same way a section of memory is accessed.
- *Clock, power, ground bus.* The clock, power, and ground bus used to supply power to other circuits and to distribute clock signals. Fig. 4-22 shows all these components connected to a bus.

BOARDS

Chips are soldered or plugged into connectors attached to flat plastic panels called boards (or cards). Connections between chips are provided by electrically conductive paths printed on the boards. The buses just mentioned are examples of these conductive paths. The boards themselves are sometimes called PC (printed circuit) boards.

A collection of electronic switching elements on a board may simply be a "logic device." Such a board may perform an intelligent control function, but the function is fixed. However, with the addition of an MPU (or CPU) chip and possibly some memory the device can perform whatever control program has been loaded into memory.

A small system may only have one board; larger systems may have several. In most cases, multiple boards are in the same enclosure; however, that is not mandatory. When more than one board is used, the chips they contain are generally grouped by function. Here are some possibilities.

CPU/memory

The CPU chip and the memory chips may be included on the same board. There may be other components, but these are the most important. The memory chips will probably be organized into ROM, PROM,

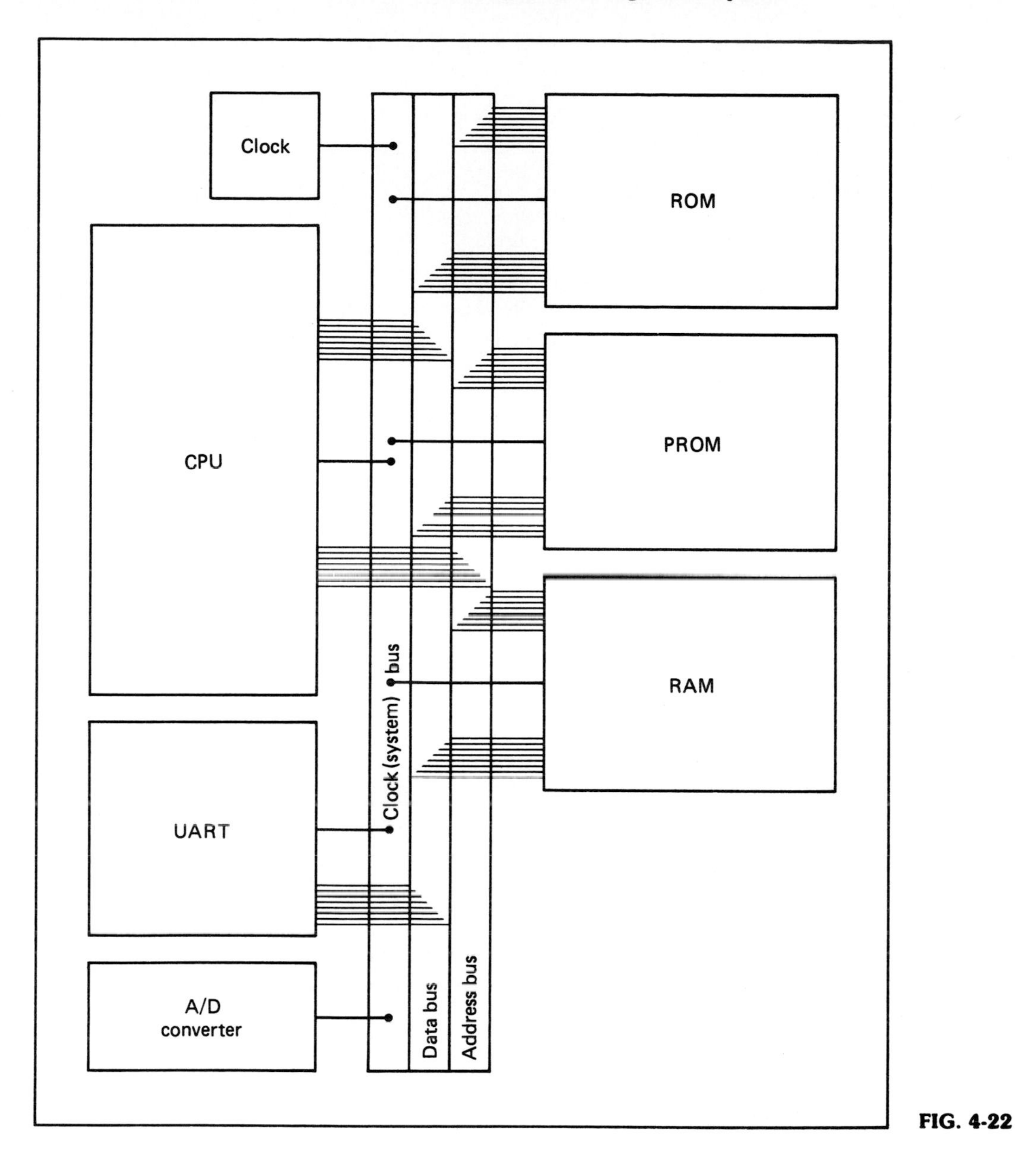

FIG. 4-22

and RAM groupings. In many cases, the PROM chip may be accessible to the technician in the local shop because the chip contains calibration settings for this vehicle and must be saved in case the computer is swapped out or sent away for repairs.

Power supply

Solid-state chips use dc power in the range of 5 to 12 volts dc. Since information is represented by small changes in voltage levels, power control is very important. In some systems, a single board is used to contain the circuitry used to manage power levels. In one system, a single power supply for multiple MPU managed devices is contained within a separate enclosure.

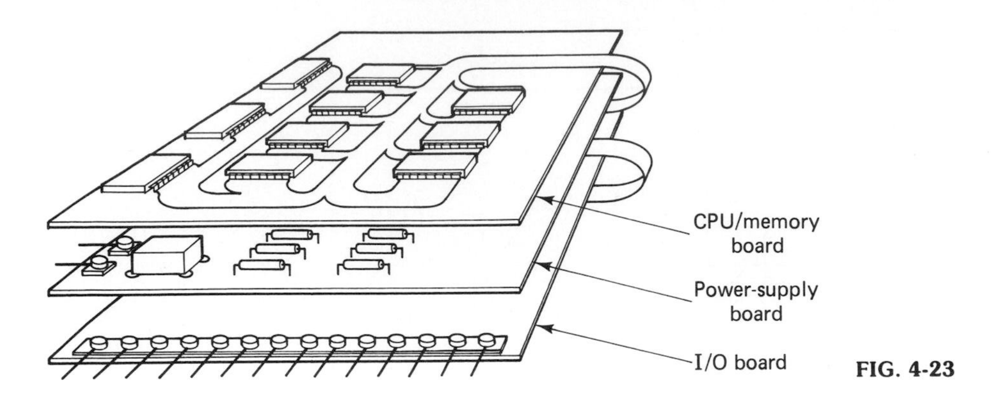

FIG. 4-23

Input/output

If the computer system has varied contacts with the outside world, it is not uncommon to put all the I/O connectors and associated circuitry on one board. Input sensors and output controls are connected here. Figure 4-23 shows a *card stack* containing the boards just described.

SOFTWARE

Just as gates are the foundation of computer hardware, the instructions available in a CPU's instruction set are the foundation of software. All programs can ultimately be reduced to a combination of these low, or machine-level instructions. However, programs do not usually start out at this level, as a pattern of bits in the CPU's instruction register. Very few people could understand what is going on. Figure 4-24 and the following paragraphs depict a typical sequence of events.

1. Programs begin with an idea or statement of capabilities—in other words, a description of the functions the computer will perform.
2. These general statements are rephrased as a set of flow charts providing a more detailed description of how information is processed.
3. A computer program is written. Using the vocabulary and "grammar" established for one of the computer languages, specific processing instructions are written. This involves rephrasing previous descriptions according to the rules specified by the language.

So far, all the descriptions and instructions have been written for human understanding. Nothing has been devised or stated that the computer can understand. In other words, nothing has been written in the language of the machine. The next steps do this:

4. The program (called the source code) is entered into another computer, which contains a special program called a compiler, interpreter, or assembler (depending on the type of language used). This program converts the human-oriented statements into a binary pattern of 1s and 0s. This pattern corresponds to

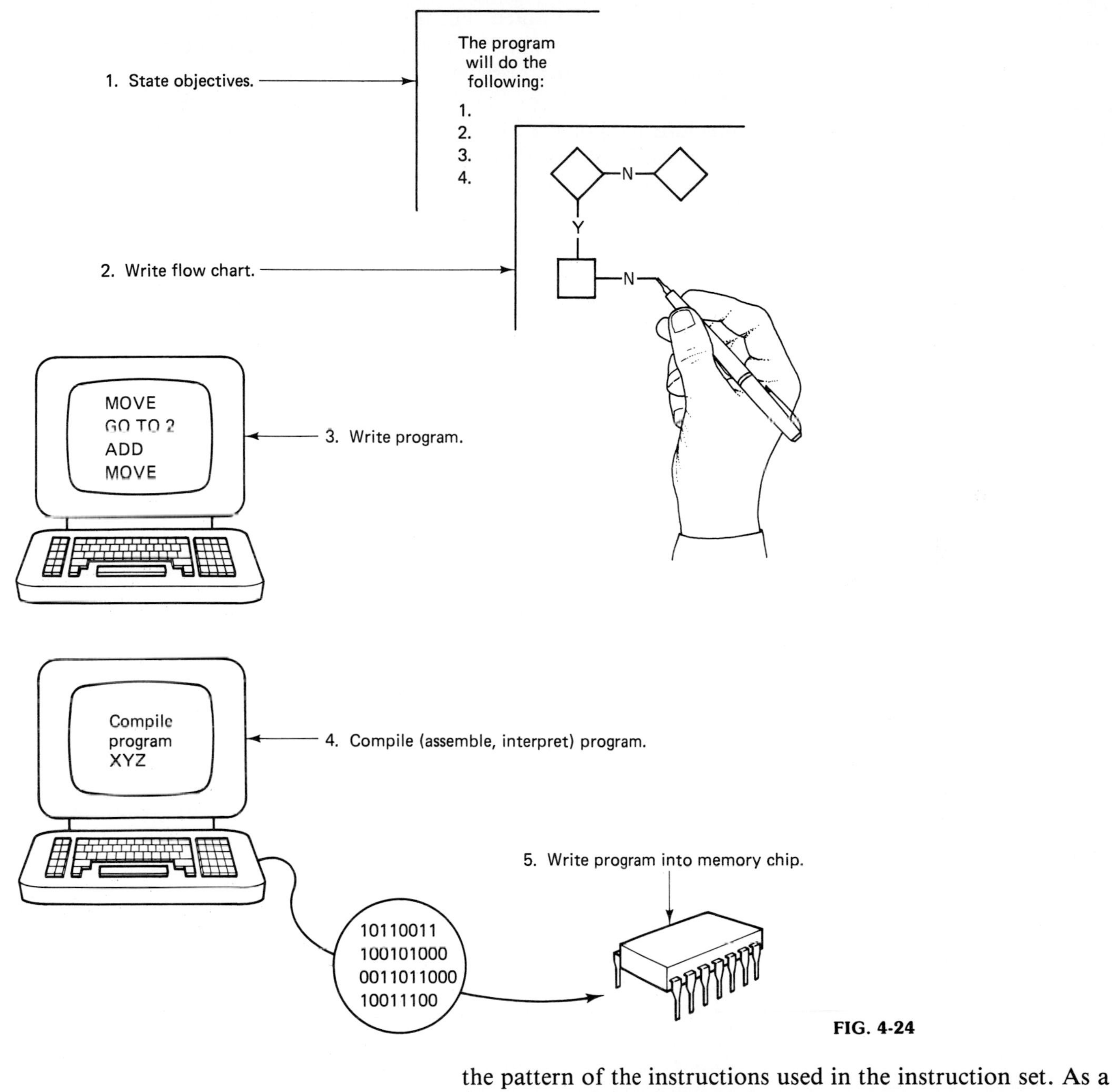
1. State objectives.
The program will do the following:
1.
2.
3.
4.
2. Write flow chart.
N
Y
N
MOVE
GO TO 2
ADD
MOVE
3. Write program.
Compile program XYZ
4. Compile (assemble, interpret) program.
5. Write program into memory chip.
10110011
100101000
0011011000
10011100

FIG. 4-24

the pattern of the instructions used in the instruction set. As a result, the human-oriented source code is reduced to a machine-comprehensible form called *object code.*

5. The object code is transferred to the memory of the computer in which it is to run. In the case of an automotive computer, the program is most likely written to a ROM or PROM memory chip.

The computer is now ready to run. The program steps are moved out of memory, one step at a time, into the CPU's instruction register. The steps, which have been rephrased into the language of the instruction set, cause the control unit to perform the actions required to execute the instruction.

It should be noted that the languages and associated compilers, interpreters, or assemblers used to create programs do not all yield the same level of performance. So-called higher-level languages, like BASIC, PASCAL, or FORTRAN, are several steps removed from the machine level of the CPU's instruction set. Each instruction written by a human in one of these languages may result in many machine-level operations. As a result, the source code is relatively easy for trained humans to read and write, but the object code may take up too much room in the computer for certain applications. This is especially true for real-time process control systems like those used in automobiles. The amount of memory available is generally limited and each instruction has to be precisely specified.

In these cases, a lower-level language is likely to be used. Called *assembly level programming,* each source code instruction results in one instruction set action. The only concession to human understanding is the way the instructions are phrased in the source code. Instead of expressing the instruction directly in 1s and 0s, mnemonics such as ADD and MOVE are used. The assembler then converts the mnemonics into 1s and 0s. Because the programmer is working directly with the CPU's registers, any wasted motions that may have resulted from a high-level compiler or assembler can be eliminated. The exact requirements of the function can be satisfied. Automotive systems do very sophisticated, real-time process control and tend to be written in assembly language code.

TYPES OF AUTOMOTIVE COMPUTERS

Digital computers like the kind introduced in this chapter can perform a variety of automotive information-management functions. Two major groups of functions are engine control and body operation. Programs running in engine-control computers examine inputs from various sensors located about the vehicle and then send out signals to regulate spark timing, fuel mixture, and various pollution-regulation operations. Body computers manage digital instrument displays, automated braking systems, cruise control, and so on.

At present, any single vehicle is likely to have multiple computers, each comprising one or more board and each performing its own individual functions. There is some degree of communication between different computers, with the output of one computer serving as the input to another, and so on. At the present, it is difficult to tell if multiple functions will continue to be performed by multiple computers or if functions will be combined into one computer. Functionally, it does not make any difference.

5

Concept of Control

GENERAL

The first automotive computers were used to control fuel mixture and spark timing. Since then, on-board computers have been used to manage pollution-control devices, automatic transmissions, brakes, suspension systems, air conditioners, intake and exhaust valves, and so on. Given the number of control functions performed by computers, it will be helpful to examine briefly the concept of control itself.

ASPECTS OF CONTROLS

Almost every mechanism, no matter how simple, requires some kind of control. Control devices existed long before computers were invented. Consider a child's seesaw (Fig. 5-1). The action of the board is man-

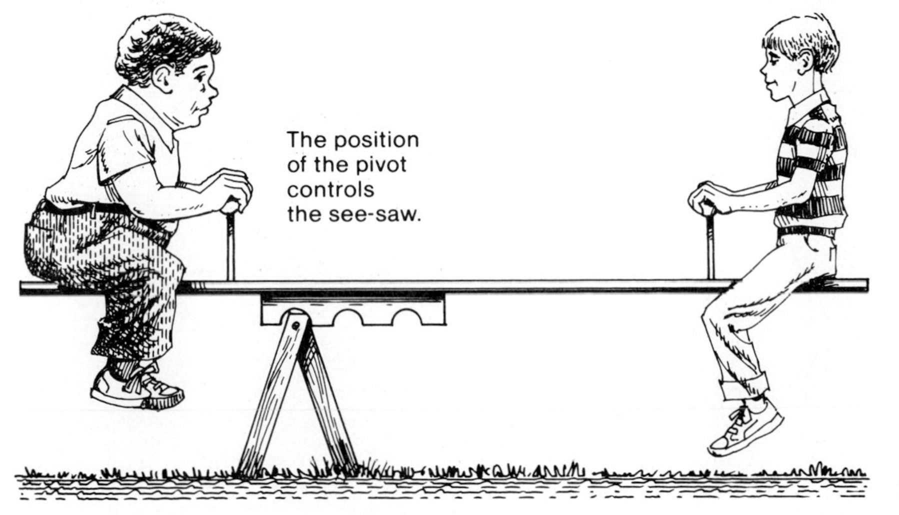
The position
of the pivot
controls
the see-saw.

FIG. 5-1 Seesaw control.

aged or *controlled* by the location of the pivot point. It is actually a control device. Children of equal weight achieve balanced operation by putting the pivot in the center of the board. Children whose weights are not the same must adjust the board back and forth to find a balance point. In either case, the operation of the seesaw must be controlled. Otherwise, it could not function in a useful manner.

The concept of control is tied in with the basic design of a machine. For instance, in an internal combustion engine, the size and location of the cylinders controls the size and operation of the pistons. In a house design, the size and location of the window controls the flow of air through the house (Fig. 5-2).

Even the way we use words points out the basic nature of control. Well-designed, properly operating machines are said to function in a *controlled* manner. They perform their job in a predictable way. On the other hand, malfunctioning machines are said to be *uncontrolled.* They do not work the way they are supposed to and are therefore considered unpredictable and not useful.

Control means intelligence

Control represents thought or intelligence. The control function is a deliberate attempt by a person to manage the behavior of a device. It is one of the ways in which we use to control nature. You might say that it is *our* nature to establish control over things.

Control is how we separate ordinary objects (such as sticks and stones) from useful tools. An ordinary object (such as a stick or stone) can become a tool when a person uses it in a controlled way. A higher degree of utility is obtained when the object is reshaped for better or more precise control (Fig. 5-3). An even higher degree of utility is achieved if the control junction is built into the device itself—if the device inherits some of the intelligence of its human creator.

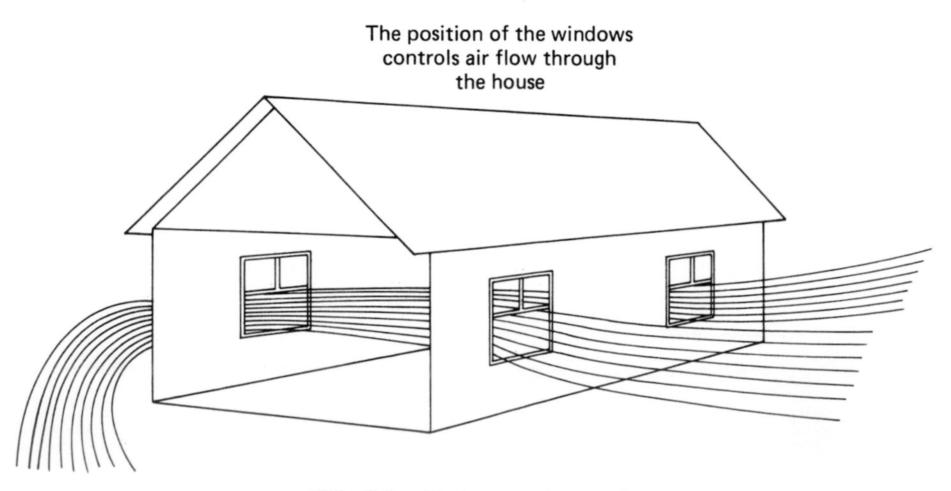

FIG. 5-2 Windows as air control.

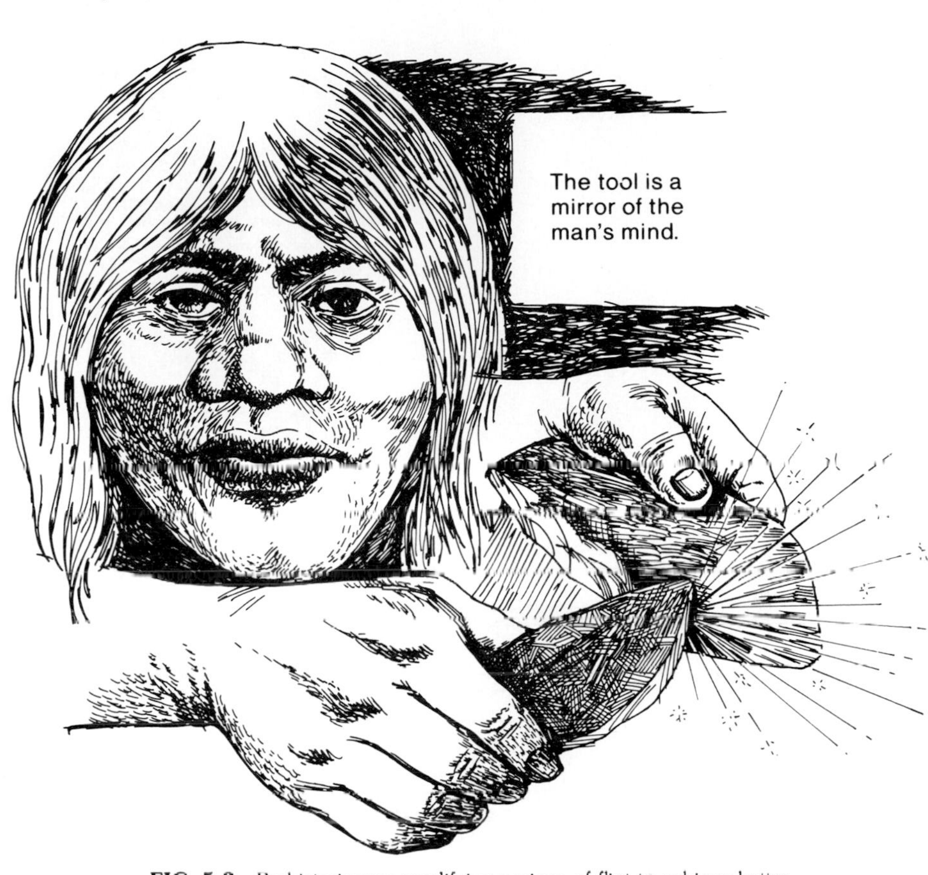

FIG. 5-3 Prehistoric man modifying a piece of flint to achieve better control.

LEVELS OF CONTROL

Control levels can be grouped into some general categories.

Static shape control

The most fundamental level of control comes from the very shape of a tool or machine. Simple tools are good examples of static shape control. Screwdrivers, hammers, pliers, and wrenches are controlled by their shape. Of course, the operator also has a hand in it. But even so, change the shape of any tool and the control requirements change.

These control devices also have another common factor. They are static with respect to time and space (Fig. 5-4). In other words, the control function does not depend on any moving parts and, unless wear or deterioration set in, it will remain the same forever. A screwdriver acts the same now as it does 1 year from now or 1000 years from now.

Dynamic shape control

Dynamic control refers to moving objects. To see what this means, think about the way a camshaft works. Its shape or contour certainly provides control: determining when and how fast valves open and close, controlling breaker-point operation in old-style distributors, and

man grows old,
the wrench still

FIG. 5-4 Static shape control is permanent.

so on. However, you would not call a camshaft a static device. The design or shape might be fixed, but its operation is not. In fact, a camshaft's control function takes place only when the cam is moving (Fig. 5-5). A stationary camshaft does not do anything. So a camshaft is dynamic with respect to space because it must move to provide control. It is dynamic with respect to time because no point on the cam surface stays in one place more than an instant.

Dynamic shape control is provided by any device whose configuration and movement control the operation of another device. Some examples are gear sets, chain and sprocket assemblies, and crankshafts.

Feedback control

Most machines have both dynamic and static control elements. We can think of the static control components as providing a framework or skeleton for the dynamic control parts.

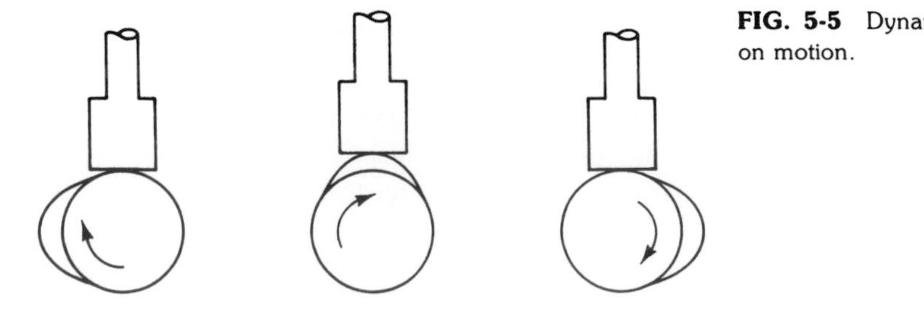

Follower goes up and down only
when cam is in motion

FIG. 5-5 Dynamic shape control depends on motion.

Many machines also need another kind of control. Take your wristwatch, for example. The configuration of the body and the design of the dial provide a certain kind of static shape control. If it is an old-style watch, the spring and timing gears provide a high degree of dynamic control. But suppose that the watch starts to run fast or slow. Then you need to be able to adjust it one way or another. This is what we call *feedback control.* It involves adjusting the operation of a machine in response to changing conditions to maintain a built-in goal or reference.

In the case of the wristwatch, the feedback control is not automatic. You are an active part of the operation. You supply the reference goal. Until several hundred years ago, this is how most feedback control was done. The machines were fairly simple, often driven by human or animal power. If any control adjustments were needed, the person attending the machine could make the required corrections.

However, the situation changed at the beginning of the Industrial Revolution with the invention of the steam engine. Short of constantly looking at dials and gauges, there was no way for a human operator to tell if a steam boiler was about to blow up or if a shaft was running too fast. Automatic feedback controls had to be provided in the form of pressure relief valves and governors. The reference goal was built into the machine—as well as the means to achieve the goal.

Today, almost all complex machinery requires some kind of automatic feedback control. Such devices represent a different order of control and information management than the control elements mentioned previously. Feedback controls come closer to duplicating the human decision-making process. In comparison, dynamic and static shape controls are inflexible. They process information in the same way regardless of the circumstances.

Automobiles possess a variety of automatic feedback controls. One example is the thermostatic control valve in the cooling system. At a present temperature, it opens to let coolant flow from the radiator through passages in the engine block. If the engine cools too much, the valve closes again. The thermostat is the designer's representative in the engine, responding as the designer would to the information presented (Fig. 5-6).

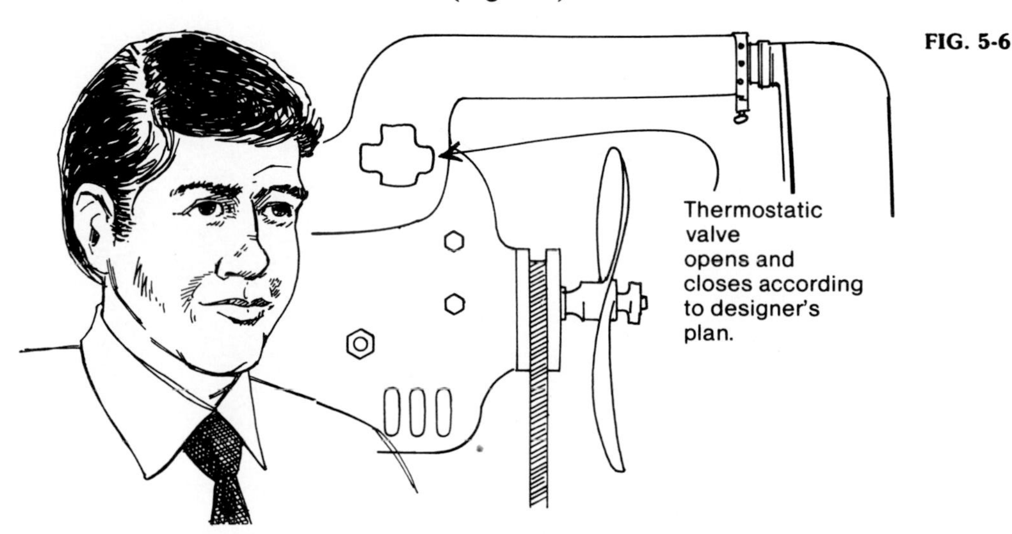
Thermostatic
valve
opens and
closes according
to designer's
plan.

FIG. 5-6

Another "representative" is the starter drive engagement mechanism. After the engine reaches a certain speed, the drive pinion is moved out of mesh with the flywheel, thus preventing damage to the starter motor. Other examples include the automatic advance mechanism in old-style distributors, carburetor control circuits, voltage regulators, and so on.

Computerized controls

As "thinking machines", it is natural that computers have become heavily involved in feedback-control systems. They perform many of the functions that were handled by mechanical devices. In most cases, you would have trouble telling the difference between results obtained by computer controls versus those obtained mechanically. The real difference lies in the variety of results available. Computers provide control over a wider range of circumstances. In other words, computers can manage more information.

Why? Unlike mechanical devices, computers do not depend on the presence of relatively large physical shapes. As noted in the last chapter, the designer's intention or program is stored as a pattern of very small electrical circuits (Fig. 5-7). Simply speaking, it is possible to pack more control information into a much smaller space (Fig. 5-8).

Another difference involves versatility. Although computers, like mechanical controls, operate according to fixed programs, the computer program can easily be changed by altering the circuit pattern. Using the same basic on-board computer, a manufacturer can write different programs for different engines and operating conditions. It is much easier and less expensive to reprogram a computer than it is to produce a new mechanical control device.

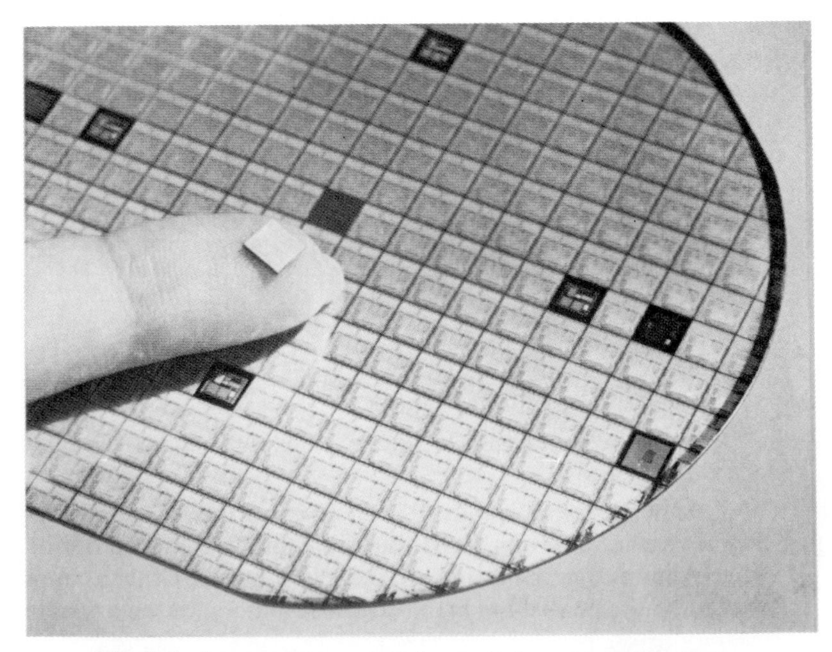

FIG. 5-7 A tremendous amount of control information is packed into tiny circuits on this silicon chip. (Courtesy of Delco Electronics Division, General Motors Corporation.)

FIG. 5-8 The actual computer is not much bigger than a paperback book. (Courtesy of Delco Electronics Division, General Motors Corporation.)

6

Overview of Engine-Control Systems

GENERAL

The first on-board automotive computers were used to control basic engine functions such as ignition timing and fuel mixture. They replaced electromechanical control systems already in place. Now, various engine operations are managed by one or more engine computers. The engine computers "talk" to instrumentation computers as well as to shop diagnostic computers.

Instrumentation computers are discussed in Section III. Diagnostic computers are described in Chapter 19.

The next group of chapters examine engine control systems. Electromechanical control systems are reviewed, and then computerized systems are examined.

The remainder of this chapter looks at some of the reasons why engine controls were automated in the first place.

REASONS FOR TECHNICAL CHANGE

Technical change, whether it relates to computers, TV sets, or whatever, comes about only for certain reasons. Two of the primary factors are need and profit. In the past, need and profit were two sides of the same coin. Innovations reached the market only if need was perceived to equal a widespread public desire to spend money. In other words, although necessity might be the mother of invention, profit is the father.

The innovations that shaped the first 50 or 60 years of automotive development followed this pattern. Electric self-starters, feedback devices of various kinds, automatic transmissions—all were produced for profit. These inventions may have provided great satisfaction for their creators; they may have also addressed real social needs; however, it was the public's willingness to buy that made them possible. Those that were not accepted fell by the wayside.

This pattern continued up until the middle to late 1960s. The emphasis was on power and drivability because that is how the manufacturers perceived the public's taste and buying habits. For instance, a typical American-made sedan weighed 2 tons and was powered by a 300-horsepower engine that got 11 miles to the gallon. An average "sporty car" of the period had a performance-modified version of the same engine, went from 0 to 60 miles per hour in 8 seconds and got only 8 miles per gallon. There were occasional attempts to produce economy cars, such as the Corvair and the early Falcons and Valiants. But they did not sell well for long periods. After all, fuel was readily available and cheap. And pollution, although recognized as a problem, was not subject to strict controls.

If you view engine operation as a four-part equation whose major factors are power, drivability, pollution control, and economy, the formula was shifted in favor of the first two factors. Designers did not have to be concerned with complicated feedback control devices. Engines operated on relatively rich mixtures that were fairly simple to maintain. The basic controls that had been used for years in the carburetor and distributor remained adequate for the job (Fig. 6-1).

LEGISLATED PRIORITIES

Then, in the late 1960s and early 1970s another factor entered the picture. The federal government, responding to increases in airborne pollution, began to pass laws that competed with the public's seeming passion for big, fast cars. As a result, manufacturers embarked on a program of federally mandated change. For the first time, car companies had to consider seriously the third part of the four-part operation equation, pollution control.

However, even after emission control became a factor, the mechanical control systems remained adequate. Power and drivability did suffer slightly because of leaner mixtures and other internal and external engine changes. Yet drivability was not all that bad and it was always possible to make engines larger to compensate for the power loss.

FIG. 6-1 1966 Volvo, 122S. A prepollution control European sedan.

And despite the fact that larger pollution control engines of the period used more fuel, the operating costs were still within the range of most people. There was no need to make any basic changes in the control systems used to monitor and adjust air/fuel mixtures and ignition timing.

Then, in the middle to late 1970s, a number of significant events took place:

1. Fuel costs rose dramatically because of political conditions in the Middle East and increased consumption plus inflation at home.
2. The federal government passed legislation requiring manufacturers to produce more-fuel-efficient vehicles.
3. Pollution controls became even more rigid.
4. The cost and size of computers decreased while their power and capacity increased.

So, for the first time, all four factors in the equation had to be given equal weight. Engineers had to design smaller engines for smaller cars and still provide adequate power, good drivability, improved economy, and reduced emissions.

Despite the usual changes in load, acceleration, and other operating conditions, engines had to stay even closer to the 14.7:1 stoichiometric fuel mixture. As a result, conventional systems were pushed to the limit of their capacity. One solution was to rely more and more on electronic, solid-state ignition systems. These devices, which became common in the middle 1970s, were designed to produce "hotter" ignition sparking, which allowed engines to run on leaner fuel mixtures.

The problem is that these hotter, solid-state ignitions still do not provide much better control than points-type systems. Slight changes in lean-burning engines can shift the equation into the misfiring range, where power, drivability, economy, and emissions all suffer. To avoid these problems, it is necessary to take into account a number of operating factors and to be able to make almost instant changes in timing and in fuel mixture.

ENTER THE ERA OF ELECTRONIC CONTROLS

At this point, the era of low-cost, high-powered computers entered the picture. It was simply not possible to pack into an analog device the amount of control information needed to maintain an engine at the stoichiometric ratio. Furthermore, the sheer mass of any strictly mechanical system prevented it from reacting fast enough to changing conditions. From our current vantage point, it seems inevitable that computerized control systems were developed.

The first computerized automotive-control system was introduced by Chrysler in 1976. Called the Lean Burn system, it was primarily an add-on device designed to control spark timing in response to changes in engine and air temperature, throttle position, and engine speed. Figure 6-2 shows some of the components in a more recent engine-control system.

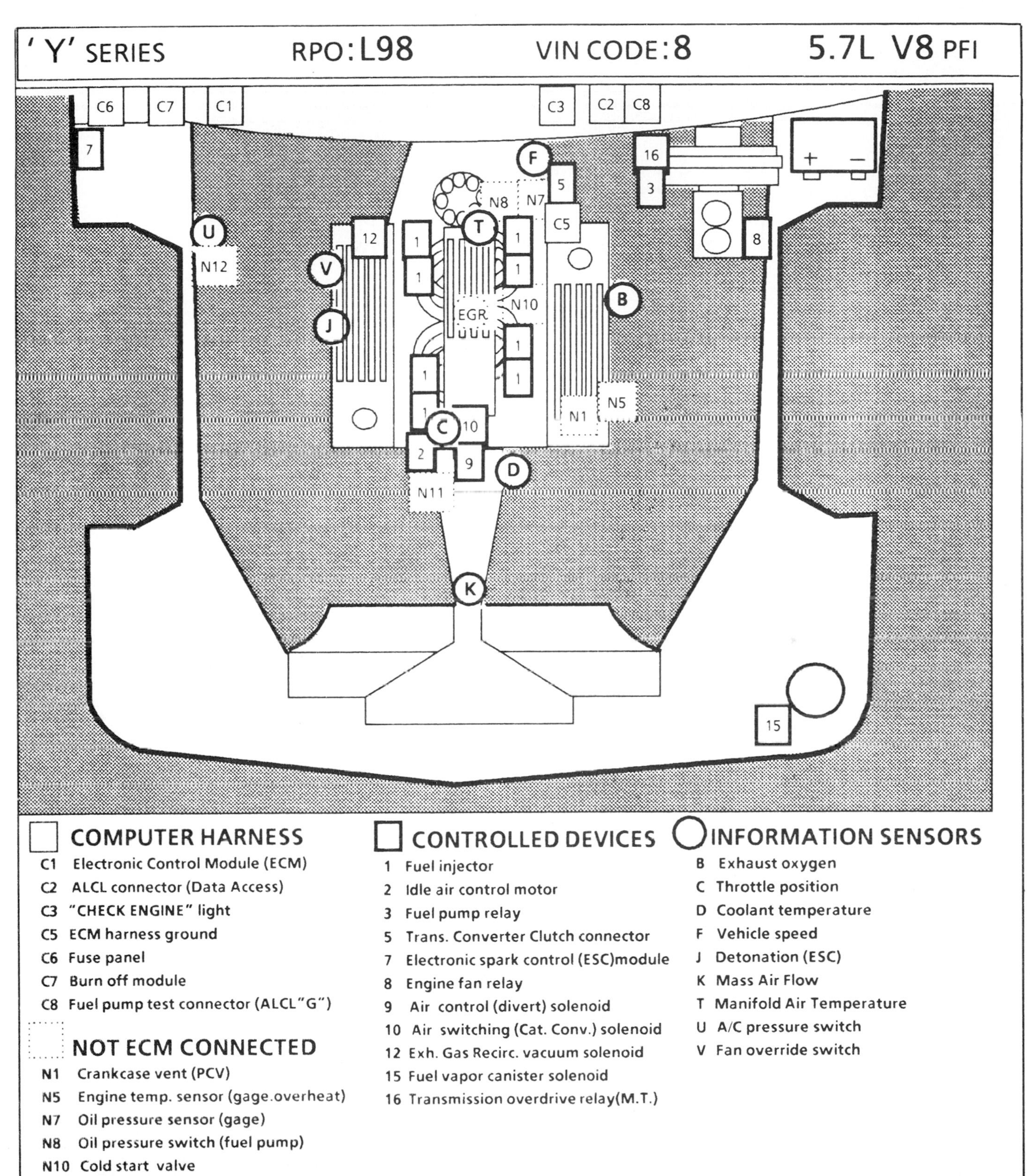
'Y' SERIES
RPO:L98
VIN CODE:8
5.7L V8 PFI
C6
C7
C1
C3
C2
C8
7
F
16
5
3
N8
N7
C5
U
T
12
1
8
N12
V
B
EGR
N10
J
N5
N1
C
10
2
9
D
N11
K
15
COMPUTER HARNESS
C1 Electronic Control Module (ECM)
C2 ALCL connector (Data Access)
C3 "CHECK ENGINE" light
C5 ECM harness ground
C6 Fuse panel
C7 Burn off module
C8 Fuel pump test connector (ALCL"G")
NOT ECM CONNECTED
N1 Crankcase vent (PCV)
N5 Engine temp. sensor (gage.overheat)
N7 Oil pressure sensor (gage)
N8 Oil pressure switch (fuel pump)
N10 Cold start valve
N11 Cold start thermal time switch
N12 A/C pressure cycling switch
CONTROLLED DEVICES
1 Fuel injector
2 Idle air control motor
3 Fuel pump relay
5 Trans. Converter Clutch connector
7 Electronic spark control (ESC)module
8 Engine fan relay
9 Air control (divert) solenoid
10 Air switching (Cat. Conv.) solenoid
12 Exh. Gas Recirc. vacuum solenoid
15 Fuel vapor canister solenoid
16 Transmission overdrive relay(M.T.)
INFORMATION SENSORS
B Exhaust oxygen
C Throttle position
D Coolant temperature
F Vehicle speed
J Detonation (ESC)
K Mass Air Flow
T Manifold Air Temperature
U A/C pressure switch
V Fan override switch
7-17-84
*5S2129-6E

FIG. 6-2 5.7-LV8 component locations.

7

Precomputerized Ignition- and Fuel-Control Systems

When all is said and done, the basic job of any engine control computerized engine-system is to deliver the right proportion of air and fuel inside the combustion chamber and then to ignite the mixture at the right time. Therefore, controlling these operations is the primary job of any engine-control system, whether mechanical, human, or electronic.

This chapter reviews basic ignition system principles and the basic control requirements of ignition and fuel systems. Although particular emphasis is placed on precomputerized systems, some of the devices described in this chapter, particularly breakerless ignitions, are still in use.

EARLY AUTOMOBILES

You often read or hear people say that new cars have gotten too complicated, that they are difficult to understand and fix. The point is hard to dispute. In some engine compartments you have trouble even seeing the block because it is covered by a maze of pipes, hoses, and wires. Earlier engines, in comparison, were a marvel of simplicity (Fig. 7-1).

However, there is another side to the coin. Although earlier cars might have been easier to fix, they were not easier to drive. That is because most of the machinery in these old cars was devoted to static or dynamic shape control (explained in Chapter 5). Very little automatic feedback was provided. The driver had to make many of the periodic adjustments necessary to keep the ignition and fuel systems in the proper operating range. The ideas of the vehicle designer might be represented in the shape of the block, or in the configuration of the camshaft. But it was up to the driver to decide exactly what the ignition

FIG. 7-1 Underhood simplicity of older car.

timing or fuel mixture ought to be. In a very real sense, the driver was an active participant in the operation of the engine.

Modern engines do not require the same degree of driver involvement. All you have to do is turn on the ignition key. Automatic feedback controls take over from there. This is one of the primary reasons for the underhood complexity of new cars (Fig. 7-2). The pipes, hoses, wires, and associated "black boxes" are devoted almost exclusively to sensing changing conditions and making corrective adjustments.

FIG. 7-2 Underhood complexity of late-model car.

Some of these controls, as we will see in the discussion of fuel and ignition systems, were designed to reduce the need for driver intervention. Others, as Chapter 8 explains, were designed to automate functions that never did belong to the driver.

IGNITION SYSTEMS

MAIN JOB OF THE IGNITION SYSTEM

All ignition systems in gasoline engines have one feature in common. They produce a high-voltage surge of electricity at a spark plug's electrode gap. The resulting arc ignites the air/fuel mixture in the combustion chamber, producing the expanding gases that push down on the piston and operate the engine.

HEART OF THE IGNITION SYSTEM (COIL)

The coil is the heart of any ignition system, old or new. It converts low-voltage battery current into the high-voltage surge needed to jump across the spark plug's electrode gap.

As shown in Fig. 7-3, a coil has two sets of wire windings, one called the *primary* and the other called the *secondary*.

The primary windings are part of the primary ignition circuit, which includes:

- The ignition switch
- The battery
- A control component to interrupt periodically the primary current flow (explained momentarily)

As pictured in Fig. 7-4, the secondary windings are part of the secondary circuit, including:

- The distributor cap and rotor assembly
- The spark plugs

FIG. 7-3 Ignition coil cutaway.

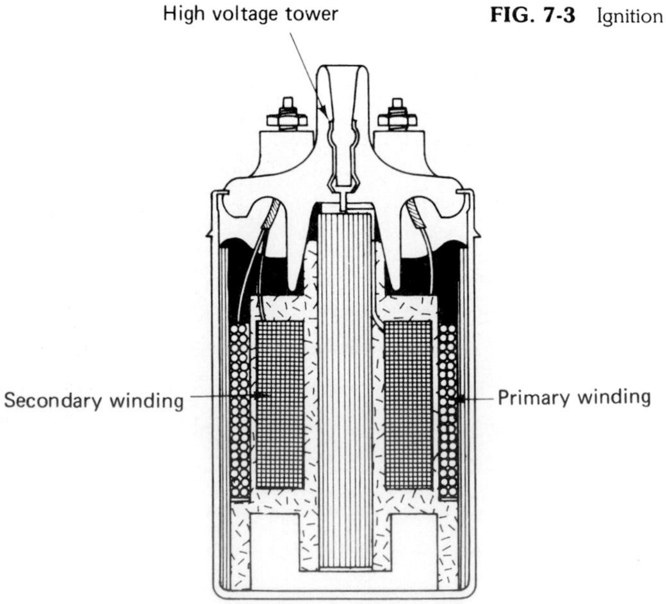

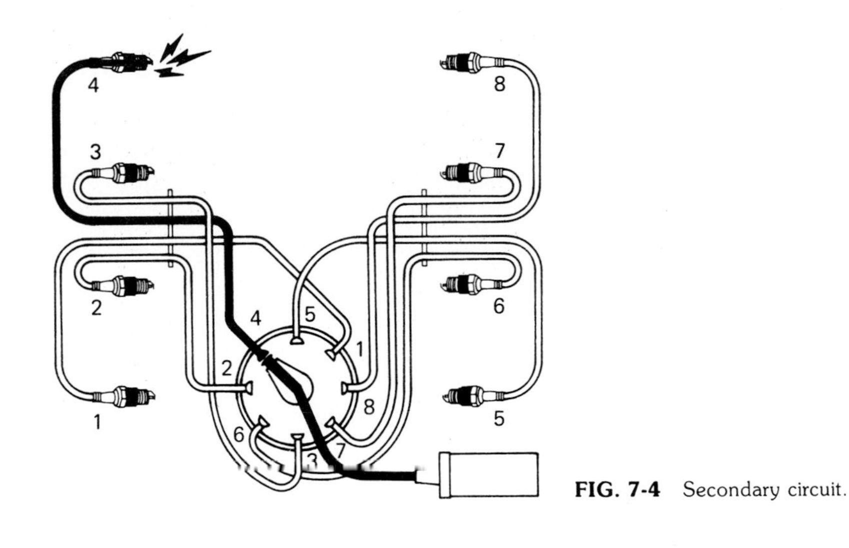

FIG. 7-4 Secondary circuit.

- All the high-voltage cables going from the coil to the distributor and from the distributor to the spark plugs

At the center of the coil is a core made from thin strips of soft iron, or a bundle of soft iron wires. The core is wrapped with electrical tape to insulate it from the secondary windings.

The secret of the coil's operation is in the windings. The outside, primary windings are made from hundreds of turns of relatively coarse wire. They are coated with varnish or shellac to provide electrical insulation. The inside secondary windings are made from many *thousands* of turns of relatively fine wire, also electrically insulated.

Low-voltage primary current flows through the primary windings. Current enters the coil at a positive terminal marked (+) or (bat) and leaves at a negative terminal marked (−) or (neg) (Fig. 7-5).

As a result of the current flow, the primary windings are surrounded by lines of magnetic force. The flow of primary current is periodically interrupted at times corresponding to the moment of spark plug ignition. When current flow ceases, the magnetic lines of force fall inward toward the center of the coil, crossing the secondary windings. The soft iron core concentrates the lines of flux.

As explained in Chapter 2, voltage is induced in a conductor whenever it is crossed by magnetic lines of force. This means that an electrical surge is produced in secondary windings. The secondary voltage is many times greater than the primary because the secondary wind-

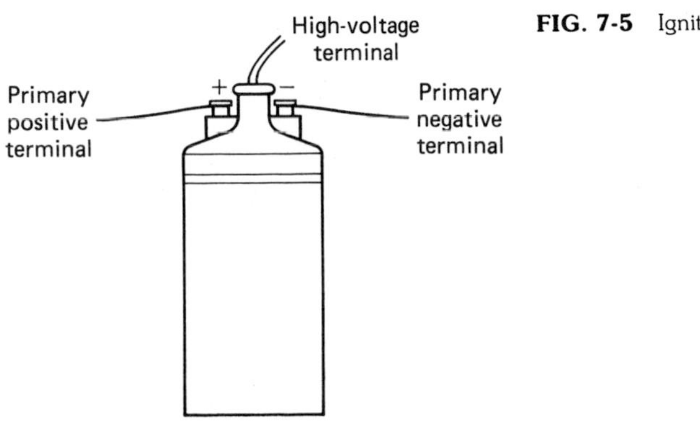

FIG. 7-5 Ignition coil terminals.

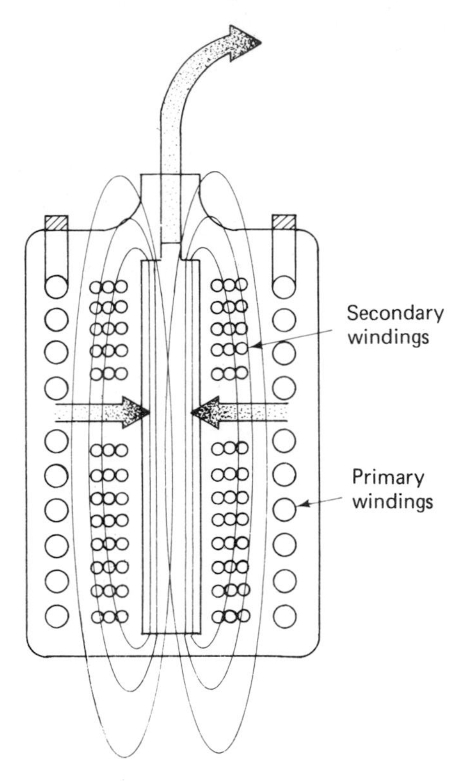

FIG. 7-6 Collapsing lines of force induce voltage in secondary coils.

ings have many more turns of wire. The electrical pressure from all the secondary windings totals 20,000 to 30,000 volts (Fig. 7-6).

One end of the secondary winding is attached to one of the primary terminals and the other end is connected to the high-voltage tower terminal. The high-voltage surge goes from the tower to the distributor. The distributor cap and rotor send the surge to the spark plug in the cylinder that is ready for ignition.

TYPES OF COILS

All coils serve the same purpose, although there are differences in appearance and location. For instance, some types, as noted in Fig. 7-7, look like transformers (*all* coils are really transformers) and are mounted inside the distributor or on the side of the engine or on the fender apron. Other coils are of the more typical cylinder shape (noted in the same illustration) and are mounted on the engine, fender apron,

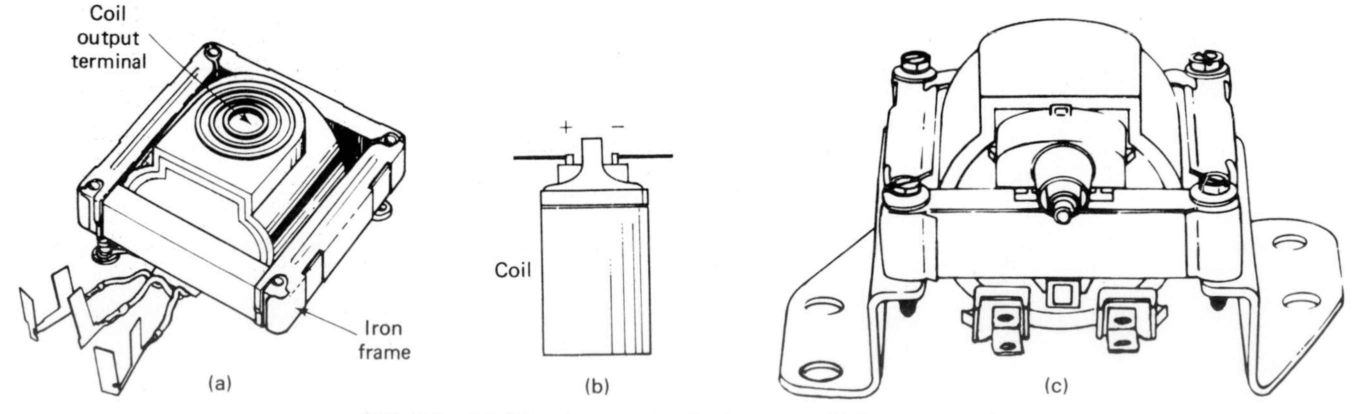

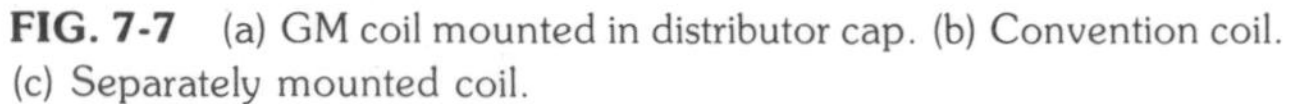

FIG. 7-7 (a) GM coil mounted in distributor cap. (b) Convention coil. (c) Separately mounted coil.

or firewall. These types of coils typically are enclosed in an insulating plastic shell and topped with an insulating cap and tower assembly. Some systems have multiple coils, as noted in the section titled "Distributorless Ignition System."

BRAIN OF THE IGNITION SYSTEM AND CONTROL INFORMATION

If the coil is the heart of the ignition system, the components that periodically interrupt the primary circuit and then distribute the resulting high-voltage surges comprise the brain.

The main control requirement is deciding when and how long to break the primary circuit. The main information needed to perform this operation is the piston's position. The interruption of primary current and distribution of secondary surges must be timed to the movement of the piston up and down in the cylinder. (Information about engine speed and load are also necessary, but we'll wait until later in the chapter for that.)

In precomputerized engines, a distributor drive shaft connected by a spiral drive gear to the camshaft was solely responsible for supplying piston-position information. The distributor drive shaft turns at one-half crankshaft speed. In computerized engines, a sensor on the end of the crankshaft may supply piston-position information. However, the distributor drive shaft also supplies information in most cases (Fig. 7-8).

Connected to the top of the distributor drive shaft are a rotor button and some kind of switch or trigger device:

- The button rotates with the distributor shaft. As it turns, high-voltage surges from the coil's secondary are routed through the button and out the distributor cap to the proper spark plug. Both precomputerized and computerized systems have rotor buttons.
- The switch or trigger element is responsible for interrupting the primary current flow. In precomputerized systems, the trigger was always located inside the distributor. In computerized sys-

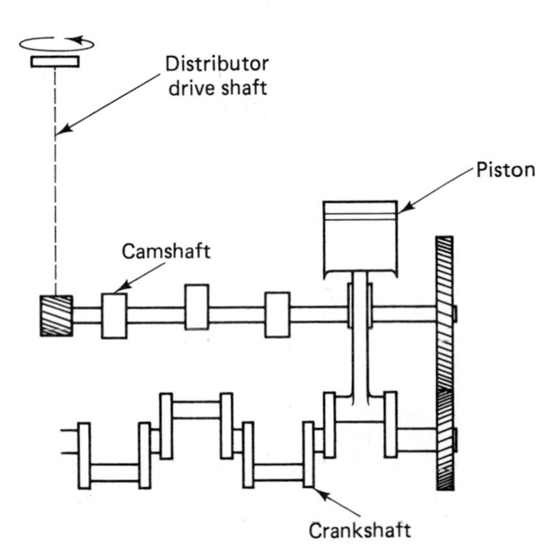

FIG. 7-8 A direct connection from the distributor drive shaft to the camshaft to the crankshaft to the piston supplies information about piston position.

tems the trigger may be located in the distributor, or it may be at the end of the crankshaft, or it may be in both places.

PRECOMPUTERIZED TRIGGER SWITCHES

Mechanical breaker points

Up until the 1970's, most ignition systems used a mechanical cam and breaker point assembly to interrupt the primary circuit.

The cam is connected to the end of the distributor drive shaft. As it rotates, the cam lobes push against a rub block to separate the contact points. Primary current flow through the points is interrupted, causing the high-voltage discharge from the secondary windings. As a cam lobe rotates to a low position, a spring tensioner closes the points. The primary circuit is once again complete, ready for the next firing cycle. Figure 7-9 pictures a typical breaker-point assembly.

Contact-controlled transistorized systems

As far back as the late 1950's, an alternative was available for points-type ignition systems. Using the recently introduced transistor as a control element, this system promised higher voltages for the spark plug and longer point life. It was the forerunner of current electronic systems.

Here is how it works: In addition to all the elements of a conventional, points-type system, a contact-controlled transistorized system contains an amplifier module. The main component in the amplifier is a transistor.

The primary circuit is divided into two parts:

1. A low-voltage control circuit goes from the points to the base of the transistor in the amplifier module.

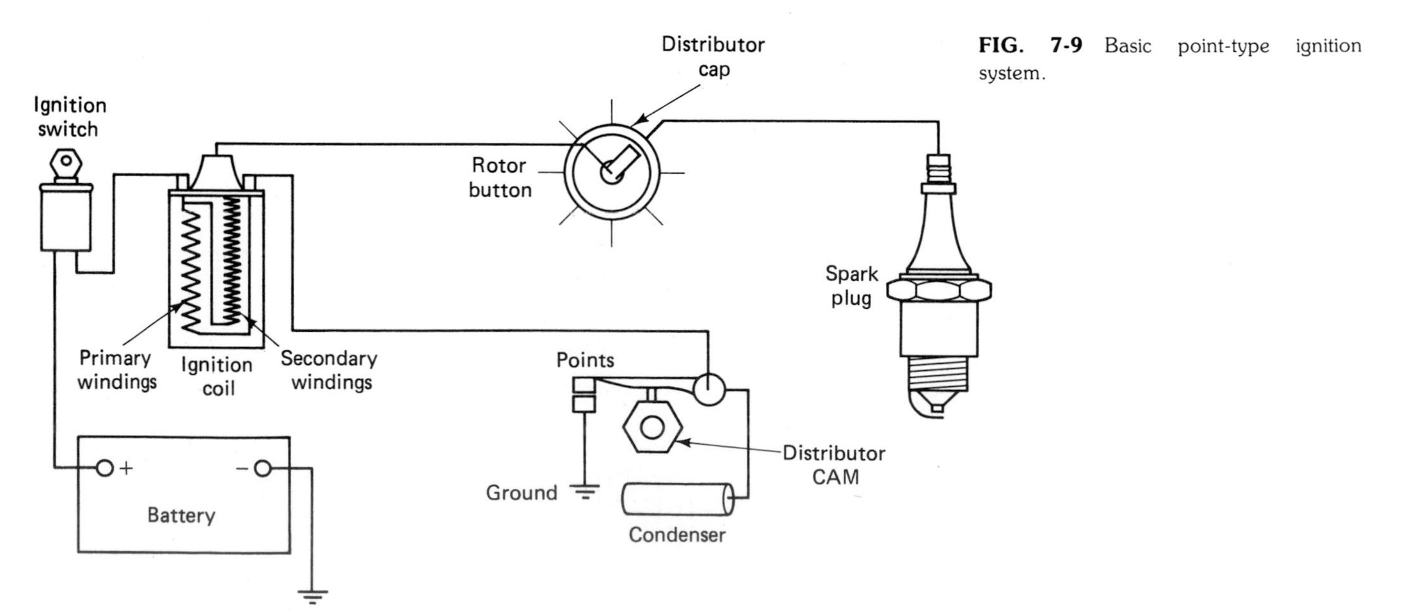

FIG. 7-9 Basic point-type ignition system.

2. Battery-level voltage goes from the battery, through the ignition switch, through the transistor's collector and emitter to the primary terminal of the coil.

When the points open the circuit to the transistor base, flow through the higher-voltage emitter-collector circuit is also interrupted. (See Chapter 4 for transistor operation.) This results in a high voltage surge from the coil's secondary. When the points close the base circuit, current flow through the primary is also resumed.

Such systems have several advantages:

- Reduced current flow in the points control circuit reduces arcing and prolongs point life.
- Higher voltage in the main part of the primary circuit results in higher secondary voltages and improved spark plug performance.

However, despite such advantages, contact-controlled transistorized systems still relied on a mechanically operating point set and distributor cam. Components in rubbing or rotating contact tend to wear out at a fairly rapid rate.

BREAKERLESS IGNITION SYSTEMS

The solution to point-related problems was to eliminate the points—in other words, to create a system that senses piston position and interrupts the primary circuit in another way. The result was breakerless ignition systems. Such systems are still used in many computer-controlled systems.

Breakerless ignitions are presently one of two types: induction coil pickup systems and Hall-effect systems. As described in Chapter 2, both of these systems look somewhat alike and operate in similar ways.

Induction coil pickup systems

Induction coil pickup systems were among the first types of breakerless systems. The parts located inside the distributor included a stationary pickup assembly and a rotating trigger wheel.

- The pickup assembly consists of a core element enclosed in a coil wrapping at one end and attached to a permanent magnet at the other. Two leads are attached to the coil.
- The trigger wheel is attached to the rotating shaft. The wheels usually have one tooth per cylinder.

The main component located outside the distributor is a unit alternatively called the amplifier module or the electronic control unit. Its main component is a transistorized switch circuit. The main elements of an induction coil pickup system are shown in Fig. 7-10.

The induction coil pickup system operates in this manner: as the trigger wheel rotates, its teeth pass by the pickup coil. When the teeth

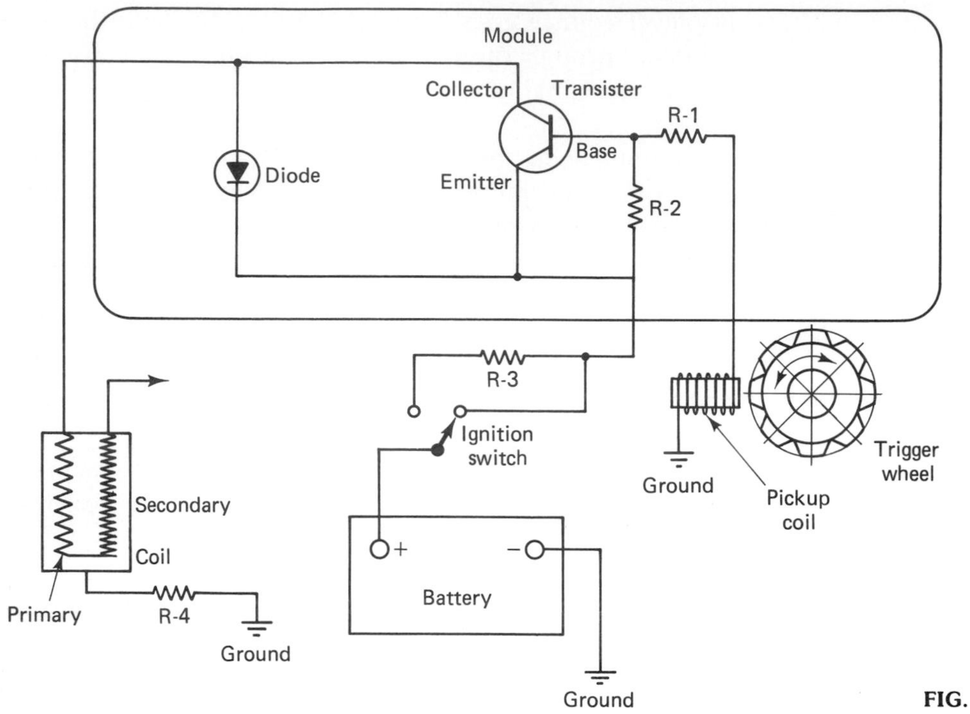

FIG. 7-10 Basic breakerless primary circuit.

and the core piece line up, the lines of force from the permanent magnet are concentrated. When the teeth and the core piece move apart, the lines of force diminish. As a result, lines of force balloon in and out across the pickup coil. This causes a pulsating dc current to be transmitted through the two leads attached to the coil. Figure 7-11 pictures the process.

The current from the two leads goes to the amplifier module. Although the current is not very strong, it is sufficient to operate the base of a transistor. The transistor acts like a solenoid switch to control current flow through the ignition coil.

Many types of induction coil pickup systems have been and are now being used:

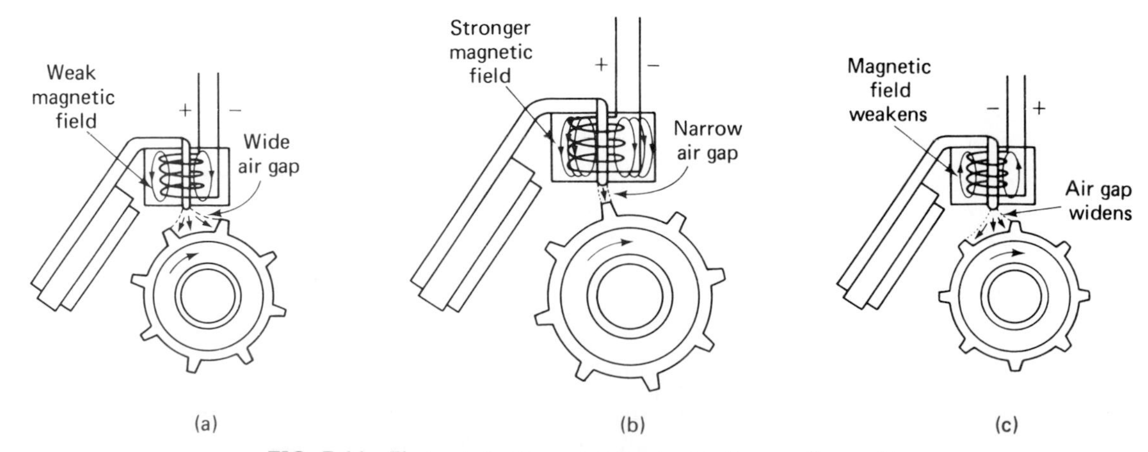

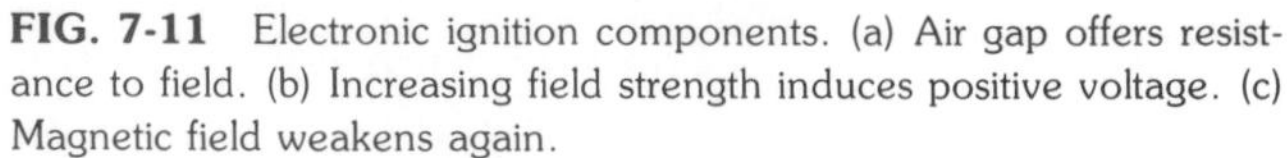
FIG. 7-11 Electronic ignition components. (a) Air gap offers resistance to field. (b) Increasing field strength induces positive voltage. (c) Magnetic field weakens again.

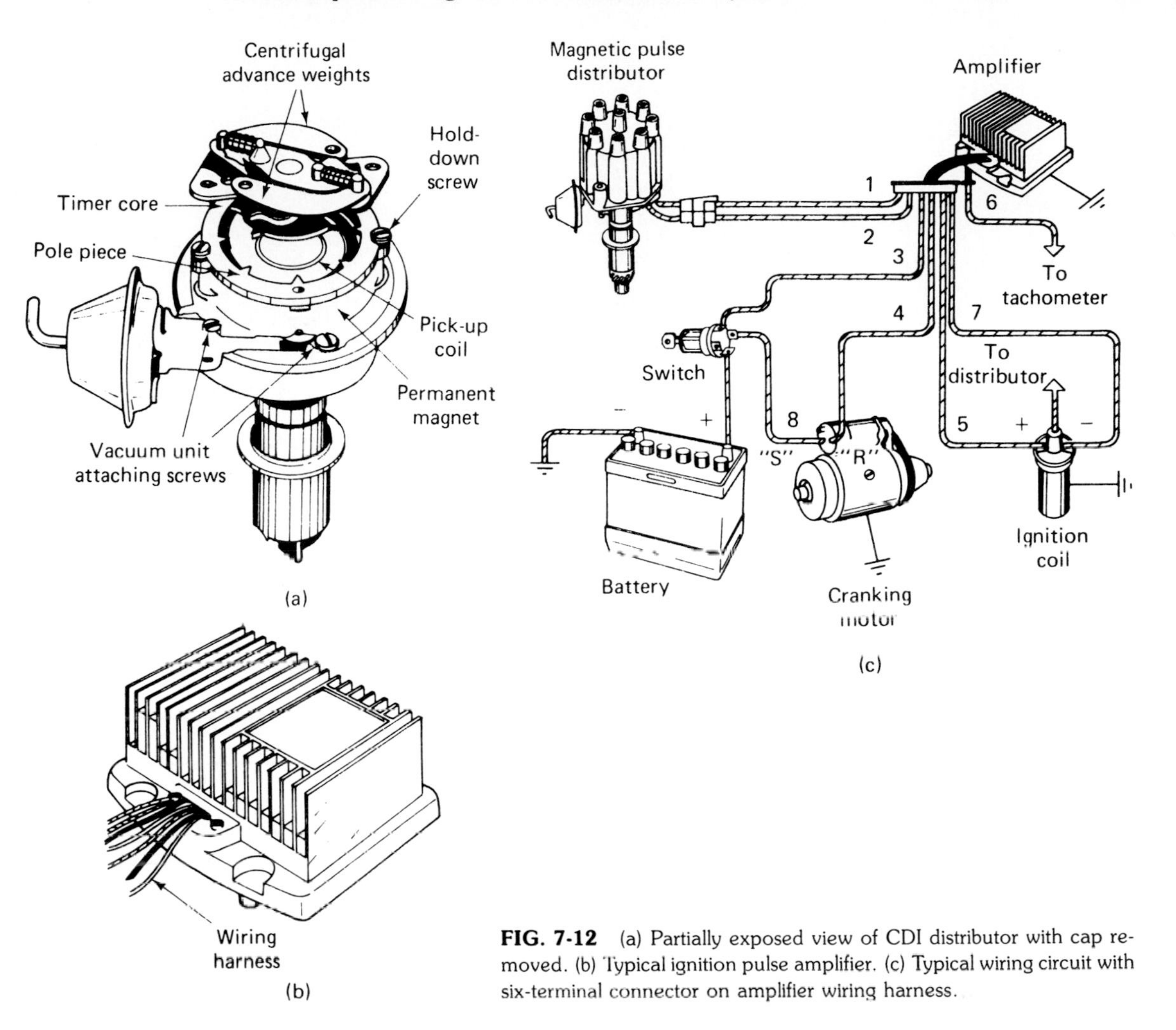

FIG. 7-12 (a) Partially exposed view of CDI distributor with cap removed. (b) Typical ignition pulse amplifier. (c) Typical wiring circuit with six-terminal connector on amplifier wiring harness.

- One of the first breakerless ignitions was Chrysler's capacitor discharge ignition (CDI). The pulse amplifier in the CDI was a collection of electronic components (diodes, transistors, thyristors, resistors and capacitors). On a signal from the distributor pickup coil, the amplifier would send a 300-volt burst of electricity into the coil's primary circuit. Figure 7-12 pictures a CDI system.
- GM HEI (high-energy ignition) distributors use a pickup assembly that surrounds the trigger wheel (timer) with an equal number of internal teeth. Ford and Chrysler systems generally only have one pickup point. Figures 7-13 and 7-14 picture GM and Chrysler systems.

Hall-effect Systems

Starting in 1977 with the Chrysler Omni, several manufacturers began using Hall-effect triggers to sense rotation of the distributor driveshaft. As described in Chapter 2, a Hall-effect system has components similar to those of an induction coil pickup system. However, instead of teeth on the trigger wheel, shutter blades are used. When a shutter blade passes between a magnet and a pickup unit (called the Hall-effect sen-

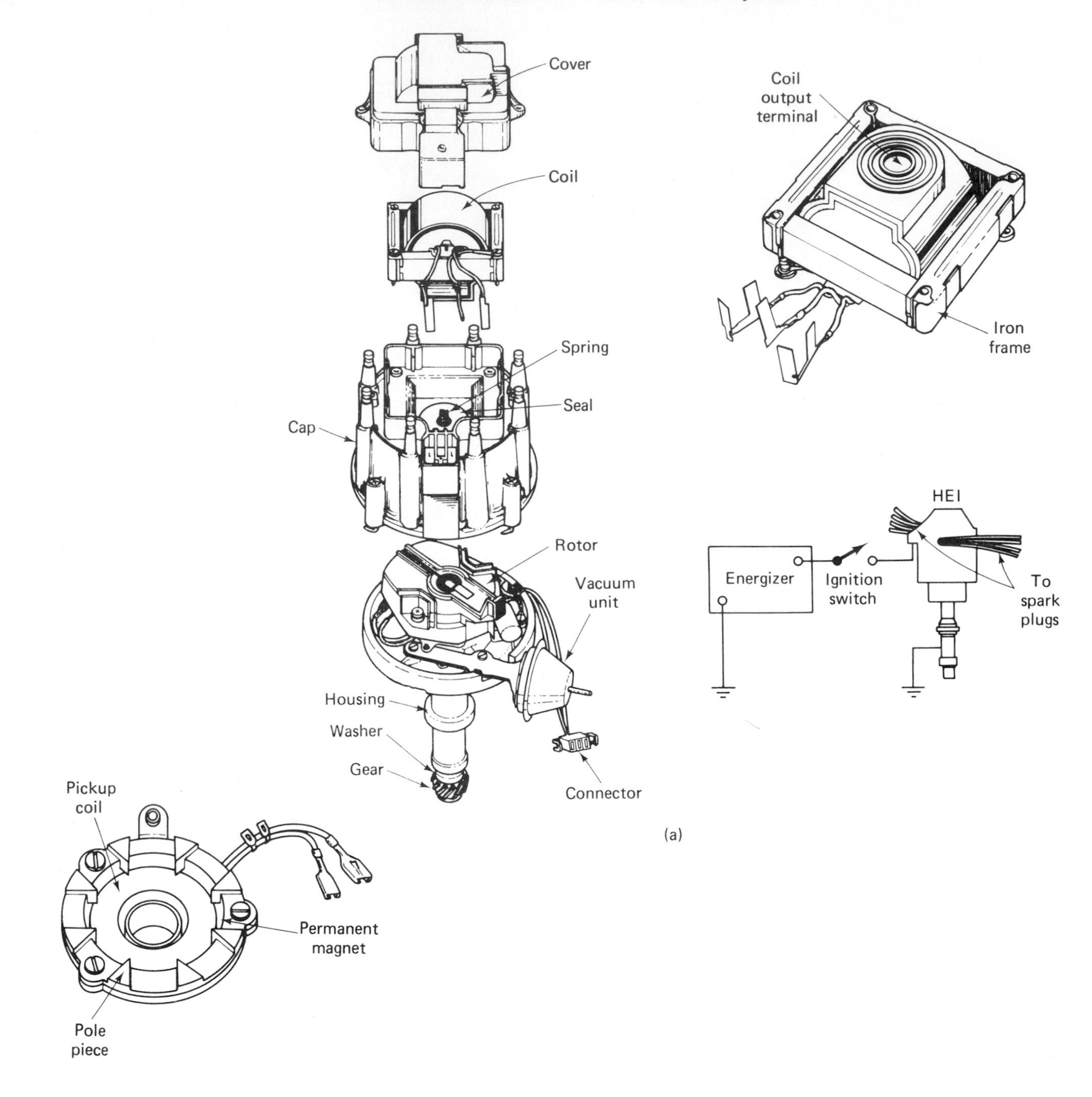

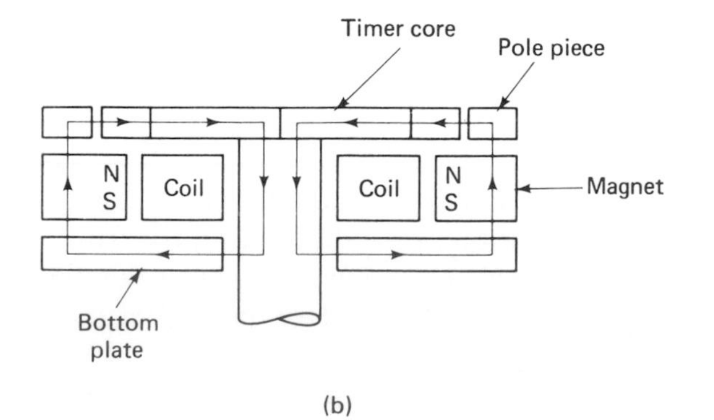

FIG. 7-13 High-energy ignition system. (a) H.E.I. distributor assembly pickup coil. (b) Magnetic lines of force through pickup coil.

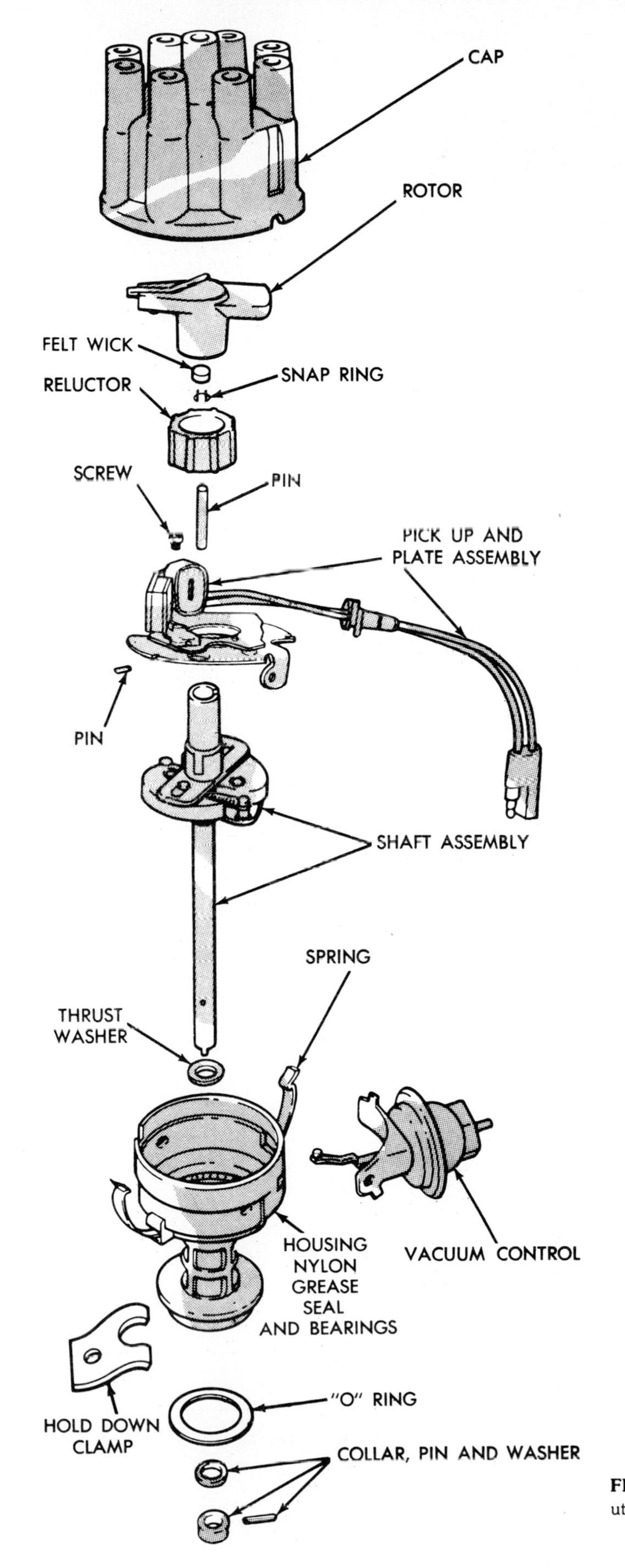

FIG. 7-14 Chrysler magnetic pickup distributor.

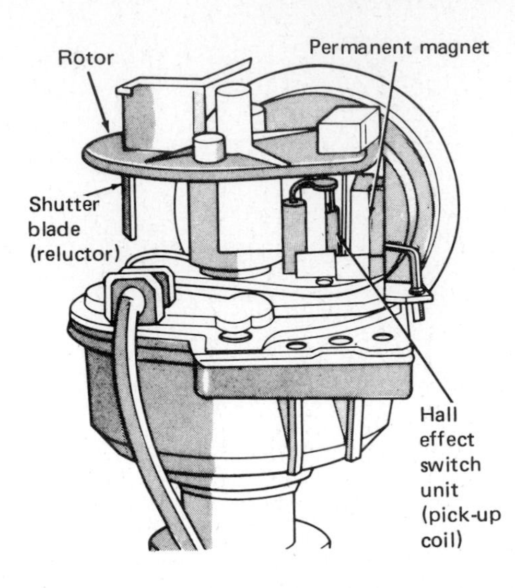

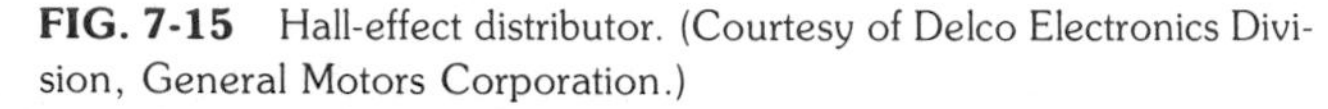

FIG. 7-15 Hall-effect distributor. (Courtesy of Delco Electronics Division, General Motors Corporation.)

sor, or trigger), the voltage level in the sensor circuit changes. Hall-effect systems are intended to reduce signal distortion that results from engine-speed changes. A GM Hall-effect unit is shown in Fig. 7-15.

DISTRIBUTORLESS IGNITION SYSTEMS

A few manufacturers have, at the time of this writing, eliminated the distributor altogether in certain models. It is possible that more manufacturers will do away with the distributor over the next several years.

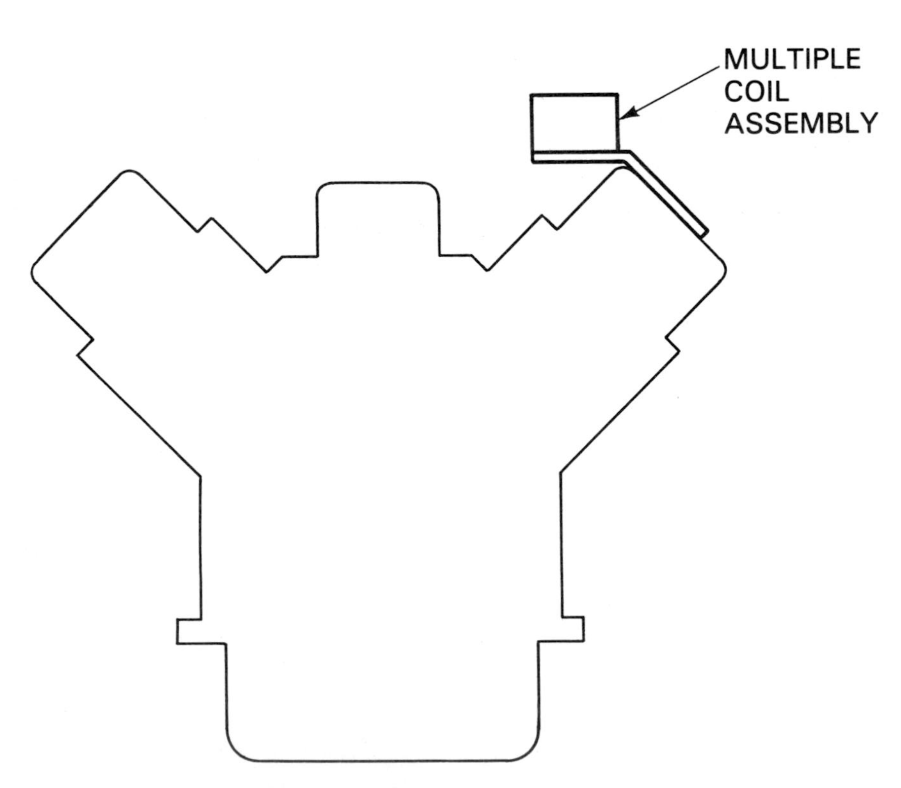

FIG. 7-16 Multiple coil system.

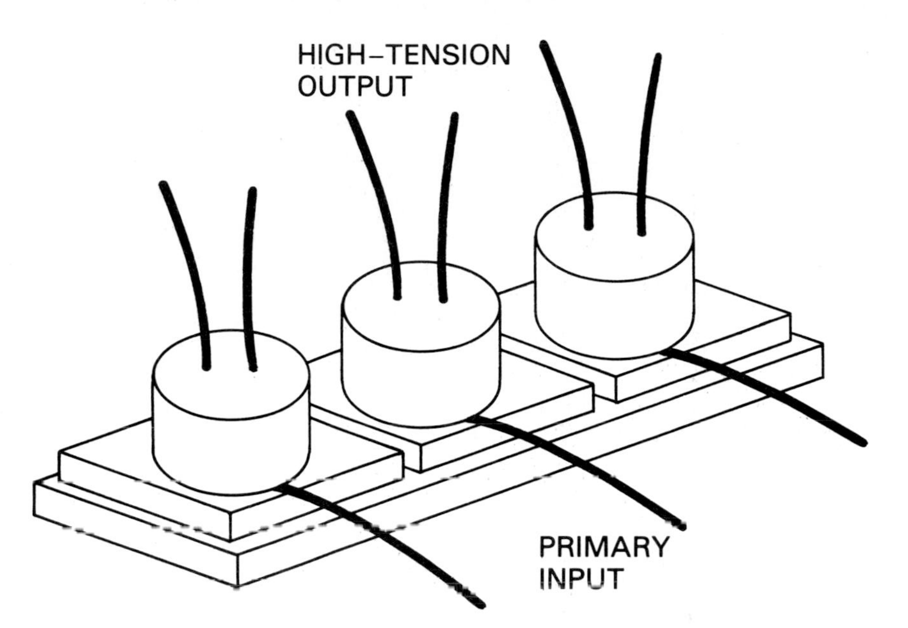

FIG. 7-17 Module and coil assembly.

In a distributorless system, all the intelligence that was once located in the distributor is now located inside a control computer. It controls the flow of low-voltage, primary current to the coil, interrupting the flow to produce the high-voltage surges from the secondary circuit.

Typically, such systems use multiple ignition coils. The system shown in Fig. 7-16 and 7-17 uses three ignition coils, one per every two cylinders in the V-6 engine being served.

The coil pictured is mounted on top of the engine. It has two high-tension terminals and one low-voltage, primary terminal. The high-tension leads are connected to the spark plugs in two paired cylinders. Both spark plugs are fired simultaneously. The primary leads are connected to the control computer. The electrical schematic in Fig. 7-18 pictures this arrangement.

When the primary circuit in one of the coils is interrupted by the computer, the high-voltage surge is directed along the two high-tension leads to the two paired cylinders. The cylinders are matched so that when one is in the compression stroke, the other is in the exhaust stroke. Very little energy is required to fire the cylinder in the exhaust stroke, so more energy is available for firing the one in the compression stroke. As the engine rotates, the circumstance is reversed and the other member of the pair moves into the compression position and is fired.

In this system, the information needed by the computer to fire the spark plugs comes from a Hall-effect sensor located in the engine front cover timing assembly. The unit is pictured in Fig. 7-19. Every time the magnetic interrupter mounted on the camshaft gear passes by the stationary Hall-sensor, a pulse is produced. The control computer uses

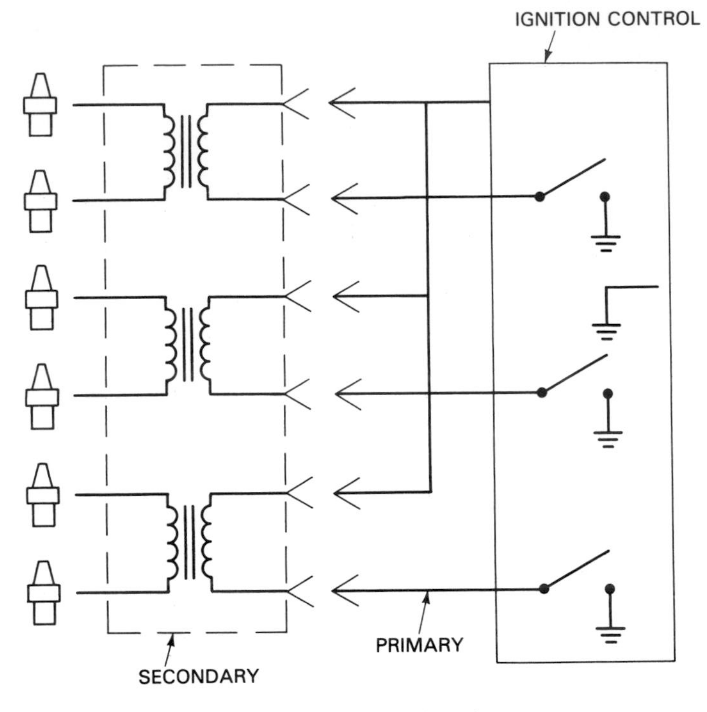

FIG. 7-18 Schematic of type II distributorless ignition system.

this signal to time ignition operation (and, in this system, to time the operation of electronic fuel injectors).

In the engine being described, another Hall-effect sensor is also mounted on the engine timing cover. The Hall-effect switch is activated by three vanes with windows spaced 120 degrees apart. The vanes, which are permanently mounted to the rear of the front crankshaft pulley, pass between the fixed-position magnet and the Hall-effect switch. The signal produced is used to provide engine rpm information. Figures 7-20 and 7-21 picture the crankshaft sensor.

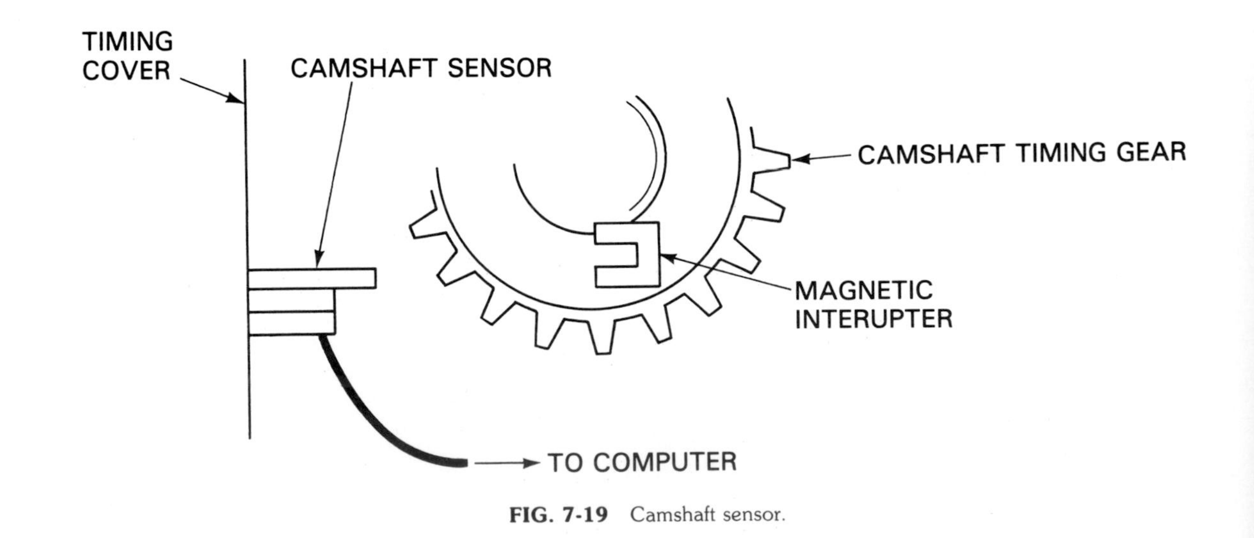

FIG. 7-19 Camshaft sensor.

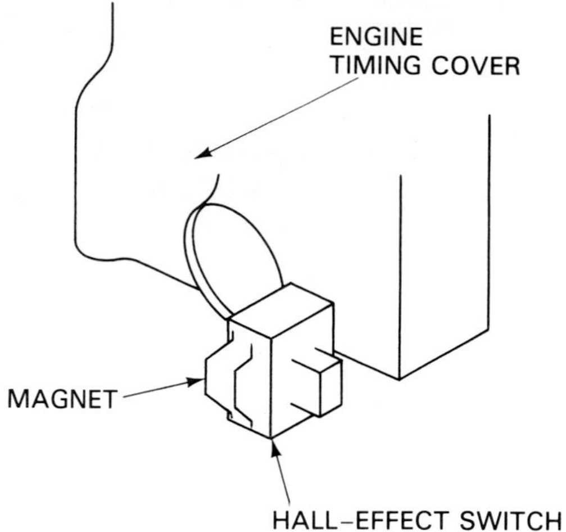

FIG. 7-20 Crankshaft sensor.

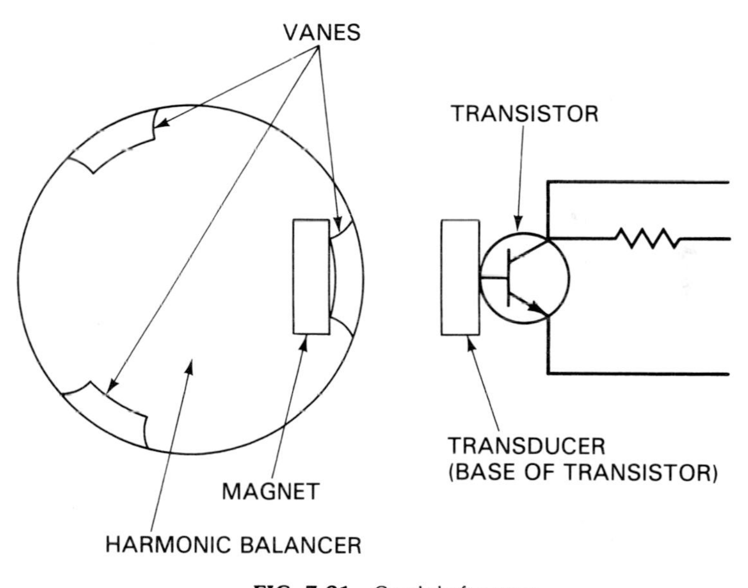

FIG. 7-21 Crankshaft sensor.

TIMING ADVANCE

In a previous section, we noted that ignition-control systems also needed to know about engine speed and load. That is because spark timing must be advanced or retarded as changes in speed and load change the air/fuel mixture going into the combustion.

Today, the on-board computer, using information from its sensors, controls ignition timing. In precomputerized systems, mechanical and pneumatic (air-operated) devices controlled timing. Following is discussion of those devices, as well as a review of reasons for having timing advances. Pay particular attention to the review, because it applies to both computerized and precomputerized systems.

REASONS FOR HAVING TIMING ADVANCE

At idle speed, the spark is timed to occur just before, just as, or just after the piston is at TDC (top dead center) in the compression stroke. However, as the engine speed increases, the spark must be advanced. That is, it must be made to occur earlier in the stroke. The time to burn a given amount of fuel remains much the same, regardless of engine speed. However, at faster engine speeds there is less time available, so the burning process must be started sooner (Fig. 7-22).

The burning process must also be started earlier when the fuel delivery system is at part throttle. At that time there will be a partial vacuum in the intake manifold and cylinders. Because of the partial vacuum, less fuel will be drawn into the cylinders during the intake stroke. The reduced amount of air/fuel mixture will not be compressed as much in the compression stroke as a full mixture. When the air/fuel mixture is less dense, it burns slower. So the burning process must be started sooner.

There are two principal ways in precomputerized systems to cause the burning process to begin earlier: the centrifugal advance and vacuum advance methods. The centrifugal advance is generally responsive to engine speed and the vacuum advance to carburetor and/or load conditions.

Centrifugal (or mechanical) advance

In the centrifugal or mechanical advance systems, the timing is advanced by changing the position of the distributor cam. The cam is attached to the end of the distributor shaft in such a way that the cam can be rotated a certain amount independently of the shaft's rotation. When rotated in the proper direction, the cam lobes will strike the points'

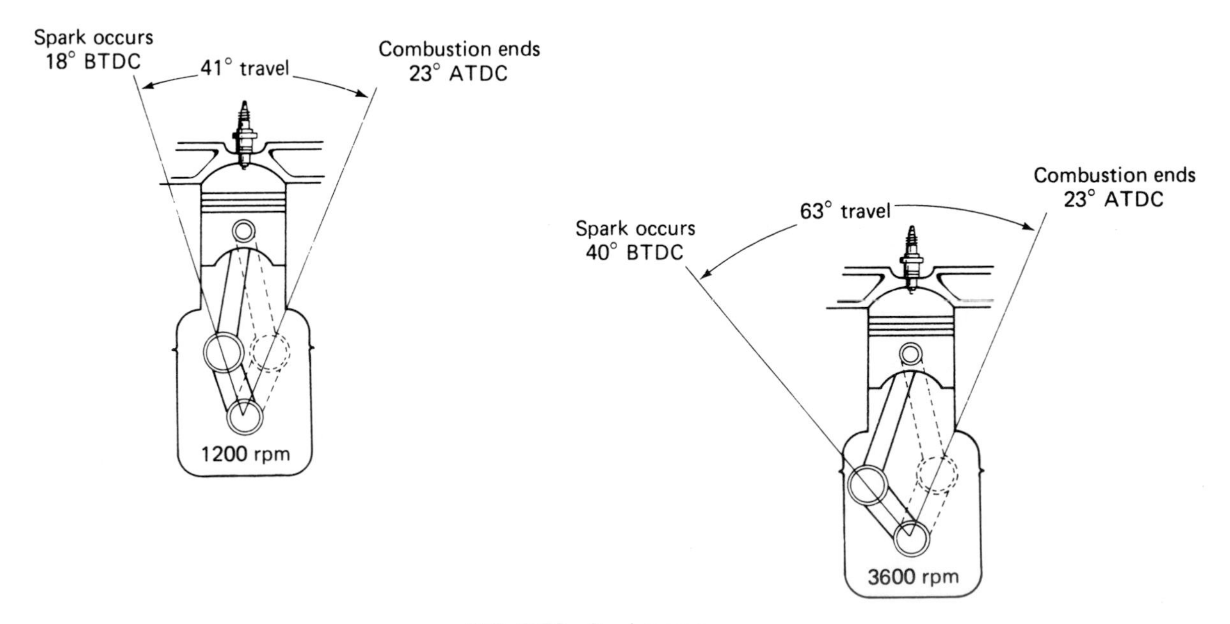

FIG. 7-22 Spark timing.

rubbing block earlier and thereby cause the points to open earlier. The faster the engine runs, the greater the advance will be.

Figure 7-23 shows a GM-type centrifugal advance mechanism, and Fig. 7-24 shows the centrifugal advance used on most other cars.

Vacuum method

Mechanical or centrifugal advance systems move the breaker cam to advance the spark. Vacuum systems move the distributor plate.

The breaker plate is a movable platform on which the points are mounted inside the distributor. The cam comes up through a hole in the center of the plate. The plate is movable so that it can be rotated around the cam.

When the plate is rotated, it changes the position of the moving arm's rubbing block with respect to the cam lobes. Depending on the position of the plate, the rubbing block will engage the lobes earlier or later. This, in turn, affects the times that the points open and close, which affects the times that the spark plugs fire.

In most vacuum systems, an air-pressure-sensitive diaphragm is used to change the position of the distributor plate. The diaphragm is located on the side of the distributor and is connected by an air line to the carburetor. The diaphragm and its linkages are arranged in such a way to move the distributor plate into an advanced position whenever there is a vacuum in the line from the carburetor.

The air line to the carburetor is connected just above the carburetor throttle plates when the plates are in the idle position, as shown in Fig. 7-25. At idle speed, the air line registers the atmospheric pressure above the throttle plates. At this time there is no movement of the distributor plate, and consequently no spark advance.

However, as the engine speeds up, the throttle begins to swing open and exposes the air line to the partial vacuum created in the cylinders. This causes the distributor plate to move into an advanced position. Then, as the throttle moves to a fully open position, the vacuum in the cylinders (and hence the air line) is reduced. The distributor plate is then returned back toward the basic timing setting by spring action.

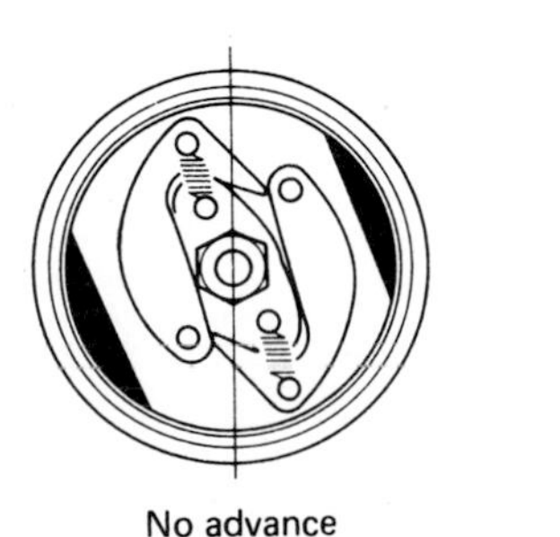

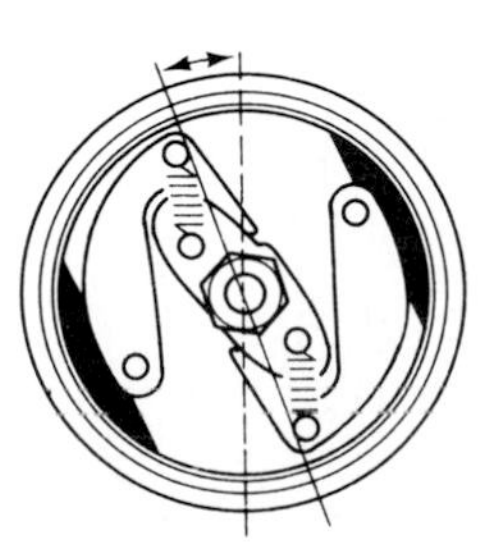

FIG. 7-23 GM centrifugal advance.

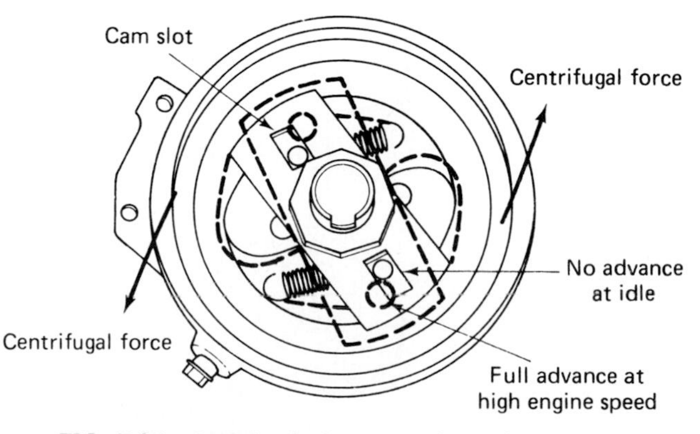

FIG. 7-24 Centrifugal advance used on other engines.

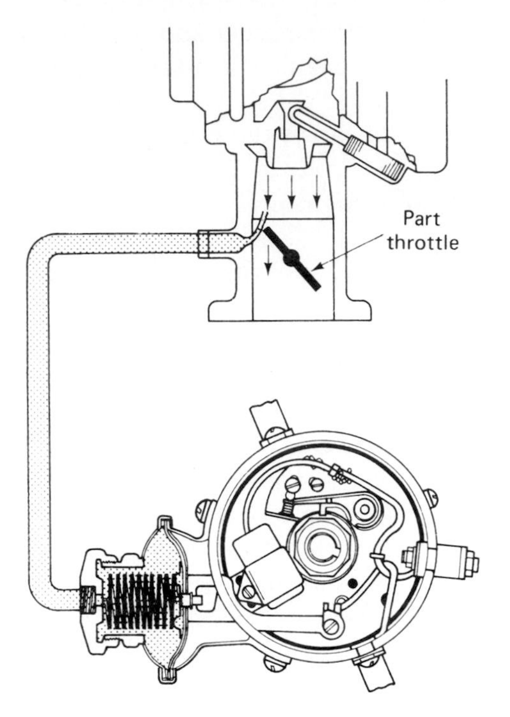

FIG. 7-25 Vacuum advance.

Many engines have both centrifugal and vacuum advance mechanisms. At any given engine speed or throttle setting, one or both systems might be in operation. See Fig. 7-26 for the operation of a typical vacuum advance system.

Basic timing

In any sort of timing advance system, there must be a starting point. This is called the *basic timing setting*. It is the position of the distributor plate and cam at idle speed when none of the advance systems are in operation. In older engines the basic timing is usually 5° to 10° before TDC. Newer pollution-control engines often operate at 0° TDC or as much as 4° ATDC (after top dead center).

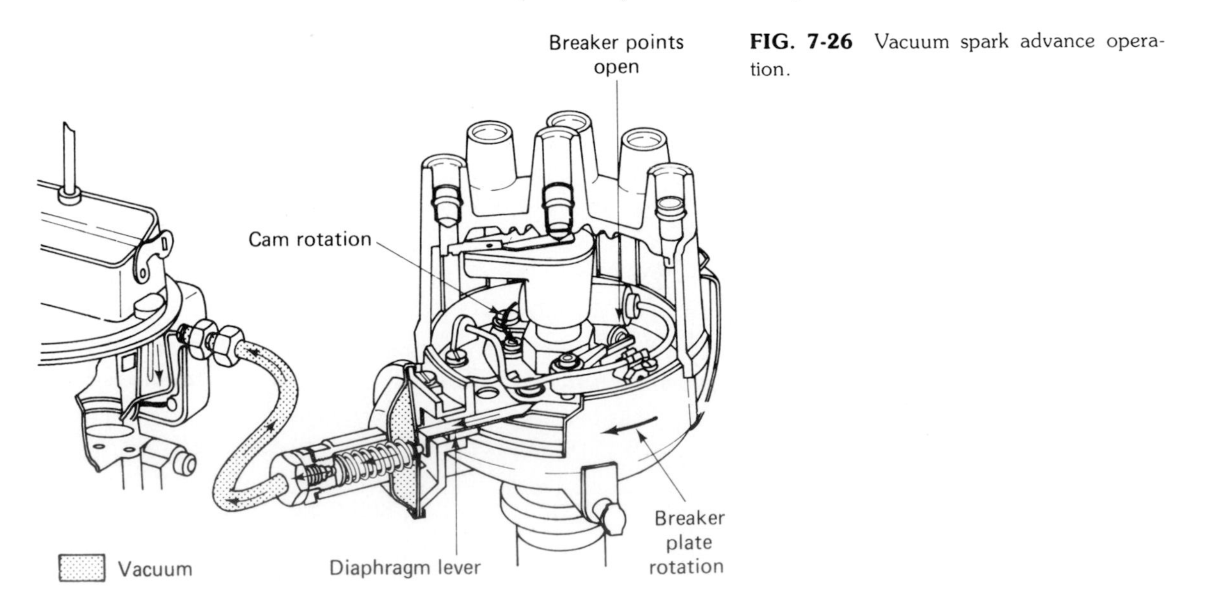

FIG. 7-26 Vacuum spark advance operation.

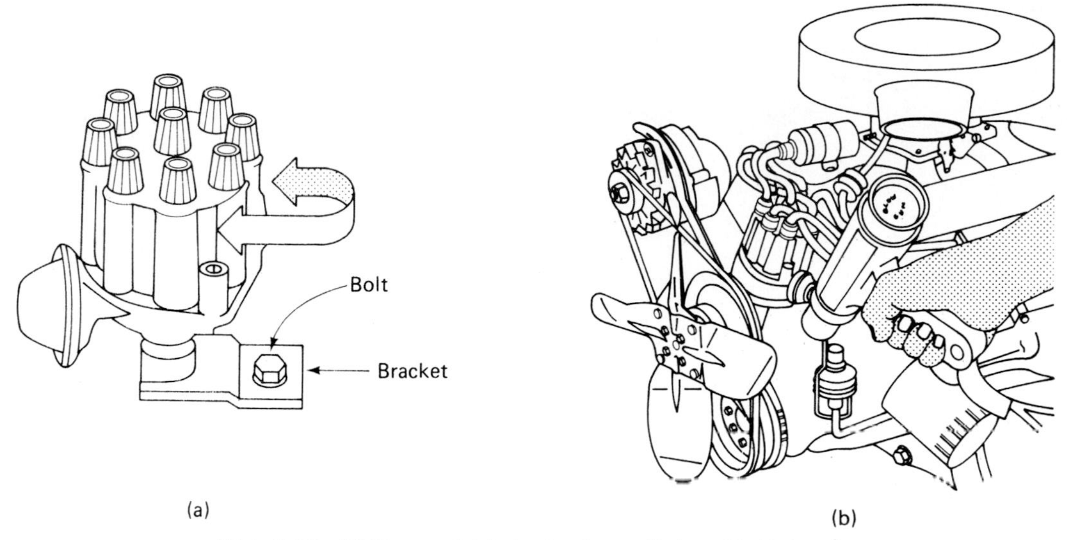

FIG. 7-27 (a) Moving distributor to change timing. (b) Aiming the timing light.

Timing advance is usually fixed, unless the mechanism breaks. However, basic timing is adjustable according to manufacturers' specifications. Basic timing is usually adjusted by loosening the distributor lock-down bolt(s) and moving the distributor housing clockwise or counterclockwise [Fig. 7-27(a)]. The correctness of the adjustment can be determined by shining a timing light on markings inscribed on one of the crankshaft pulleys or on a plate alongside the pulley [Fig. 7-27(b)].

FUEL SYSTEM CONTROLS

Stoichiometric ratio

The ideal fuel mixture contains 14.7 parts of air for each part of fuel. This so-called stoichiometric ratio results in the least pollution, the most power, and the best economy. However, in order to achieve a true stoichiometric ratio, these proportions must exist throughout the fuel mixture, down to individual molecules (because that is where the actual burning takes place). Even though the correct amounts of air and fuel are in the same general vicinity, a number of factors can affect the mixing process and thus the actual mixture available for burning.

It is the function of the fuel-control system, whether computer or mechanically operated, to adjust the mixture according to changes in operating conditions.

Carburetor operation

Today, most fuel systems are based on fuel injectors and computers. Not long ago, they relied on carburetors and mechanical controls. Later chapters in the book discuss computer-controlled injection systems. The remainder of this chapter briefly reviews carburetor-based control functions.

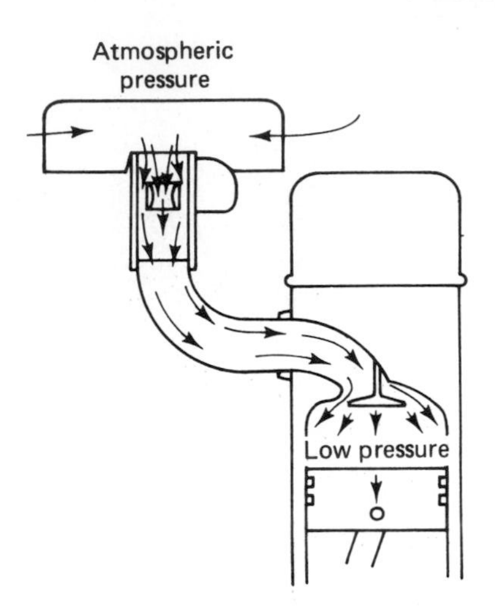

(1) Low pressure in cylinder "pulls" air through intake system (in other words, higher atmospheric pressure "pushes" air to lower pressure region)

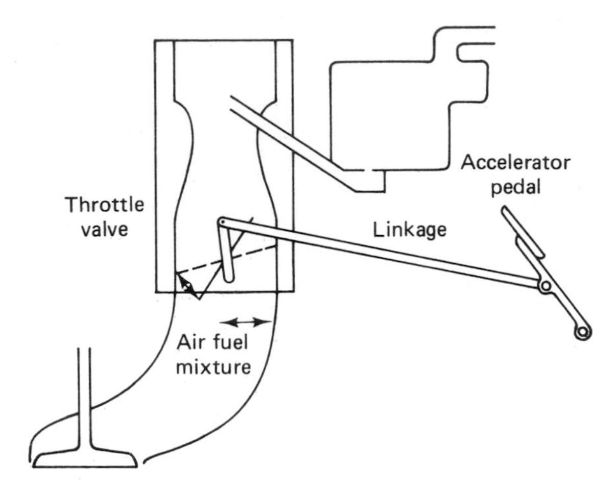

(2) Accelerator pedal controls position of throttle

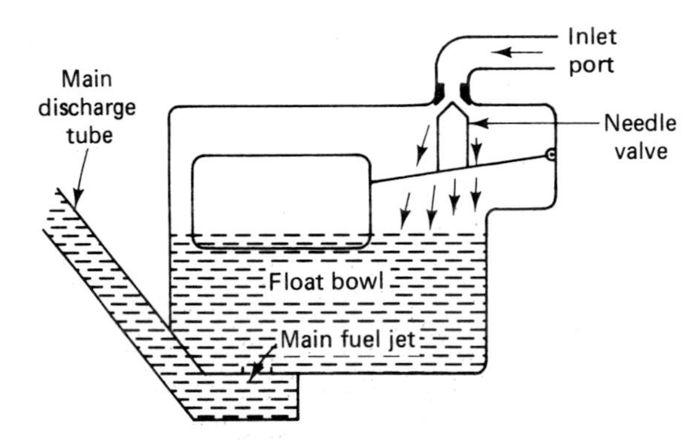

(3) Float controls needle valve and fuel delivery into bowl

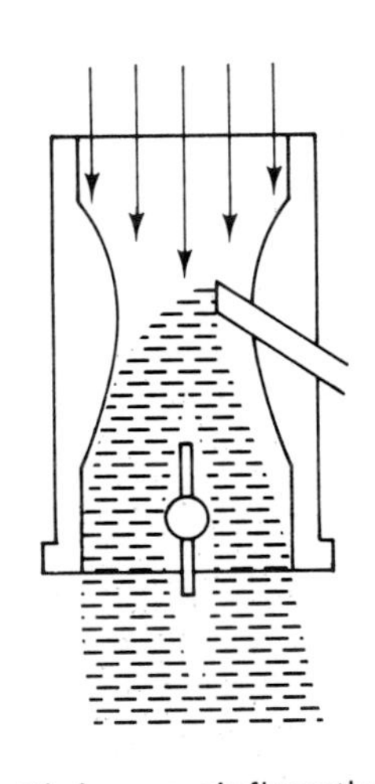

(4) When throttle is open, air flows through venturi (and manifold vacuum is low)

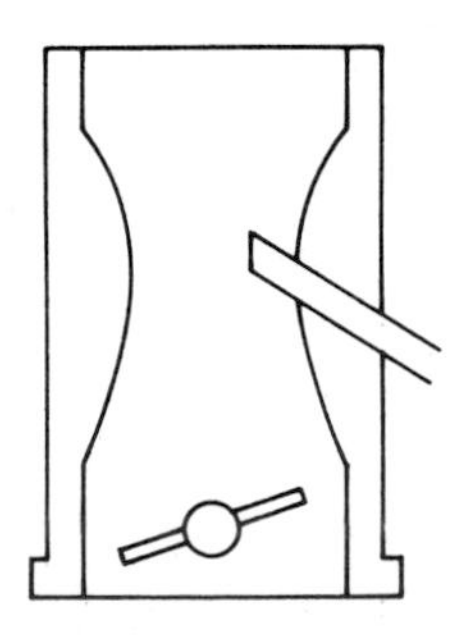

(5) When throttle closes, airflow is greatly reduced (and manifold vacuum is high)

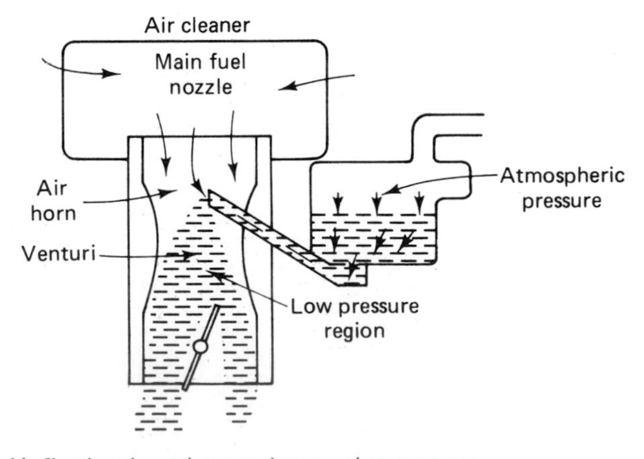

(6) Air flowing through venturi creates low pressure region which "pulls" fuel from float bowl

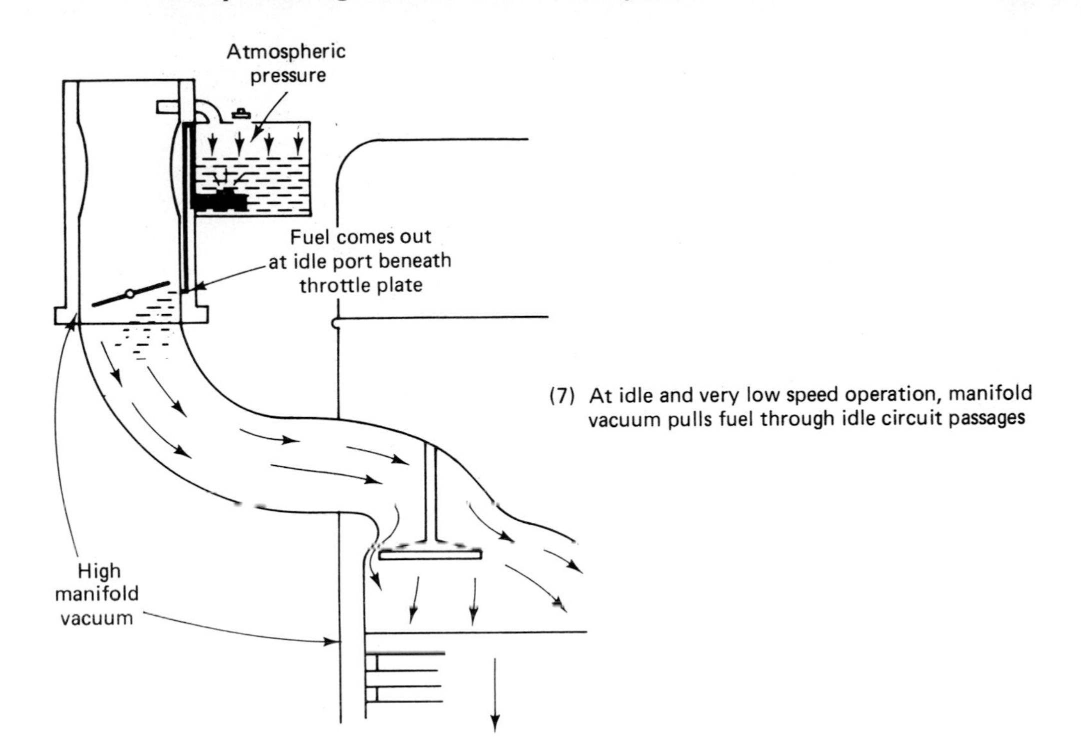
Atmospheric pressure
Fuel comes out at idle port beneath throttle plate
(7) At idle and very low speed operation, manifold vacuum pulls fuel through idle circuit passages
High manifold vacuum

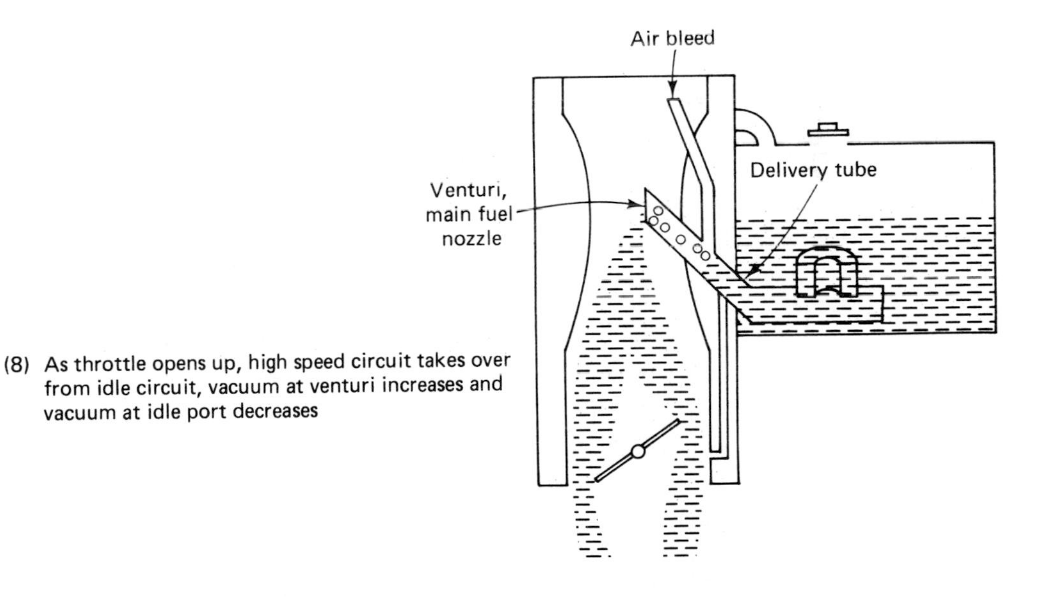
Air bleed
Delivery tube
Venturi, main fuel nozzle
(8) As throttle opens up, high speed circuit takes over from idle circuit, vacuum at venturi increases and vacuum at idle port decreases

FIG. 7-28 Fuel system review.

NORMAL CARBURETOR FUNCTIONS

Figure 7-28 shows the main steps in normal carburetor operation. If you trace the steps, you will notice that air pressure, operating through various passages in the carburetor, controls the flow of fuel through the system.

CHOKE FUNCTIONS

The choke is responsible for adding fuel to the mixture when the engine is first cranked during cold temperatures (when the fuel does not vaporize well and does not mix readily with air).

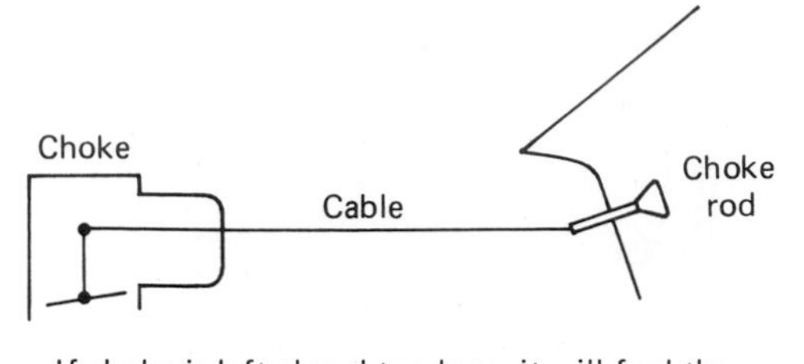

FIG. 7-29 Manual choke.

Chokes were first operated manually according to control programs running in the driver's head. Figure 7-29 shows some of the elements in a manual choke. The main component is a butterfly valve that is located near the top of the carburetor. When closed by the operator for cold-weather cranking, the choke restricts air flow, which in turn, subjects the entire throat of the carburetor to low manifold pressure. This vacuum pulls fuel from every available orifice, creating a very rich mixture (although the amount of air and fuel effectively combined and available for burning is much less).

Because of problems with the control programs contained in many driver's heads, manual chokes were replaced for the most part with automated, mechanical control systems. Figure 7-30 shows an early model automatic choke. It has a bimetal leaf spring combined with a small electric motor. The bimetal spring is mounted on the exhaust

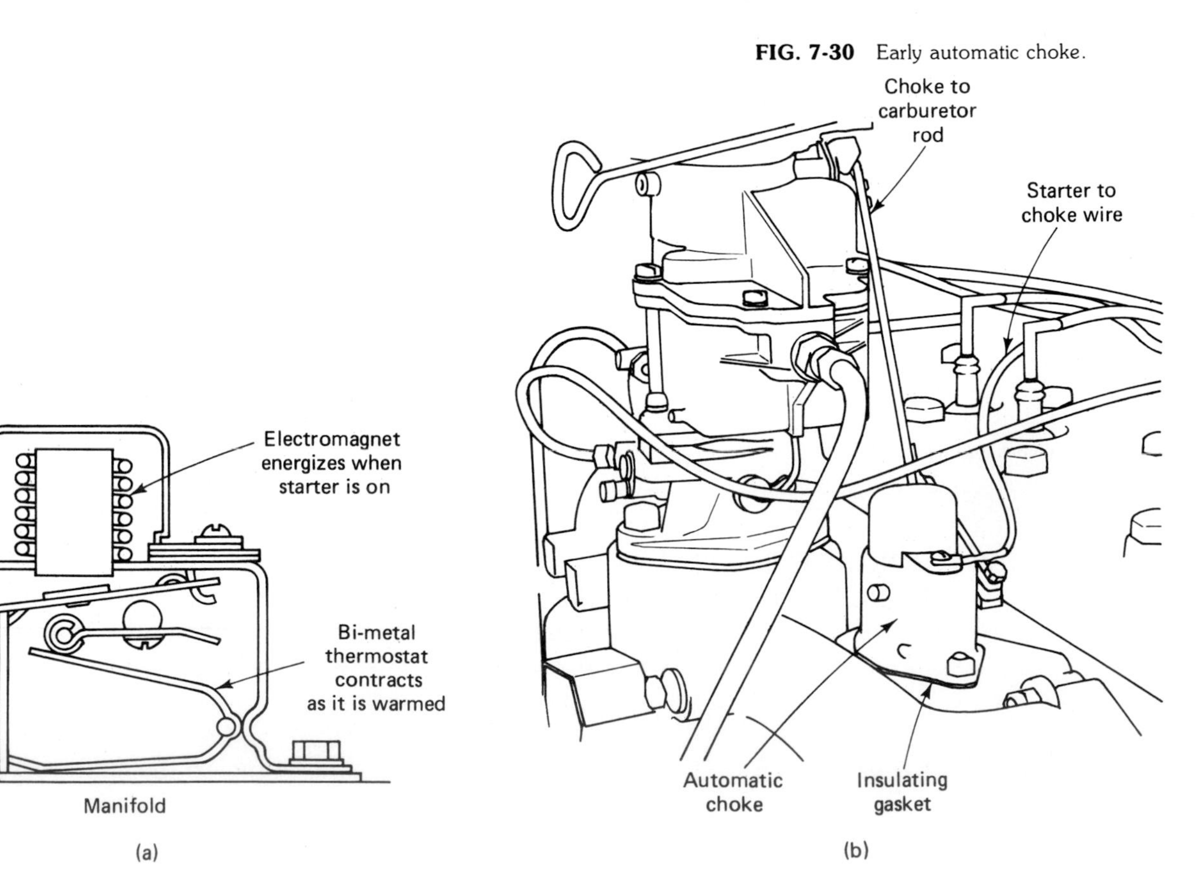

FIG. 7-30 Early automatic choke.

manifold and the solenoid is attached to the choke plate by an arm and rod assembly. When the cranking motor switch is closed, the solenoid is activated and the choke is closed. As the engine warms up, the bimetal spring responds to the heat by bending. Eventually, it bends enough to break the solenoid circuit and let the choke open.

An automatic choke like those used in later-model cars is shown in Fig. 7-31. The bimetal spring is in the shape of a coil and is altogether responsible for closing and opening the choke plate in response to temperature changes.

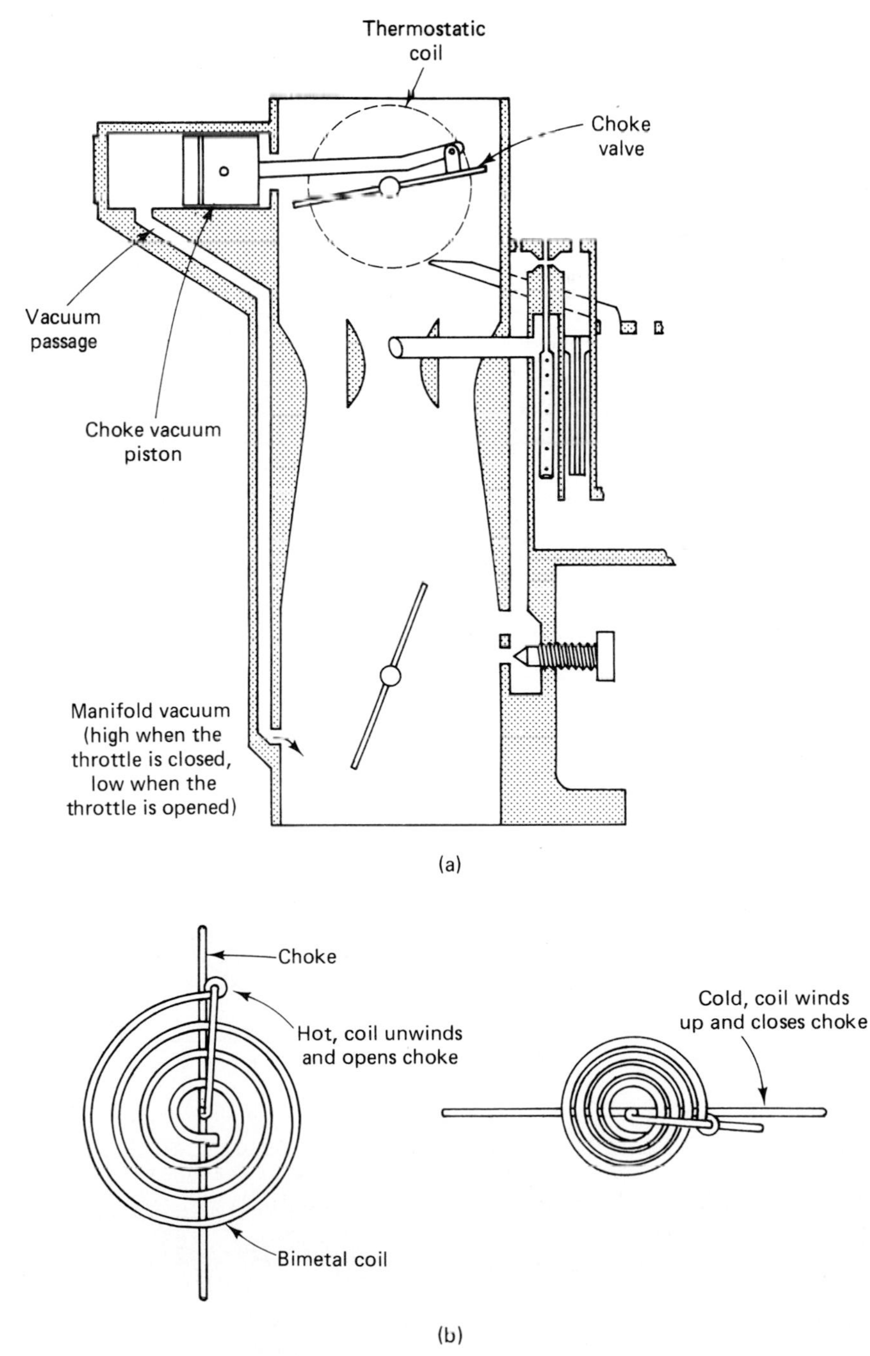

FIG. 7-31 Later-model automatic choke.

OPTIMIZING FUEL-METERING FUNCTIONS

After the engine is started, new sets of conditions affect the fuel-mixing process. For instance, sudden acceleration reduces the manifold vacuum, causing the mixture to become lean. During high-speed operation, the air moves faster than the fuel through the intake manifold passages, which tends to upset the effective mixture available for burning. In addition, heavy loading alters the intake manifold pressure for a given throttle setting, which disrupts the balance of air and fuel.

For optimum operation, it is necessary to make compensating adjustments in the amount of fuel delivered. As in the case of the manual choke, the first systems relied on control programs running in the driver's head. One example is the system used in Model T Fords manufactured in the 1920's. The system used a connection running from a dash-mounted knob to the carburetor needle-valve assembly. By adjusting the position of the knob, the driver could vary the position of the needle valve with respect to its seat. This controlled the amount of fuel delivered, thus altering the mixture, making it richer or leaner as the operator considered necessary.

Model A Fords, manufactured in the late 1920's and early 1930's, combined the mixture control and the manual choke in one knob, as shown in Fig. 7-32. The knob was pulled out for choking and twisted to adjust the mixture. The problem with these and similar systems is that they rely on driver skill, i.e., on the quality of the control programs contained in the driver's head. Skilled drivers could exercise a high degree of control over engine operation. However, ordinary operators often made damaging mistakes. Also, many engine events happen too fast for human response or cannot be detected by human sensory equipment.

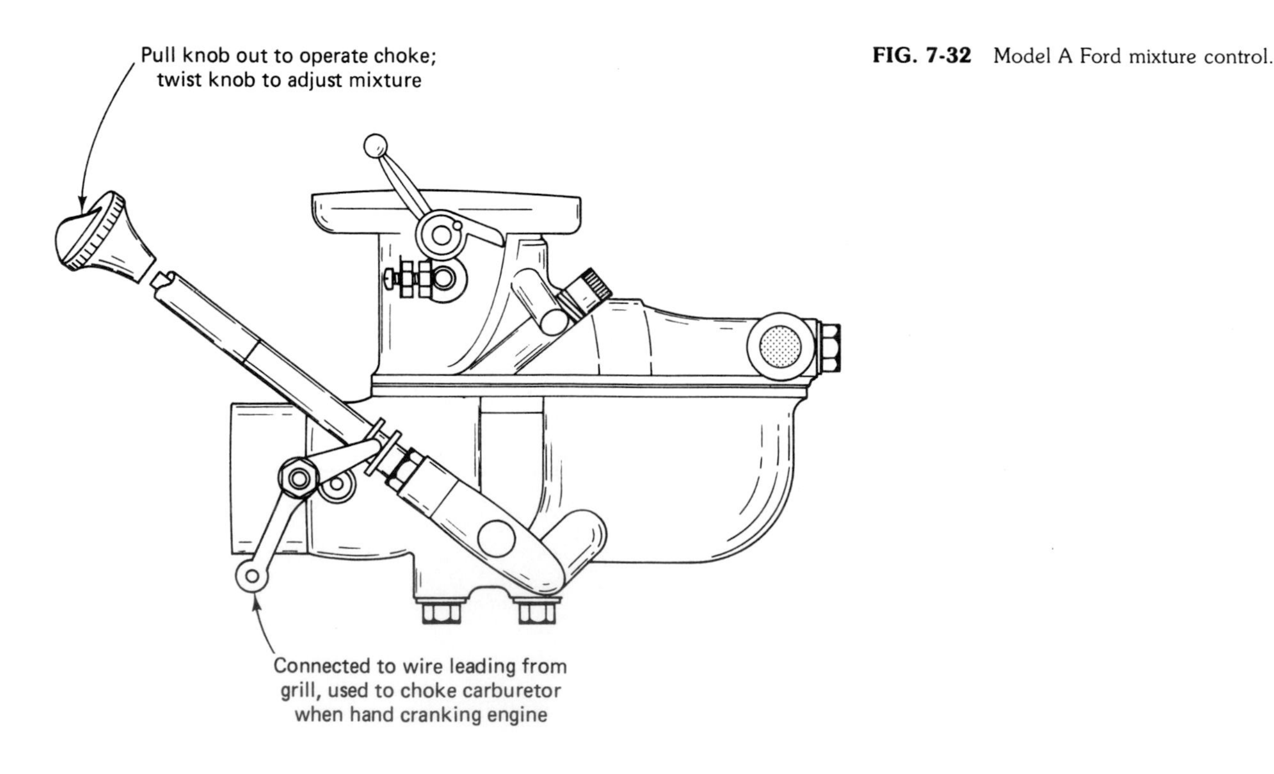
Pull knob out to operate choke; twist knob to adjust mixture
Connected to wire leading from grill, used to choke carburetor when hand cranking engine

FIG. 7-32 Model A Ford mixture control.

Consequently, fuel system controls were developed that did not rely on the operator. The automatic choke, as we have already seen, is one example. Figure 7-33 shows another, a power-metering rod that supplies extra fuel for sudden acceleration or increased load.

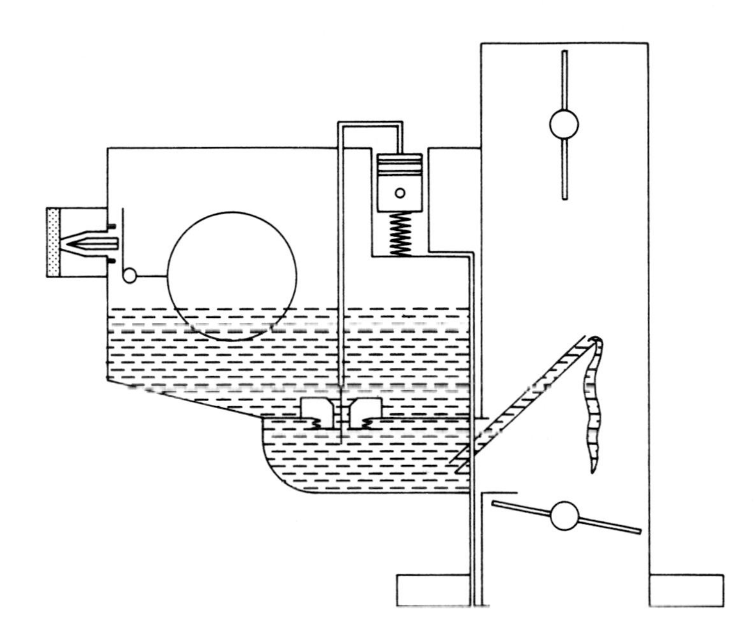

Power metering rod is held down by high vacuum at low engine speeds. When vacuum drops . . . because of sudden throttle opening . . . the spring pushes the rod up which enlarges the opening and lets more fuel flow through the jet

FIG. 7-33 Power metering rod.

8

Precomputerized Pollution-Control Systems

The term *pollution* covers a host of twentieth-century sins. We have beer cans in parks, chemical poisons seeping into drinking water, rivers that burn, air that chokes, and microwaves that rearrange our chromosomes. Sometimes the situation seems pretty grim. However, there is one relatively bright spot. Automakers, more than most other industries, have gone a long way toward cleaning up their products. From 1960 to 1973, they reduced unburned hydrocarbon emissions by 85% and carbon monoxide by 70%.

Responding to pressures from the federal government, the car companies have created a virtual new branch of technology. Computer controls are part of this technology. They were developed because the last vestiges of automotive pollution have been considerably more difficult to clean up than the first 70 to 80%. The feedback requirements are more exacting and comprehensive. Therefore, one of the major themes underlying this book is pollution control.

The present chapter reviews some of the antipollution measures that preceded on-board computers. We will look at the three main sources of pollution, how they have been controlled, and the combustion factors that affect pollution.

SOURCES OF POLLUTION

Automotive pollution (with the exception of NO_x) comes from gasoline or its by-products. Three main sources are the (1) crankcase, (2) fuel tank and carburetor, and (3) exhaust. The first two sources contribute 20% each or 40% of the total hydrocarbon pollution produced. The exhaust accounts for the remaining 60% (Figs. 8-1 and 8-2).

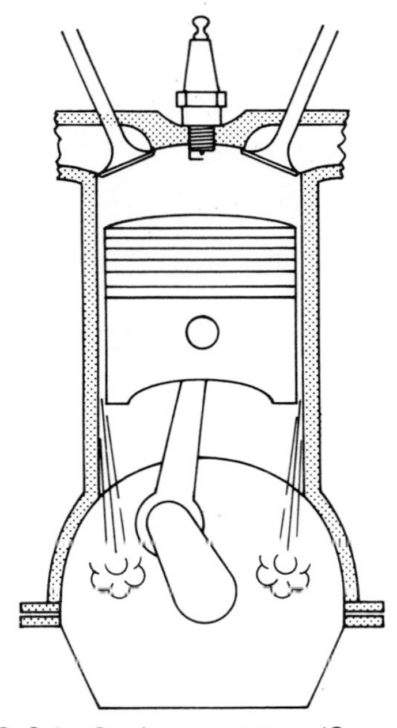

FIG. 8-1 Crankcase emissions. (Courtesy of the Chrysler Corporation.)

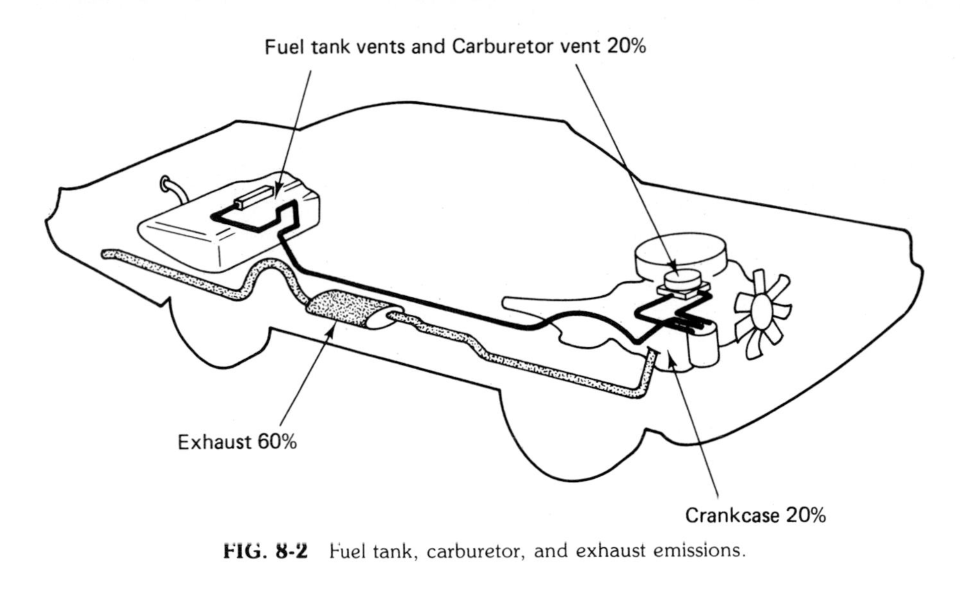

FIG. 8-2 Fuel tank, carburetor, and exhaust emissions.

CRANKCASE EMISSIONS

Causes

Crankcase emission starts with *blow-by,* which occurs when unburned fuel and combustion by-products escape past the piston rings during the compression and power strokes. These blow-by gases fill the crankcase with contaminants. If allowed to remain, they can turn lubricating oil into sludge.

Before pollution controls, the principal means for getting rid of crankcase vapors was the road draft tube (Fig. 8-3). It was simply a pipe that vented the crankcase to the atmosphere beneath the vehicle. Air moving past the outlet created a low-pressure region, which helped force vapors out of the crankcase.

Control

Dumping raw contaminants out of the road tube was never very appealing. Some older readers may remember the days when black streaks

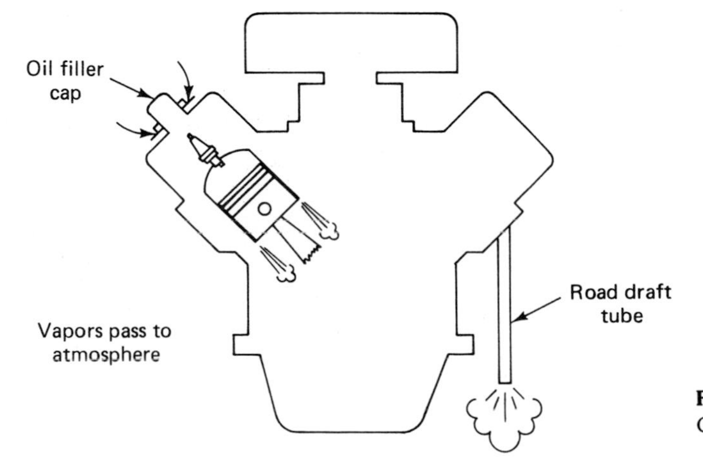

FIG. 8-3 Road draft tube. (Courtesy of the Chrysler Corporation.)

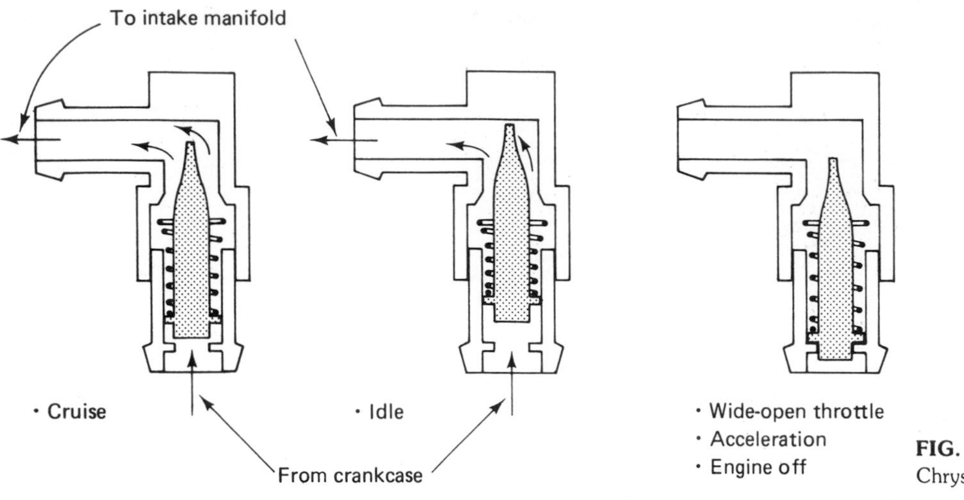

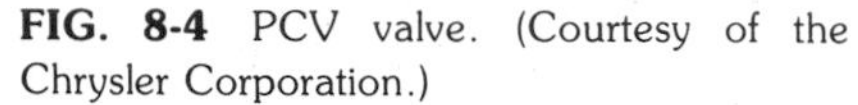
FIG. 8-4 PCV valve. (Courtesy of the Chrysler Corporation.)

were deposited between the wheel paths on all major roads. One of the first targets of antipollution legislation was crankcase emission.

In 1963, all U.S. manufacturers installed PCV (positive crankcase ventilation) valves on their cars (Fig. 8-4). These systems operate in the following manner:

1. During periods of high intake manifold vacuum, the PCV valve opens.
2. Responding to the reduced pressure, fresh air is pulled from an intake at the oil filler cap. Then, the air goes:
   a. Through the crankcase where it is mixed with contaminants.
   b. Through the open PCV valve.
   c. Into the intake manifold and then to the combustion chamber.

After 1968, the "open" PCV valves were replaced by closed systems (Fig. 8-5). They are designed to eliminate vapors pushed out of

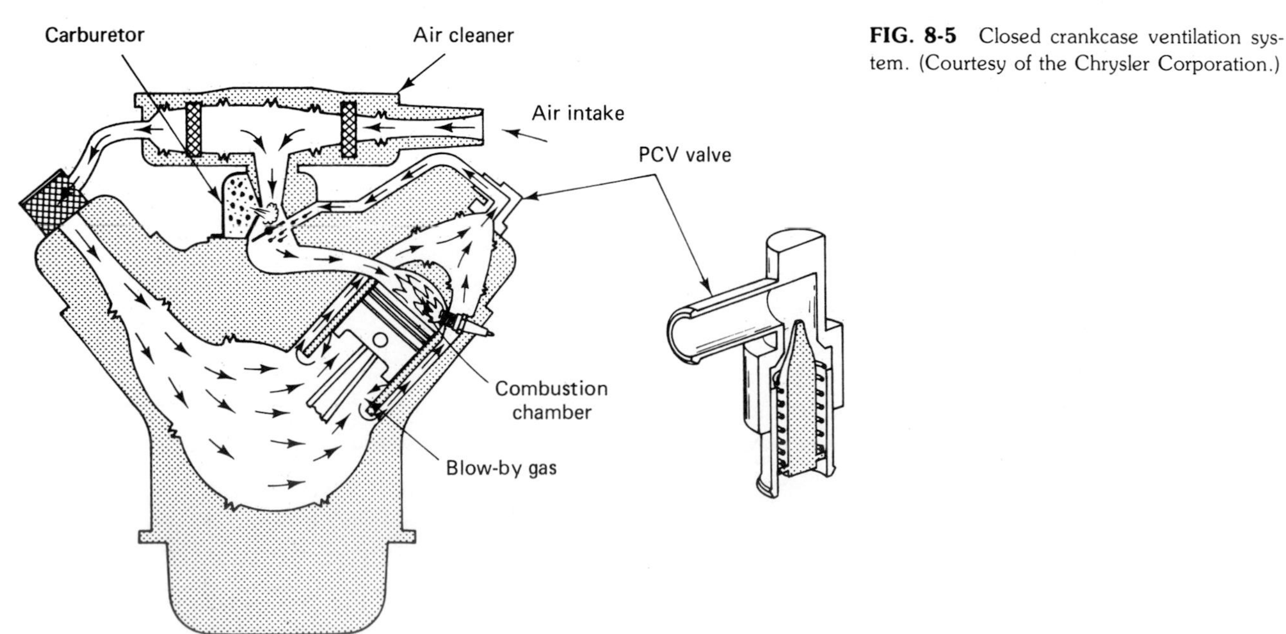

FIG. 8-5 Closed crankcase ventilation system. (Courtesy of the Chrysler Corporation.)

the oil filler cap during periods of heavy acceleration. Instead of drawing fresh air directly from the filler cap, a supply tube goes from the cap to the air cleaner. That way, excess vapors coming from the cap are vented into the carburetor. At other times, the system works the same as before (except that fresh air is always drawn through the air cleaner). Closed PCV systems are still used. However, since they are automatic feedback devices, responding to changing conditions, their operation is now subject to computer control.

EVAPORATIVE EMISSIONS

Gasoline is a very volatile liquid. Place an uncovered container in the open and before long the entire contents will evaporate. The resulting vapor consists mostly of hydrocarbons, a pollutant. In an automobile, two sources of evaporative pollution are the fuel tank and carburetor float bowl.

One obvious answer for this kind of pollution is to seal off the sources of evaporation (Fig. 8-6). That is exactly what the manufacturers have done. Gas tank caps no longer allow vapors to escape into the atmosphere, and float bowls are vented into the carburetor intake or into a special storage canister.

However, just sealing the sources of evaporation is not enough. Various kinds of vents and pressure relief valves must be provided. That is because fuel flows from the tank to the pump in response to pressure differentials between the pump and the air in the tank. If the tank is completely sealed off, a vacuum will build up as the fuel is withdrawn. Such a situation would be like trying to pour liquid from a can with only one small hole. Pretty soon, the pressure inside would be less than the pressure outside and the flow would stop. Therefore, gas caps must (and do) perform two functions. They seal closed when the pressure inside the tank increases and let air enter when the pressure inside the tank is reduced.

Another problem is the vapor that builds up inside closed spaces. Unless controlled, dangerous pressures and concentrations of explosive

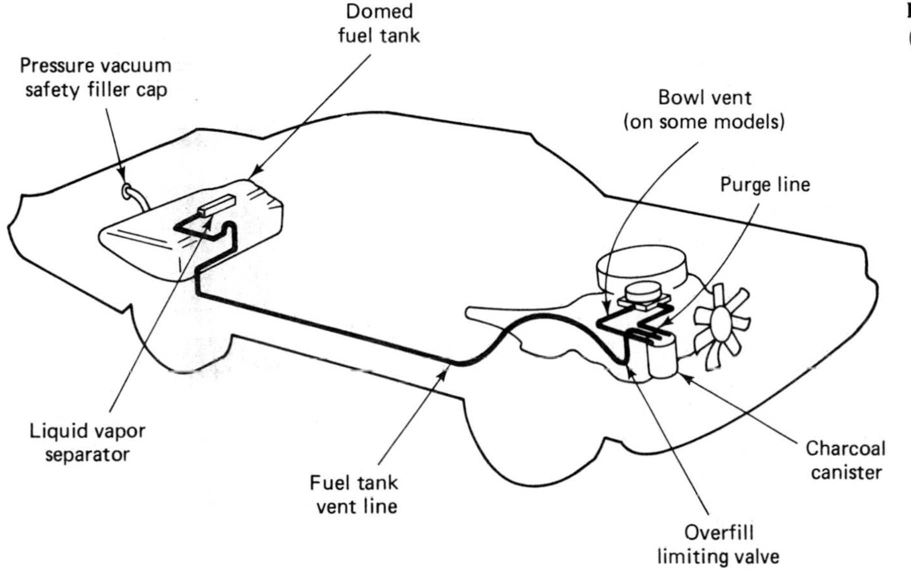

FIG. 8-6 Controlling evaporative emissions. (Courtesy of the Chrysler Corporation.)

gases can result. One solution adopted by most manufacturers is a charcoal canister connected to the source of evaporation. Fuel vapors are trapped in the canister when the engine is not running. Then, when the engine is started, a fresh air purge is pulled through the canister. The air mixes with the evaporated fuel, carrying it into the engine to be burned.

Manufacturers employ a number of other devices and techniques to help control evaporative emissions. However, it is the canister-purging operation that is most likely to come under computer control. Sensitive, late-model engines cannot tolerate unplanned additions of raw, unburned vapors into the air/fuel mixture.

EXHAUST EMISSIONS

With 60% of the total, the exhaust is responsible for most automotive pollution. If you think about it, this is natural since a majority of the combustion by-products end up in the exhaust system. Consequently, considerable time and money have been spent cleaning up the exhaust. However, before looking at particular devices and techniques, it will be helpful to review some basic facts about the combustion process and the kinds of contaminants that are produced.

REVIEW OF THE COMBUSTION PROCESS

Four ingredients are necessary for combustion: air, fuel, heat, and time. Under ideal conditions, exactly 14.7 parts of air for each part of fuel will result in complete combustion with no harmful by-products left over. (Recall that this is called a stoichiometric air/fuel ratio.) Problems occur if the conditions are not ideal, which they never are in the real world in which engines must operate.

Although any number of factors can affect the combustion process, the results usually fall into a limited number of categories. One common problem is partial mixing of air and fuel. For complete combustion to occur, the mixture must be uniform. Each molecule of fuel must be surrounded by 14.7 molecules of air. Naturally the process must start with the correct amounts of air and fuel. However, the temperature and pressure must also stay within narrow bounds. If the temperature of the air and fuel is too low, the fuel molecules will not bounce around enough to stay in a gaseous state. They will turn back into a liquid, which means that the mixture will be too lean since not enough fuel is available in a usable form. That is why engines have chokes and heat control devices.

Variations in the pressure of the air/fuel mixture can have similar effects. Generally, the lower the pressure, the easier it is for fuel to stay in a vapor state. There is less resistance to molecular movement. However, if the pressure suddenly increases, the fuel is likely to turn back into liquid droplets. This happens when the manifold vacuum drops (or the pressure increases) under load or sudden acceleration. To compensate, extra fuel is usually added during these lean periods. As a result, the overall mixture might be richer than 14.7:1, even though the effective mixture is the same or less.

Another common problem is burning time. A given quantity of a certain fuel mixture requires a definite interval for complete combustion. Therefore, when the engine speeds up, spark timing must be advanced so that the fuel will have enough time to be burned. Timing must also be advanced if the fuel mixture changes suddenly—for example, if the manifold pressure increases.

THREE PRINCIPAL EXHAUST POLLUTANTS

The targets of all engine emission control systems are three principal pollutants: hydrocarbons, carbon monoxide, and nitrogen oxide.

Hydrocarbons (HC). Hydrocarbons are the basic ingredients of petroleum products. When combined with oxygen in the burning process, hydrocarbons provide the energy to operate internal combustion engines. However, the HC molecules that do not burn and escape into the atmosphere are regarded as pollution. Any condition that affects combustion can cause HC emission. These conditions include overly rich carburetor or injector settings, dirty air cleaners, misfiring spark plugs, and improper timing.

Carbon Monoxide (CO). This gas has been referred to as the quiet killer. People who commit suicide by operating their cars in closed garages are victims of carbon monoxide poisoning. The resulting corpses usually have cherry lips and rosy complexions. Carbon monoxide generally occurs when there is insufficient air in the air/fuel mixture. Any condition that reduces airflow such as a dirty air cleaner can cause CO formation.

Nitrogen Oxide (NO_x). Nitrogen oxide is not a by-product of the fuel-burning process. It is formed anywhere that very high temperatures and pressures occur, such as the combustion chamber of a late-model automotive engine. Oxygen and nitrogen, two of the most common substances in the air, combine to form various oxides of nitrogen. As combustion temperatures and pressures go up, NO_x formation increases.

CONTROLLING EXHAUST EMISSION

A number of approaches have been employed to clean up exhaust emissions. They can be grouped into three main categories:

1. Air injection system
2. Engine modification system
3. Catalytic reactor system

At different times, most manufacturers have used a combination of all three techniques.

Air injection systems

The air injection method features an air pump driven by a V-belt connected to a pulley at the end of the crankshaft (Fig. 8-7). High-pressure air is injected into the exhaust stream near the root of the exhaust mani-

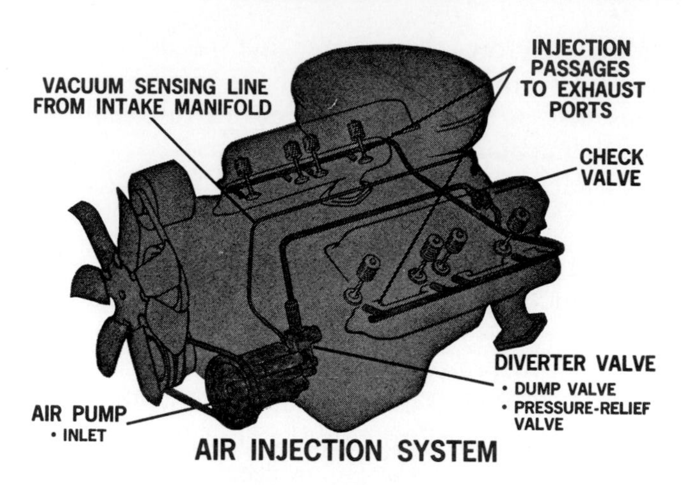

FIG. 8-7 Air injection system. (Courtesy of the Chrysler Corporation.)

fold. The extra air helps burn away HC vapors and eliminates some of the CO contained in the exhaust. The system acts like an afterburner. In earlier versions, air was supplied constantly to the exhaust. Current systems operate on a periodic basis, the exact intervals controlled, in some cases, by an on-board computer. All systems contain a variety of check and/or bypass valves to protect the components from backfiring, excessive pressure, failure of critical parts, and so on.

Modified engine system

A number of devices and techniques fall into this category (Fig. 8-8). All manufacturers have used one variation or another of the modified engine approach. Some of the trade names are:

1. ENGINE MOD, American Motors
2. IMCO (Improved Combustion System), Ford
3. CCS (Combustion Control System), GM
4. CAS (Clean Air System), Chrysler

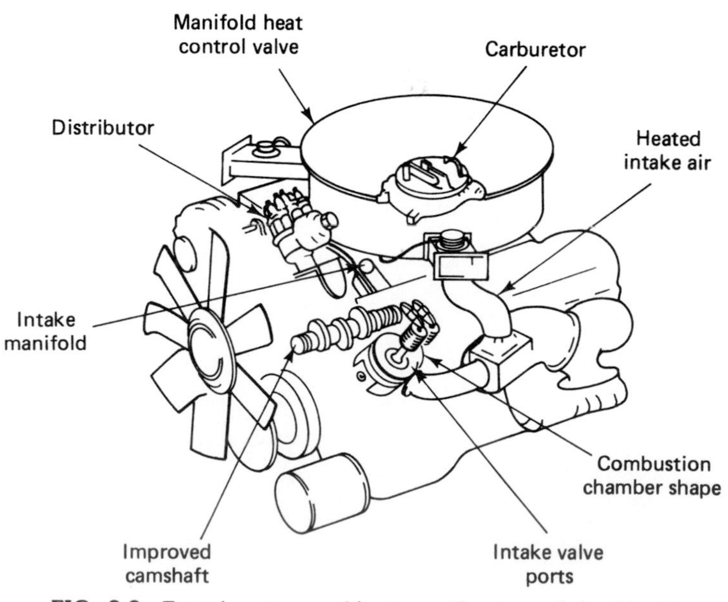

FIG. 8-8 Typical engine modifications. (Courtesy of the Chrysler Corporation.)

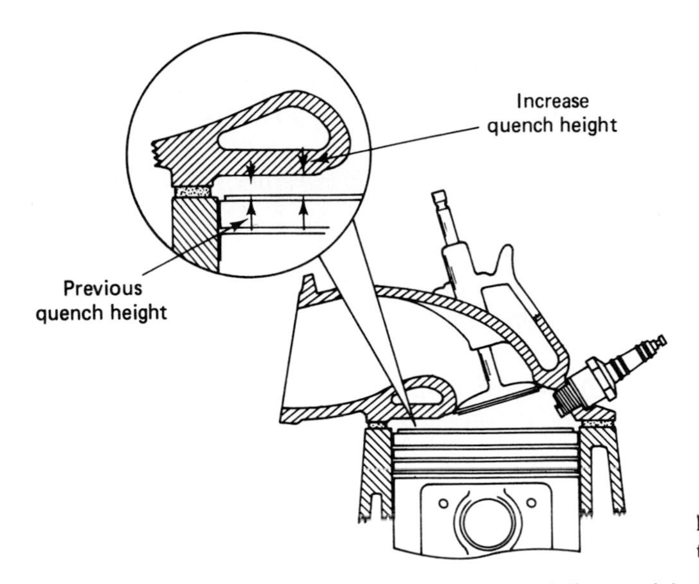

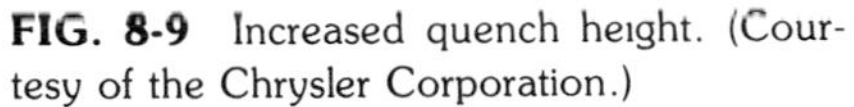
FIG. 8-9 Increased quench height. (Courtesy of the Chrysler Corporation.)

Following is a partial list and brief description of some modified engine functions.

Reduced Quench Area. *Quench* refers to the close spaces inside the combustion chamber that tend to snuff out the flame spread by burning air and fuel. Reducing the quench areas (or increasing the height of the quench zones) prolongs combustion, which reduces emission (Fig. 8-9).

Increased Valve Overlap. Camshaft lobes have been redesigned so that the intake and exhaust valve opening overlaps—in other words, so that both valves are open at the transition between the exhaust stroke and the intake stroke (Fig. 8-10). This helps scavenge or wash out exhaust gases, which improves combustion and reduces emission.

Improved Intake Manifolds. The air/fuel mixture can be affected by the flow characteristics inside the intake manifold. Rough surfaces will slow the movement of the mixture, which in turn, will cause some of the fuel to condense into a liquid. Uneven passages or ports can also affect velocity and pressure. The same is true for unequal-length passages between the carburetor and the intake manifold. Since pollution control became a factor, more attention has been paid to intake manifold design.

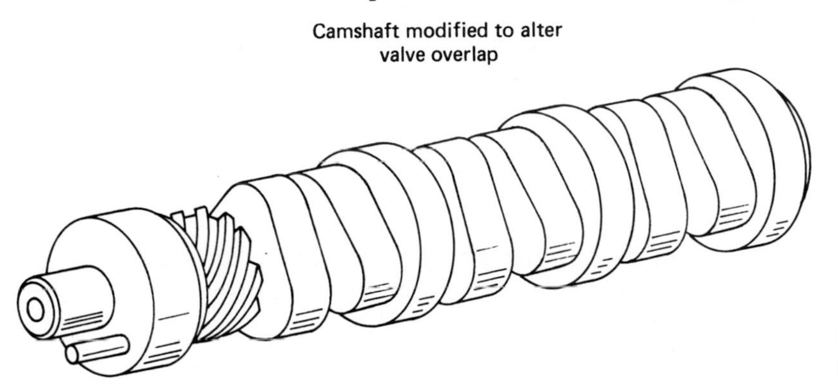

FIG. 8-10 Camshaft modifications. (Courtesy of the Chrysler Corporation.)

Reduced Compression Ratios. Compression ratios were reduced in the early 1970's when many engines were designed to run on lower-octane fuel. This change also resulted in lower levels of HC and NO_x.

Lean Calibrated Carburetors. Most prepollution engines ran on slightly rich fuel mixtures. Such carburetor settings are generally associated with improved performance and drivability. However, since rich mixtures also lead to HC and CO formation, fuel systems are now designed to operate toward the lean end of the scale.

Improved Cold-starting Procedures. Cold starting is a special operation requiring a very rich fuel mixture. If not precisely controlled, it can result in incomplete burning and increased pollution. Manufacturers have provided a number of devices to warm the engine faster and to make the choke more sensitive.

Modified Pistons. Pistons have been given special crowns and contours to help force the air and fuel into flow patterns that promote mixing and complete combustion.

MODIFIED DISTRIBUTOR OPERATION

The distributor in a precomputerized engine was an integral part of the combustion process, since it was the distributor camshaft that determined the moment of ignition. Advancing or retarding the spark, then as now, has a major effect on the pollutants produced.

Listed next are several distributor modifications that were made to reduce pollution. Variations on these and other related functions are still performed, except that now the controlling intelligence lies in a digital computer rather than in a mechanical system.

Dual-diaphragm advance

Precomputerized engines used control systems to advance the spark during light-load, part-throttle conditions. The main control element was a pressure diaphragm that was operated by changes in manifold vacuum. In some pollution-control engines, a second diaphragm was added. It retarded the spark when the throttle was fully closed (at idle speeds and during periods of deceleration). This helped reduce engine speed, giving more time for complete combustion. A dual-diaphragm distributor is shown in Fig. 8-11.

Devices such as dashpots and solenoids, which help keep the throttle open during deceleration, perform similar functions. However, all these operations are now managed by an on-board control computer.

Temperature-sensitive advance

Lean mixtures and retarded spark at idle speeds can cause engines to run hotter. This tends to promote the formation of NO_x. One solution is to advance the spark during certain warm engine conditions. The engine operating temperature is subsequently reduced. Before computer controls (but during the era of pollution controls), temperature-

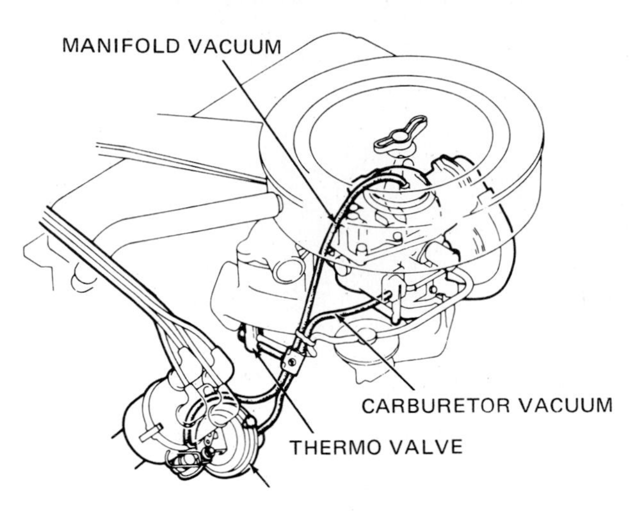

FIG. 8-11 Dual-diaphragm and temperature-sensitive advance. (Courtesy of the Chrysler Corporation.)

sensitive valves were used to control distributor diaphragms and thus spark timing. The Thermo Valve in Fig. 8-11 is part of such a system. Now as you might imagine, this feedback function is managed by a routine running in the on-board computer.

Delayed advance

There is some evidence that delaying (retarding) the spark advance during acceleration above certain engine temperatures will also reduce the formation of NO_x. An early mechanical implementation of this control function is the Chrysler OSAC (orifice spark-advance control) shown in Fig. 8-12. Located on top of the air cleaner, this valve delays advance from idle to part throttle for about 17 seconds. It operates only when the underhood temperature is above 60°F.

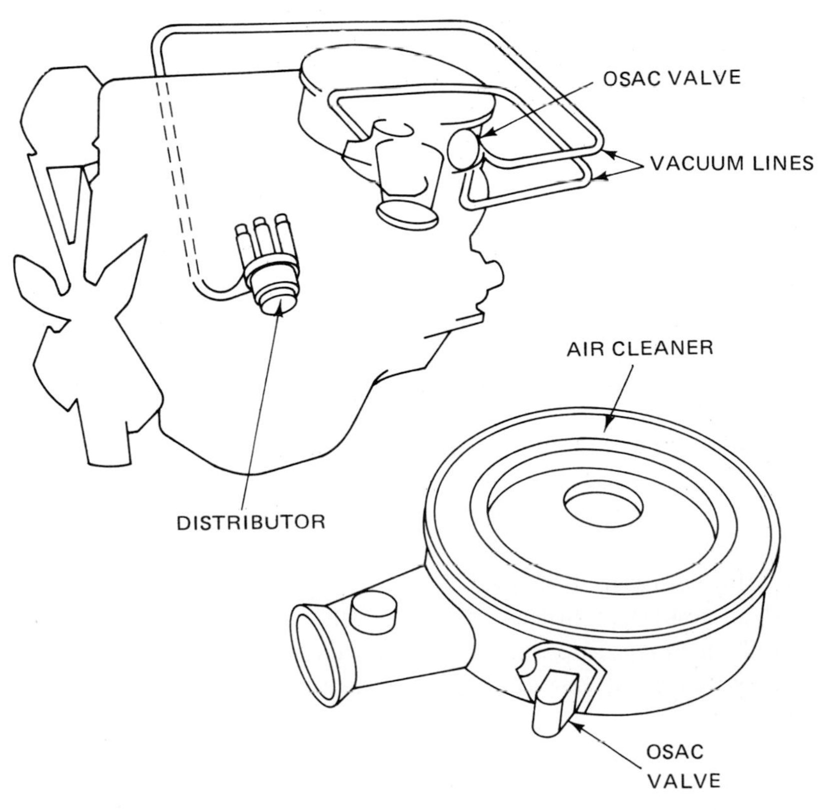

FIG. 8-12 Chrysler OSAC system. (Courtesy of the Chrysler Corporation.)

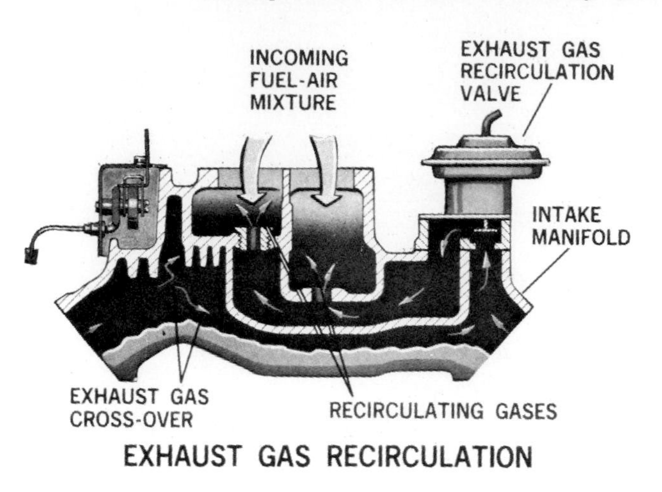

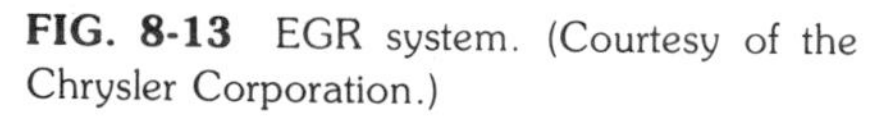
FIG. 8-13 EGR system. (Courtesy of the Chrysler Corporation.)

Exhaust Gas Recirculation (EGR). Introduced in 1973 the EGR system has been one of the primary methods for controlling NO_x emission (Fig. 8-13). The system consists basically of small, connecting passages between the intake and exhaust manifolds and temperature- and/or pressure-sensitive valves. When the appropriate conditions are reached, the system routes a small quantity of exhaust gas to the air/fuel mixture. Since the exhaust has already been burned, it will support little if any further combustion. Diluting the air/fuel mixture with an effectively inert gas reduces the temperature of the combustion process, which in turn, reduces the formation of NO_x.

Catalytic Reactors

As federal pollution standards became progressively tighter, manufacturers have had to find newer and more thorough ways to control exhaust emissions. One of the most widely adopted devices is the catalytic reactor or convertor. It was introduced in 1975 on all American-made cars and many imports. That was the same year in which permissible levels of HC and CO dropped respectively from 3.4 and 39.0 grams per vehicle per mile to 0.46 gram and 4.7 grams per vehicle per mile.

Located in the exhaust system, usually upstream from the muffler, a catalytic reactor looks like a small muffler or resonator (Fig. 8-14). Inside however, the resemblance ends (Fig. 8-15). The reactor (of certain models) is filled with a honeycomb-like ceramic core. Surrounding this fragile, clay structure is steel mesh, which offers protection from road shocks and jolts.

In precomputerized engines, the core was created with a mixture of palladium and platinum. These are the active ingredients of the system, the catalytic material. When exposed to hot exhaust gases, the catalyst gets even hotter, reaching temperatures between 1300° and 1600°F in normal operation. The catalyst itself is not changed except in the presence of very rich exhaust vapors or when leaded fuel is used.

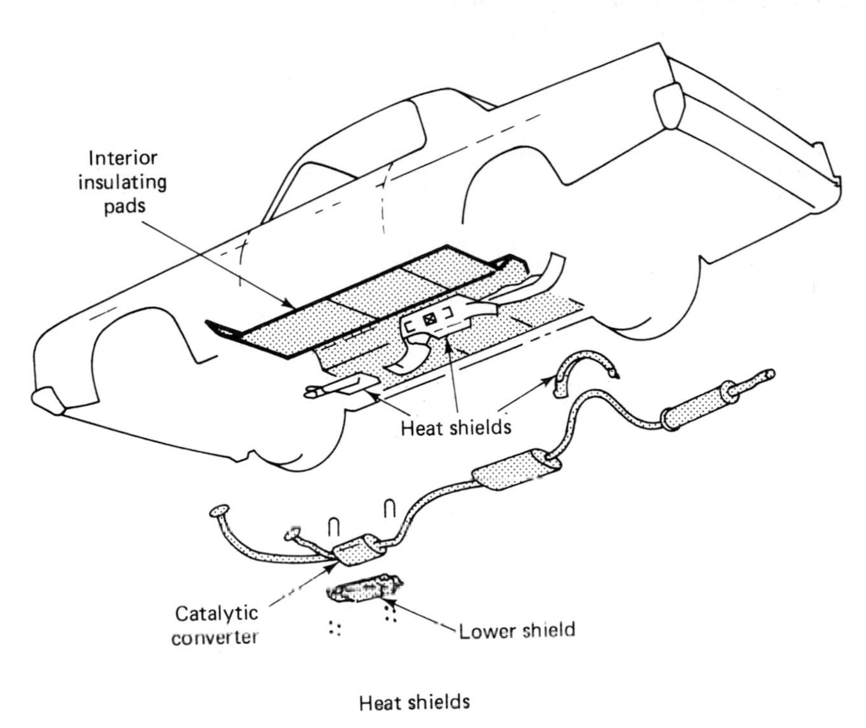

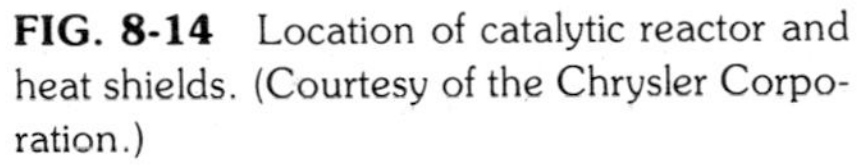
FIG. 8-14 Location of catalytic reactor and heat shields. (Courtesy of the Chrysler Corporation.)

However, the catalyst promotes a reaction between HC, oxygen (O_2), and CO. As a result, the two pollutants are changed into water (H_2O) and carbon dioxide (CO_2). The latter is a harmless substance used to provide the fizz in carbonated drinks.

In 1980, a third ingredient to control NO_x was added to many catalytic reactors. As we will see in other chapters, controlling NO_x was one of the reasons for turning to computer-managed systems.

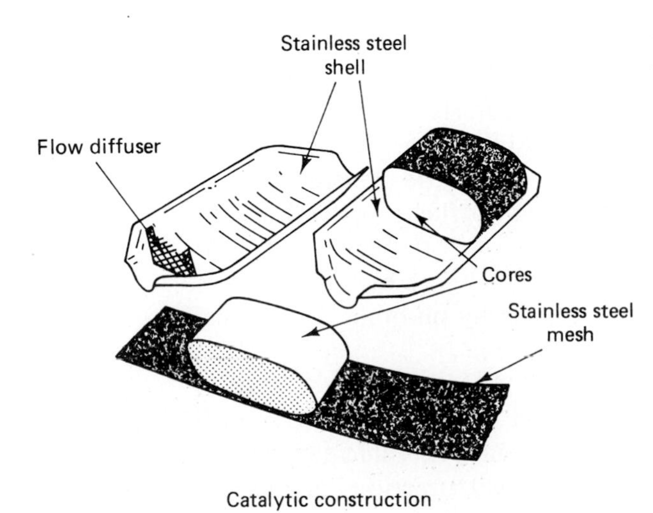

FIG. 8-15 Inside catalytic reactor. (Courtesy of the Chrysler Corporation.)

9

Computerize Engine-Control Functions

INTRODUCTION

Engine-control functions are actions taken by an on-board computer to manage the operation of an engine and its related components. Each function uses input information from one or more sensors and sends control signals to one or more output devices.

Some of the engine components that may be the object of control functions include:

- Fuel system
- Ignition system
- Idle speed control
- Exhaust gas recirculation
- Cannister purge
- Air management pump
- Automatic transmission clutch
- Early fuel evaporation

As time goes by, more and more engine-control functions will be performed by on-board computers.

TYPICAL COMPUTER FUNCTIONS

Although implemented differently, most engine-control functions are similar from one vehicle to the next. Following is a discussion of how some typical functions are implemented on a typical vehicle. In the illustrations that accompany this representative system, the control computer is referred to as the ECM (electronic control module).

FUEL SYSTEM FUNCTIONS

The amount of air and fuel that reach the combustion chamber, ready to be burned, should be at or near the ideal stoichiometric ratio of 14.7 parts air to 1 part of fuel. This mixture should be the same for all conditions. Unfortunately, this condition does not always exist.

The actual quantity of air and fuel needed to achieve this ideal mixture varies according to engine load, acceleration, temperature, and other conditions. In the past, mechanical and pneumatic control systems examined information from two or three sources to regulate the amount of fuel delivered. Today, computerized control systems look at information from a dozen or more sources.

In the typical system examined next, fuel system functions are provided for various operating modes. There modes are also two main operating states: the closed-loop State and the open-loop State.

Closed-loop state

The term *closed-loop state* refers to a feedback relationship between the amount of fuel in the air fuel mixture and the amount of oxygen detected by an O_2 sensor in the exhaust. The O_2 sensor tells the on-board computer how much oxygen is in the exhaust. The computer is programed to maintain the O_2 content of the exhaust at a level that corresponds to the desired fuel mixture. If the O_2 content moves outside the allowed range, the computer makes an adjustment in the fuel supply. That is why it is called a closed-loop system: input information controls output actions, which control input information, which control output actions, and so on.

Computer-managed closed-loop systems were developed in conjunction with three-way catalytic exhaust reactors. The first catalytic reactors, introduced in 1975, used two active elements, platinum and palladium, to reduce the amount of hydrocarbon and carbon monoxide pollutants contained in the exhaust. In order for these reactors to work effectively, the air fuel mixture had to be 14.7:1 or leaner. Mechanical and pnuematic control systems of that era could maintain mixtures at these levels.

Three-way reactors, introduced later, added a third active ingredient, zirconium. It was used to reduce nitric oxide in the exhaust. Previously, nitric oxide was controlled primarily by the EGR (exhaust gas recirculation) valve.

The problem was that zirconium requires a mixture of 14.7:1 or richer to be effective. Therefore, in order for all three catalytic elements to work satisfactorily, the fuel mixture must be maintained at or very near the stoichiometric ratio, as pictured in Fig. 9-1. The only way to achieve such precise control over a variety of changing conditions was to offload control intelligence in the form of a feedback program.

In the representative system discussed here (and in most other systems), closed-loop control occurs when these conditions have been met:

- The O_2 sensor sends signals indicating that it is working properly.
- The engine coolant temperature is above a specified value.

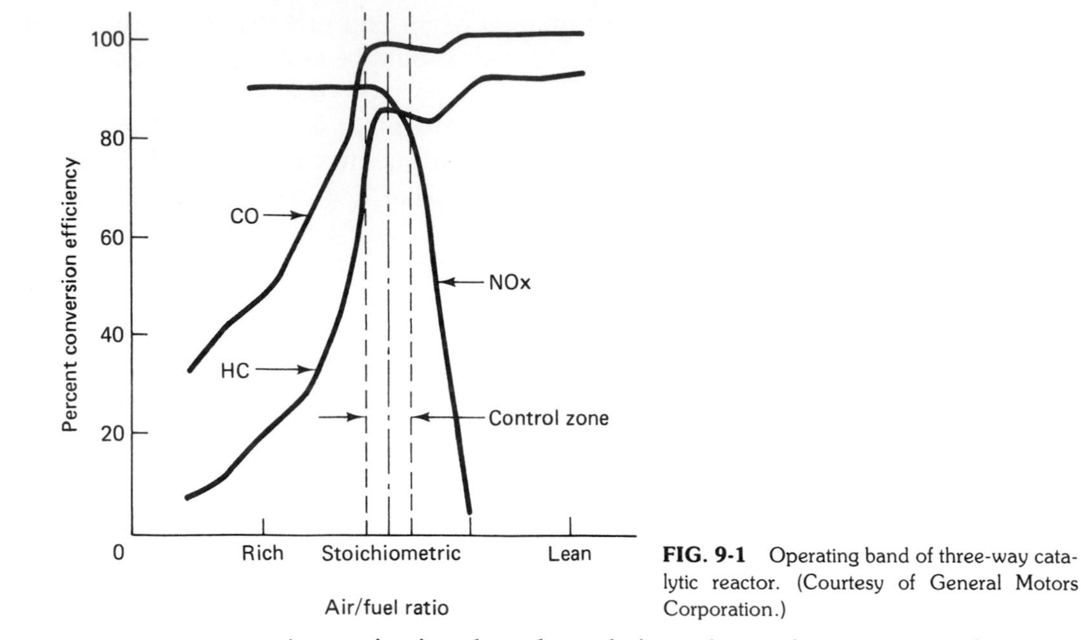

FIG. 9-1 Operating band of three-way catalytic reactor. (Courtesy of General Motors Corporation.)

- A certain time has elapsed since the engine was started.

Closed-loop operation is usually associated with the normal running mode, when no input other than the signals from the O_2 sensor must be accounted for.

Open-loop state

When conditions are not right for closed-loop operation, fixed programs take over the operation of the fuel system. This is called the *open-loop state.* Although input information triggers output actions in open-loop operation, the output action does not have the same direct feedback affect on the input data.

Some typical, open-loop operating modes include:

- Starting mode
- Clear-flood mode
- Battery voltage correction mode
- Cold-engine mode
- Acceleration mode
- Deceleration mode
- Altitude compensation

Fuel system operating modes

Following is a discussion of various fuel system operating modes as they are handled in a representative system. The object of the system is to control the amount of fuel delivered by two solenoid operated fuel injectors. In this system, the fuel injectors are located in a throttle body, which is located over the inlet to the intake manifold.

Clear-flood mode

If the engine becomes flooded during an attempt at starting, the operator is directed by the owner's manual to press the throttle all the way to the floor. When the computer senses that the throttle is open all the way and the engine is turning less than 400 rpm (revolutions per minute), it will shut off the injectors. This allows the engine to clear itself of excess fuel. If the throttle is allowed to open to about 80% of the fully open position, the computer will open the injectors and resume normal starting functions (which means that if the engine has not started yet, it will become even more flooded).

Battery-voltage correction mode

When the computer senses that the battery voltage has dropped below a certain level, it can compensate for a weak spark by:

- Increasing the amount of fuel delivered
- Increasing the idle rpm
- Increasing the ignition dwell time

Starting mode

As soon as the ignition switch is turned from the *off* to the *on* position, the following sequence of events takes place:

1. Power is supplied to the computer.
2. The computer's internal clock is reset to zero.
3. The starting programs start to cycle through the computer proccssor.

As noted in the flowchart pictured in Fig. 9-2, one of the first steps is to turn on the fuel pump so that the fuel system will become pressur-

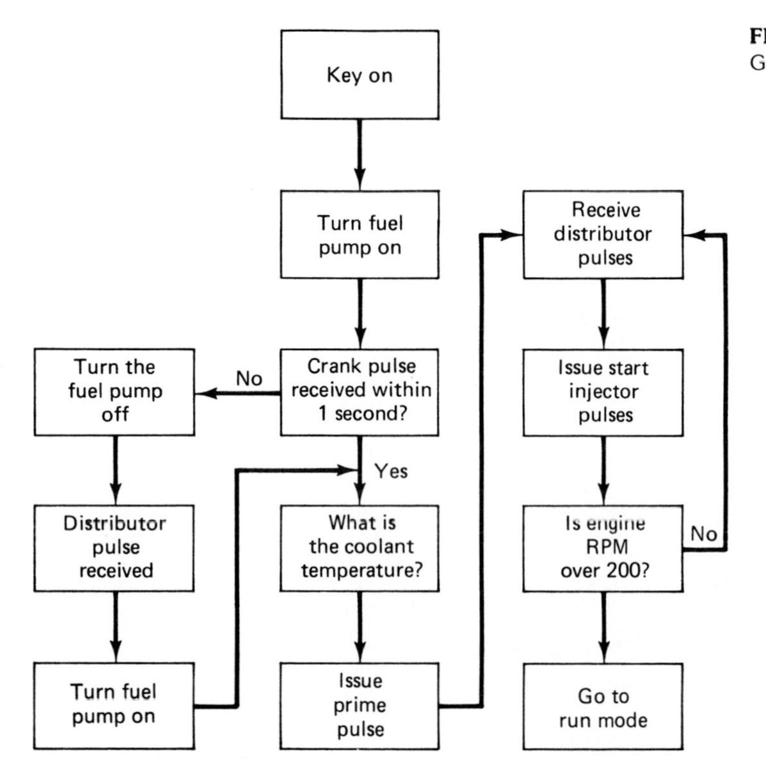

FIG. 9-2 Starting flowchart. (Courtesy of General Motors Corporation.)

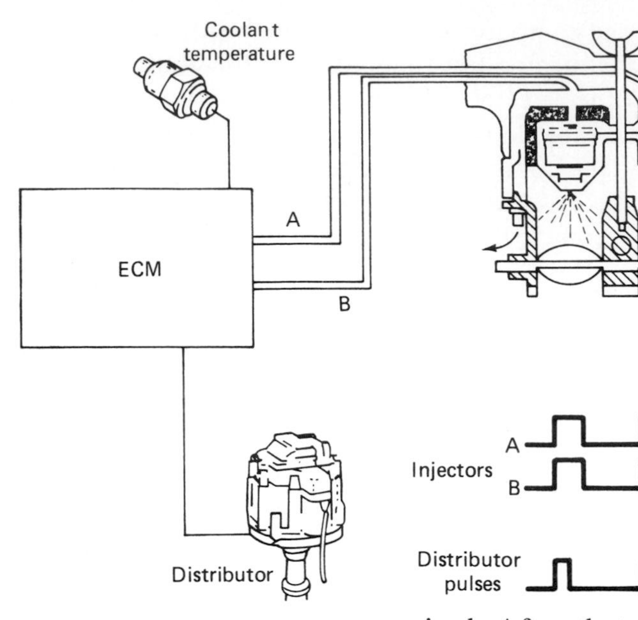

FIG. 9-3 Cranking fuel control. (Courtesy of General Motors Corporation.)

ized. After that, the computer checks for a signal indicating that the key switch has been turned to the *crank* position. If a signal is not received within 1 second, the pump is turned off until the engine signal is received.

After the crank signal arrives at the computer, the computer sends a primary pulse to the injectors. Upon receipt of the impulse, the injectors spray fuel into the intake manifold. The duration of the signal, and hence the amount of fuel sprayed, depends on the reading from the coolant sensor. Cold engines receive sprays lasting up to 170 milliseconds; warm engines get sprays as short as 10 milliseconds.

After the initial priming pulse, injector operation is by reference pulses from the distributor (Fig. 9-3). These reference pulses are associated with spark timing period.

During the crank period, both injectors spray fuel with each distributor reference pulse. The duration of the injector signal is again determined by the temperature of the coolant (Fig. 9-4). The lower the

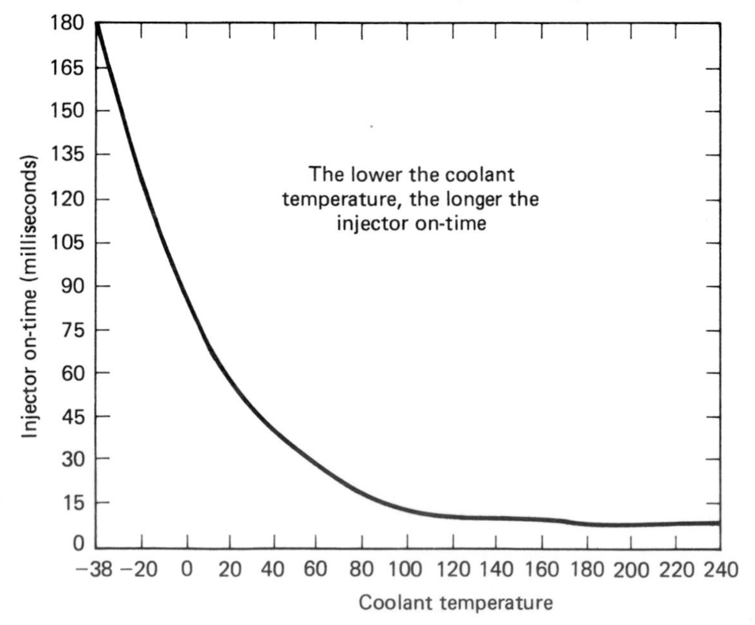

FIG. 9-4 Injector on-time. (Courtesy of General Motors Corporation.)

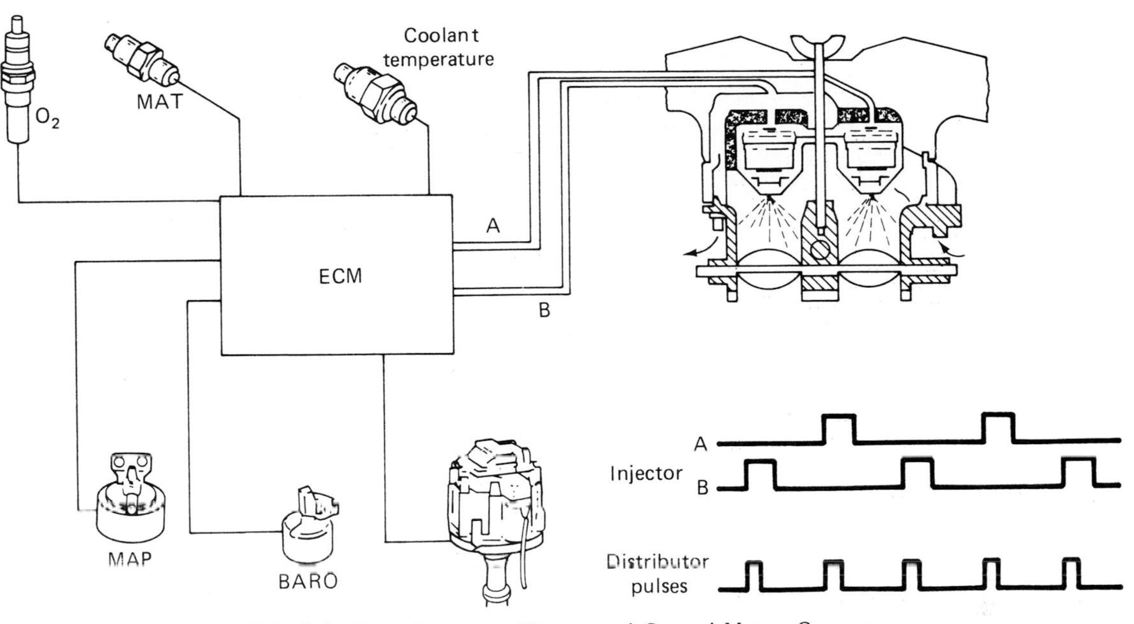

FIG. 9-5 Normal running. (Courtesy of General Motors Corporation.)

temperature, the longer the injectors remain open. (Of course, the injectors cannot remain open longer than the interval between distributor reference pulses. If that happened, the injectors would stay open all the time.) During the entire cranking period, engine speed is monitored. When it exceeds a preset PROM value, usually 200 rpm, the next operating mode begins.

Normal running mode

During normal running, the system operates in closed-loop fashion (Fig. 9-5). The injector solenoids are now energized alternately. With each distributor reference pulse, one injector opens; on the next pulse, the other injector opens. The mixture is controlled at the ideal 14.7:1 ratio by varying the pulse duration. Input factors affecting the injector on-period during normal running include (1) manifold air temperature, (2) manifold air pressure, (3) fuel pressure, and (4) O_2 content in the exhaust. Once inside the computer, this information is compared to preset PROM values. Injector operation depends on the result.

Cold-engine running mode

As in a precomputerized engine, a choke factor is needed to supply extra fuel when the temperature drops below a certain point (Fig. 9-6). Otherwise, not enough fuel will be vaporized, resulting in a mixture leaner than the desired 14.7:1 stoichiometric ratio.

The choke program examines inputs from the coolant sensor, the manifold air-pressure gauge, and an elapsed-time counter built into the computer. Until the engine reaches its normal operating temperature, the choke program causes the injectors to remain on longer. Like precomputerized choke systems, the program modifies the choke signal in response to changes in manifold vacuum. The timer insures that the choke does not remain on too long.

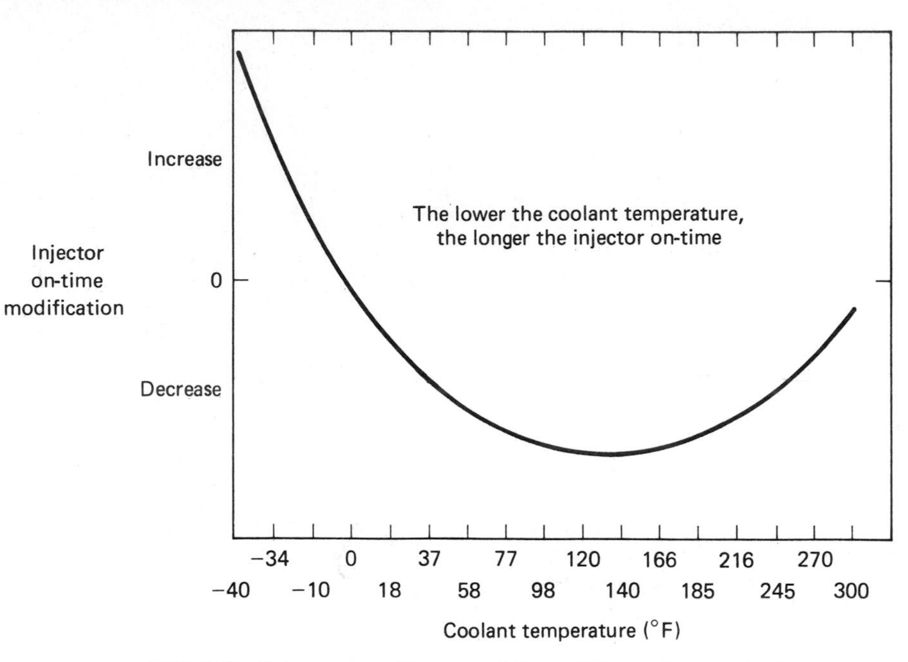

FIG. 9-6 Cold running. (Courtesy of General Motors Corporation.)

Altitude-compensation mode

At higher altitudes, fuel does not vaporize as readily when the engine load is increased (and manifold vacuum drops). Unless adjustments are made, the mixture becomes too lean. Consequently, the computer constantly checks altitude (as indicated by the barometric pressure sensor) and engine load (as indicated by the manifold pressure gauge). When the difference between the two inputs equals a preset PROM value, the program increases the injector ontime and thereby enriches the mixture (Fig. 9-7).

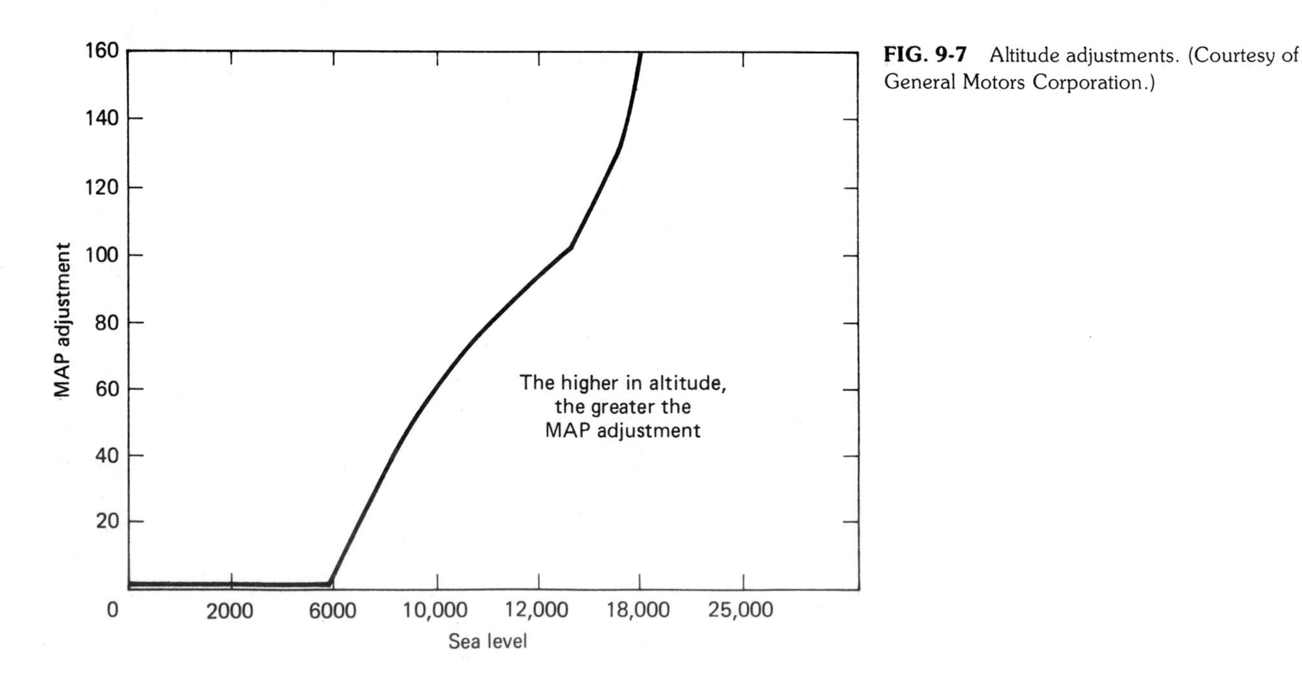

FIG. 9-7 Altitude adjustments. (Courtesy of General Motors Corporation.)

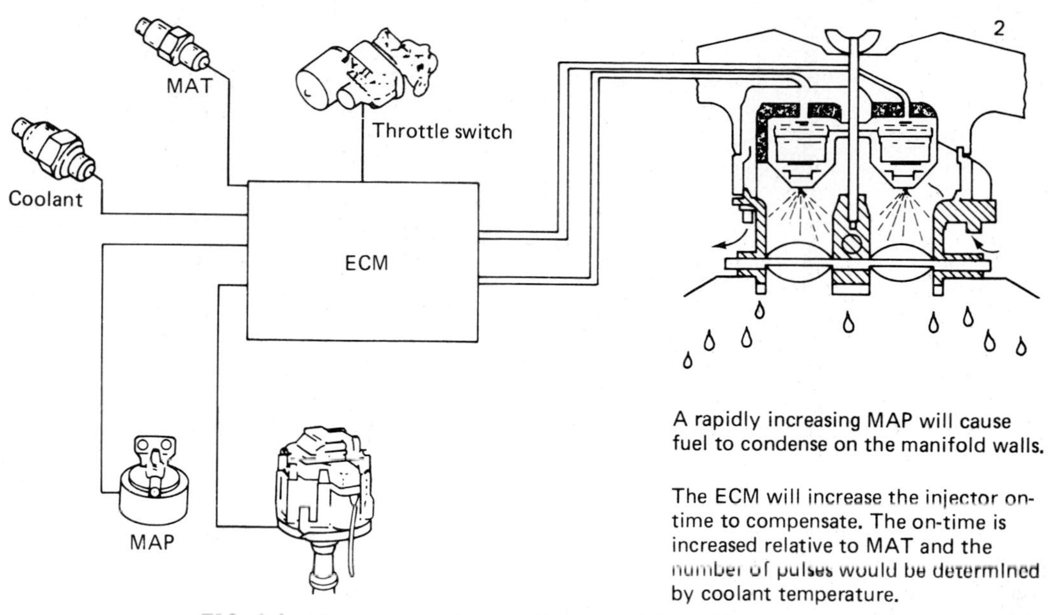
2
MAT
Throttle switch
Coolant
ECM
MAP
A rapidly increasing MAP will cause fuel to condense on the manifold walls.
The ECM will increase the injector on-time to compensate. The on-time is increased relative to MAT and the number of pulses would be determined by coolant temperature.

FIG. 9-8 Idle speed enrichment. (Courtesy of General Motors Corporation.)

Idle throttle compensation mode

When the throttle suddenly opens from an idle position, manifold air pressure increases and fuel tends to condense on the walls in the intake manifold. Extra fuel must be added or the mixture will be too lean.

The computer checks engine rpm and the position of the throttle (Fig. 9-8). If the throttle opens up from an idle position when the engine is below a certain PROM value, longer pulse signals are sent to the injectors. The exact duration of the pulses is influenced by inputs from the manifold air-temperature sensors. Since the fuel does not vaporize as well as low temperatures, the lower the temperature, the longer the pulses.

Acceleration mode

Above an idle rpm rate, the computer regards changes in manifold vacuum as being due to acceleration. However, no matter when or why it occurs, a sudden decrease in manifold vacuum (increase in pressure) causes fuel to condense and the mixture to become lean. Therefore, increases of manifold air pressure above PROM values will result in longer injector pulses (Fig. 9-9). As the pressure decreases, the pulses return to normal duration. To ensure smooth operation, the pulse change may not occur as fast as the pressure changes. In other words, there may be a lag between the two.

Wide-open throttle mode

When the difference between barometric pressure and manifold air pressure reaches a certain PROM value, the computer assumes that the throttle is wide open. As long as this condition remains in effect, the

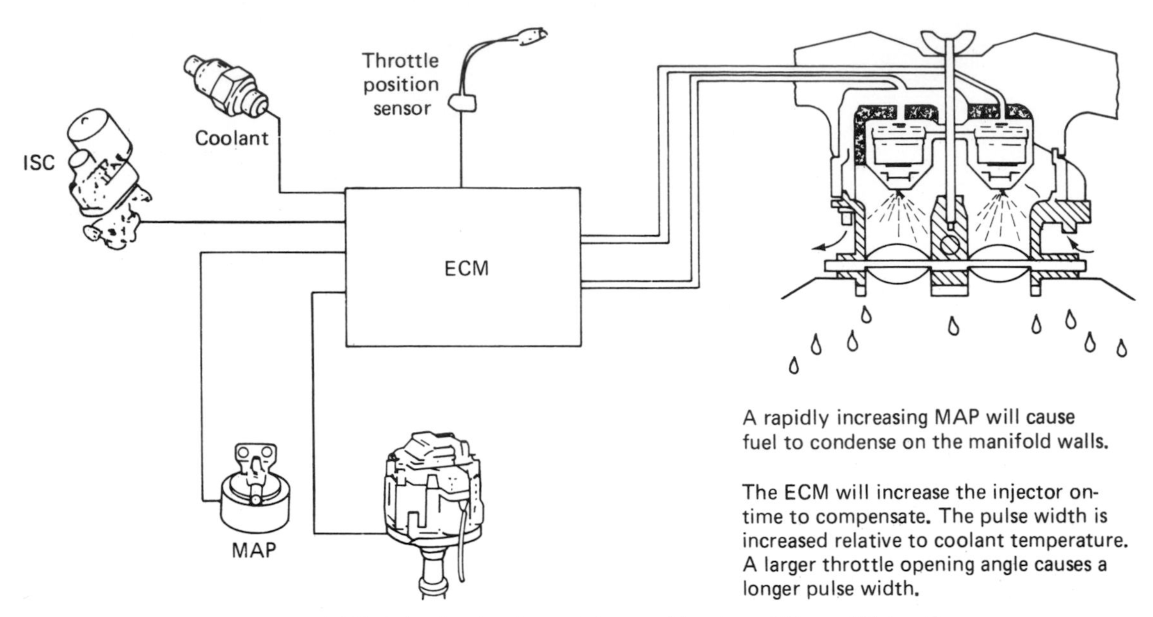

Throttle
position
sensor
Coolant
ISC
ECM
MAP
A rapidly increasing MAP will cause
fuel to condense on the manifold walls.
The ECM will increase the injector on-
time to compensate. The pulse width is
increased relative to coolant temperature.
A larger throttle opening angle causes a
longer pulse width.

FIG. 9-9 Acceleration enrichment. (Courtesy of General Motors Corporation.)

"normal running" inputs are ignored. Injector on-pulses are calculated from the barometric air-pressure readings (Fig. 9-10).

Deceleration mode

When the throttle plates are closed and the engine slows down, manifold air pressure drops (vacuum increases). This has the opposite effect of the situations previously described, which result in increased pressure (reduced vacuum). Now, any fuel that has condensed in the intake manifold suddenly vaporizes in the presence of the reduced air pres-

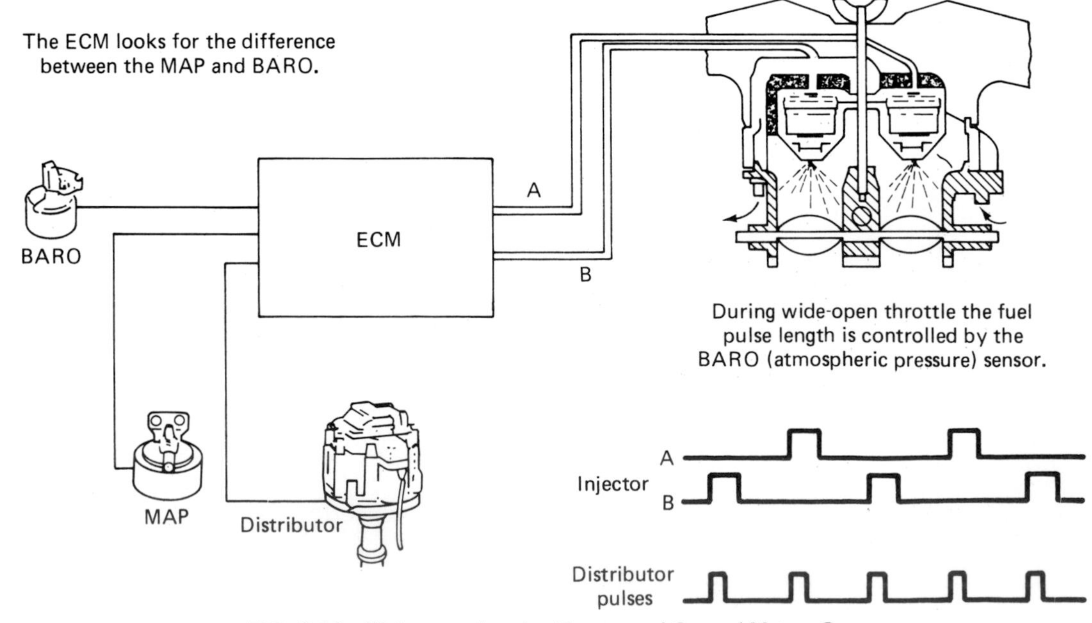

between the MAP and BARO.

FIG. 9-10 Wide-open throttle. (Courtesy of General Motors Corporation.)

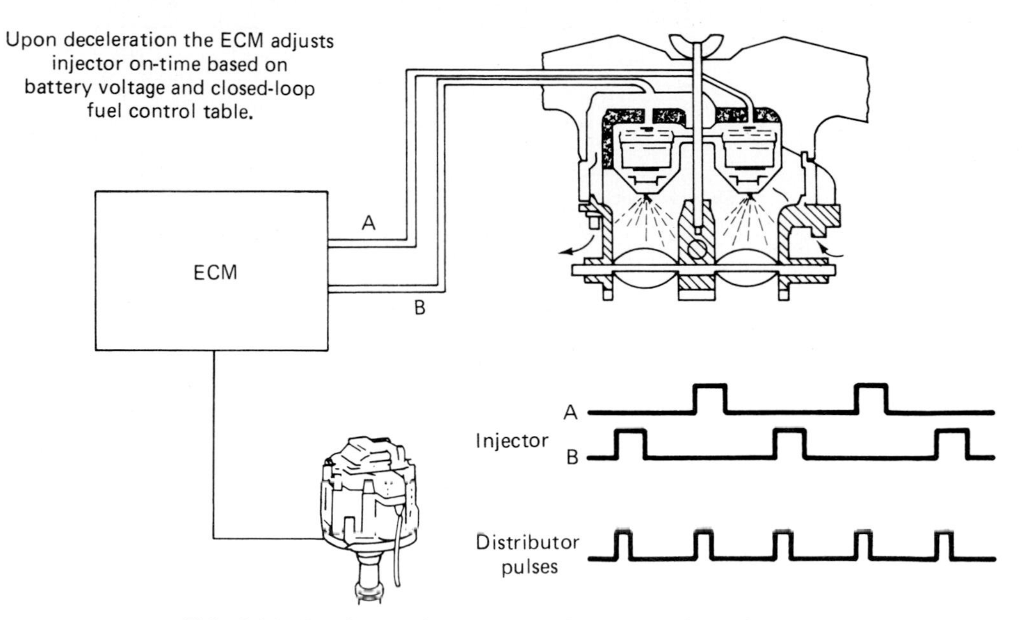

FIG. 9-11 Deceleration lean mixture. (Courtesy of General Motors Corporation.)

sure. The mixture becomes richer. To compensate for this pollution-producing condition, the ECM shortens the injector pulse signals. (Fig. 9-11) As a result, less fuel is delivered. However, if the condition occurs within a brief (preset) time of a previous acceleration or idle enrichment operation, the deceleration program will not go into effect. Under those circumstances, the manifold walls are not wet enough to require compensating adjustments.

IGNITION FUNCTIONS

Similar to the control of the injectors, the computer also determines the moment of ignition and the coil saturation period (dwell). The components are shown in Fig. 9-12. As in a precomputerized system, the

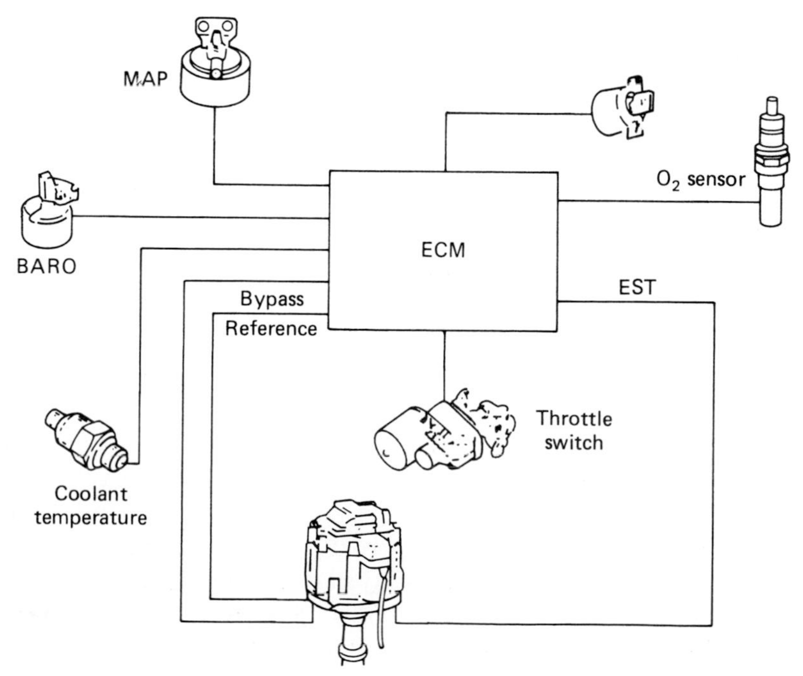

FIG. 9-12 Spark timing components. (Courtesy of General Motors Corporation.)

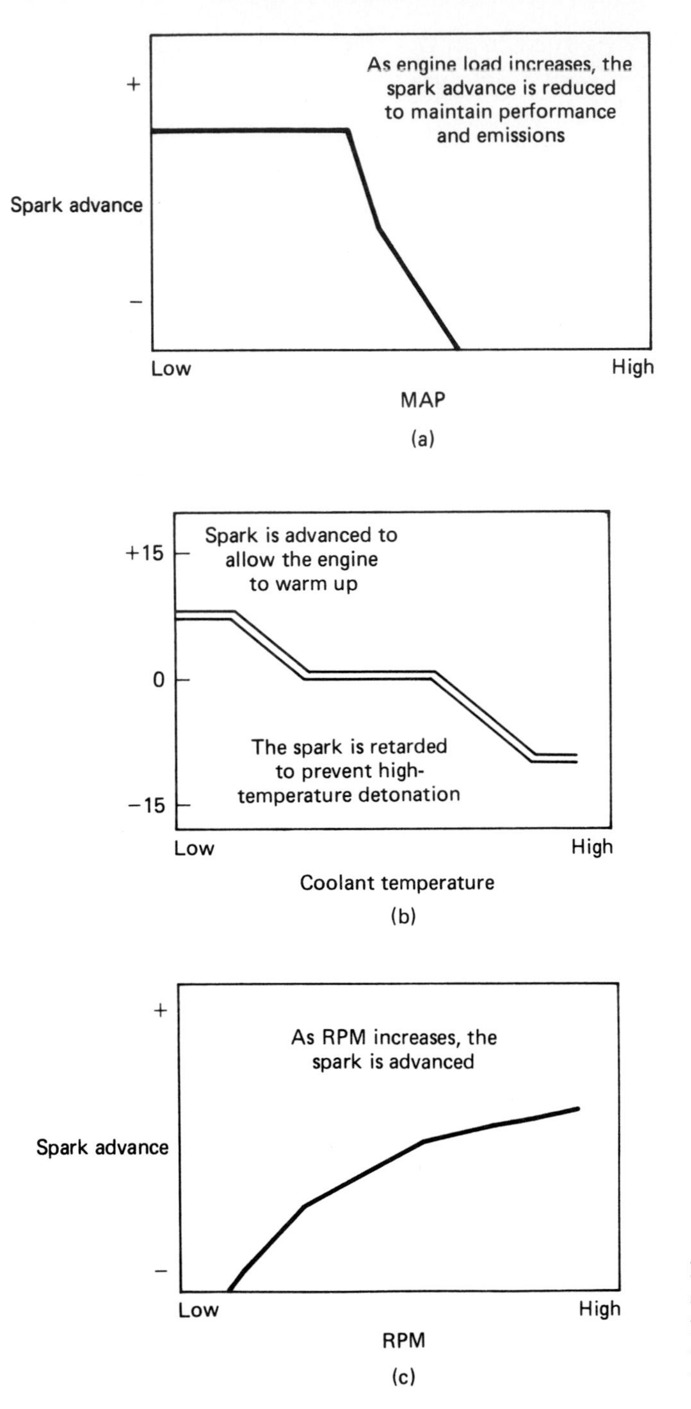

FIG. 9-13 Three conditions that modify spark timing: (a) MAP modification; (b) temperature modification; (c) RPM spark modification. (Courtesy of General Motors Corporation.)

spark is advanced as the engine speeds up. However, instead of relying on centrifugal weights, the basic advance curve is stored in a calibrated PROM (Fig. 9-13).

The timing curve is modified (again, as in a precomputerized system) as manifold pressure fluctuates. The timing curve is also modified in response to inputs from the engine-coolant sensor and the barometric-pressure sensor. Information from all these sources is converted into certain values based on calibration data contained in a computer storage area called the *look-up tables.* The look-up values are added together to get a combined figure which is then used to determine timing.

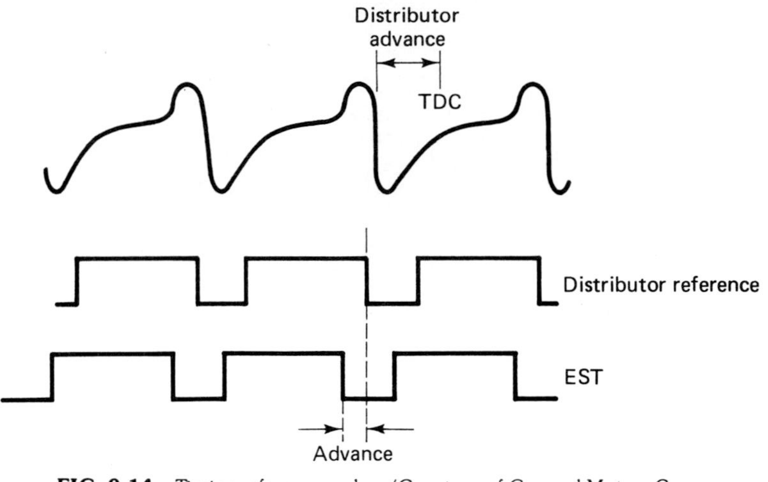

FIG. 9-14 Timing reference pulse. (Courtesy of General Motors Corporation.)

Of course, these calculations must occur many times a second to ensure that timing adjustments occur in a smooth, imperceptible manner.

In addition to advancing the ignition, controls are provided to retard timing whenever the engine knocks or detonates. The input for this control is a device located on the intake manifold or motor block. The device senses those particular kinds of vibrations which are characteristic by-products of detonation.

The basic reference pulse for the timing operations comes from a pickup coil located on the end of the crankshaft. Using a core and pole piece (similar to the core and pole assembly of the solid-state distributors covered in Chapter 7) the unit produces a fluctuating (analog) voltage signal (Fig. 9-14). This signal, once converted into digital form, is used to determine engine rpm and piston position.

The only times that the crankshaft does not determine the reference signal to the computer for spark timing are when the engine is being cranked or when the timing is being set. During cranking, a bypass line from the ECM to the distributor drops to 0 volts (Fig. 9-15). This signals the electronic distributor to ignore pulses coming from the computer and rely instead on signals generated within the ignition system itself for base ignition timing. When setting or checking basic timing, the reference signal is disconnected.

IDLE-SPEED CONTROL FUNCTION

The main purpose of this function is to control the engine speed during closed-throttle operation (Fig. 9-16). The factors that affect idle speed include:

1. *Engine temperature.* When the engine is cold, idle speed is adjusted to approximately 1200 rpm. Then as the engine warms up, the idle speed is gradually decreased, to about 450 rpm. If the engine temperature exceeds a preset PROM value, the computer assumes that the engine is about to overheat. It raises idle speed to increase the coolant flow and reduce the temperature.
2. *Battery voltage.* If the battery output falls below a certain level,

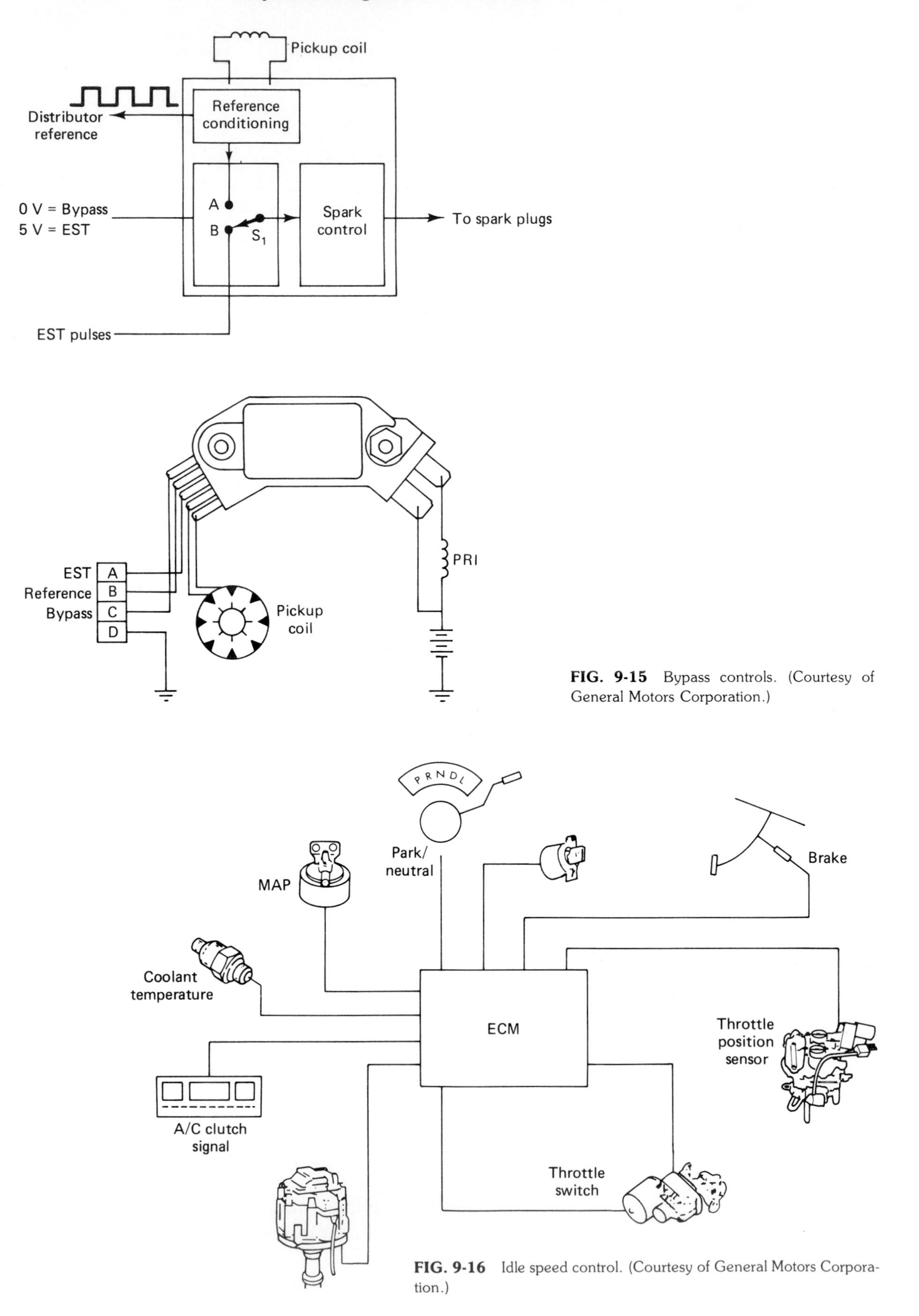

FIG. 9-15 Bypass controls. (Courtesy of General Motors Corporation.)

FIG. 9-16 Idle speed control. (Courtesy of General Motors Corporation.)

the idle speed will be increased to help in the recharging process.

3. *Transmission selector.* The throttle is opened when the transmission is shifted to drive or reverse and closed when the selector is moved to park or neutral.
4. *Air-conditioning compressor.* The throttle is also opened slightly when the air-conditioning compressor is engaged. The controls described in items 3 and 4 are primarily to even out idle-speed fluctuations rather than raise or lower idle speed.

The output of the idle-speed control program is a fluctuating voltage signal sent to a motor-driven worm and gear assembly connected to the throttle plates (Fig. 9-17).

The output device is also used to adjust the throttle position during periods of deceleration. This aspect of the idle-speed program responds solely to manifold pressure during a certain rpm range. As the manifold pressure drops, the throttle plate is opened slightly. Opening the throttle plates reduces manifold vacuum, which decreases vaporization and keeps the mixture from becoming excessively rich. Consequently, pollution is reduced. However, opening the throttle also reduces engine braking. So if the brake is pressed more than 4 seconds, the computer closes the throttle plates, regardless of inputs from the manifold air-pressure sensor.

EARLY FUEL-EVAPORATION FUNCTION

Responding to inputs from the temperature sensor, the computer also controls the operation of the exhaust heat riser valve. As the engine warms up, the position of the valve is adjusted for the best combination of emission control and vehicle performance.

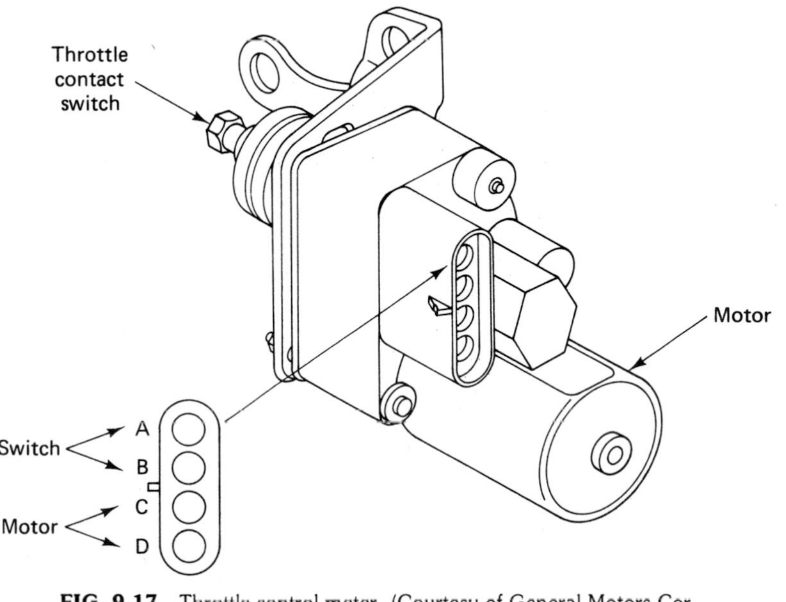

FIG. 9-17 Throttle control meter. (Courtesy of General Motors Corporation.)

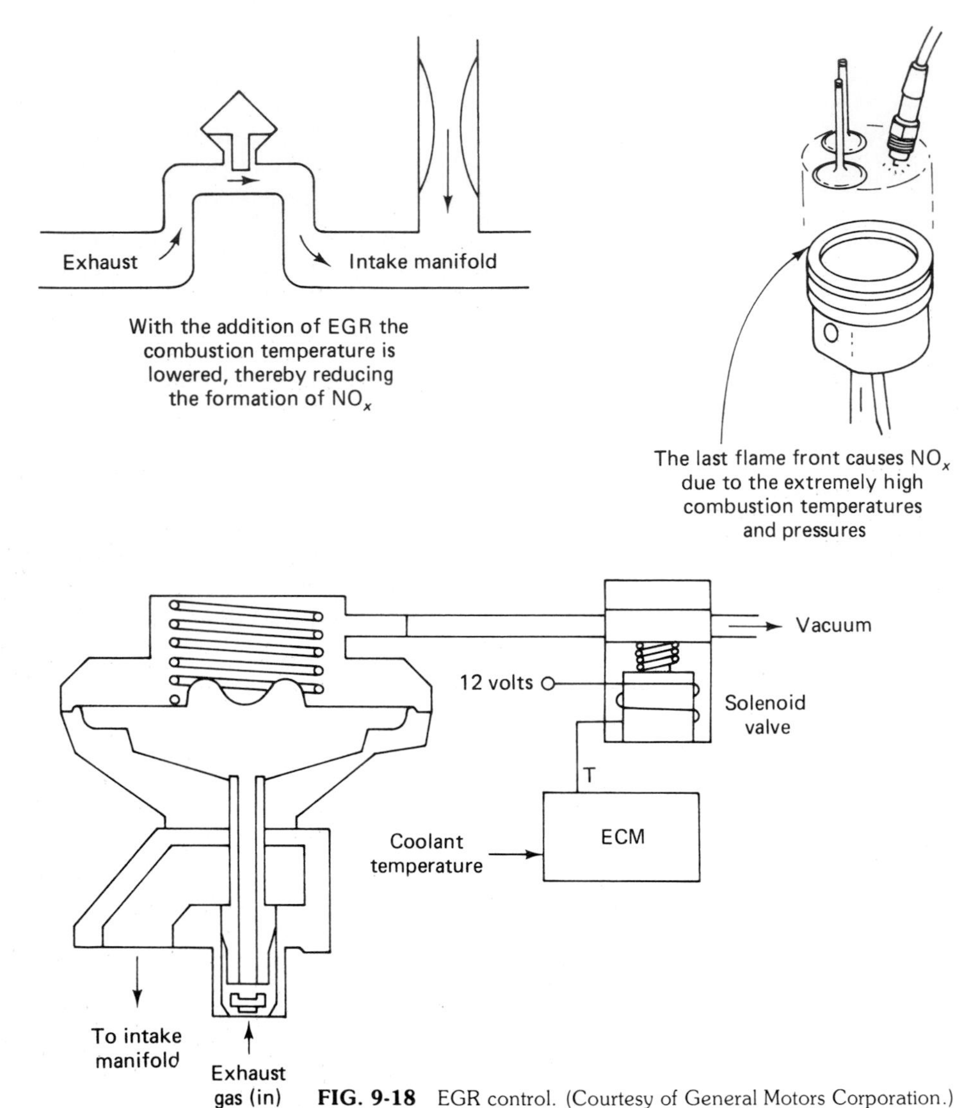
Exhaust
Intake manifold
With the addition of EGR the combustion temperature is lowered, thereby reducing the formation of NO_x
The last flame front causes NO_x due to the extremely high combustion temperatures and pressures
Vacuum
12 volts
Solenoid valve
T
ECM
Coolant temperature
To intake manifold
Exhaust gas (in)

FIG. 9-18 EGR control. (Courtesy of General Motors Corporation.)

EXHAUST-GAS RECIRCULATION (EGR) FUNCTION

The EGR valve supplies small amounts of exhaust gas to the intake system whenever manifold vacuum is high enough (Fig. 9-18). The essentially inert exhaust gas does not burn, thereby reducing the combustion-chamber temperature and nitric oxide formation.

The EGR passage to the intake manifold is controlled by a solenoid-operated valve. During engine starting and warm-up, a blocking signal is sent from the computer to the solenoid to close the valve. As a result, exhaust-gas recirculation does not take place. However, at all other times, the valve is open, allowing the exhaust gas to be recirculated in response to manifold vacuum.

CANISTER PURGE FUNCTION

As in precomputerized systems, a charcoal canister is used to store excess vapors from the fuel tank and fuel system. A solenoid valve controls the purging operation (Fig. 9-19). When the valve is open, trapped gases can be pulled from the canister to the lower-pressure region inside the intake manifold. The solenoid is energized (opened) when the engine is in the closed-loop (normal running) mode and the temperature

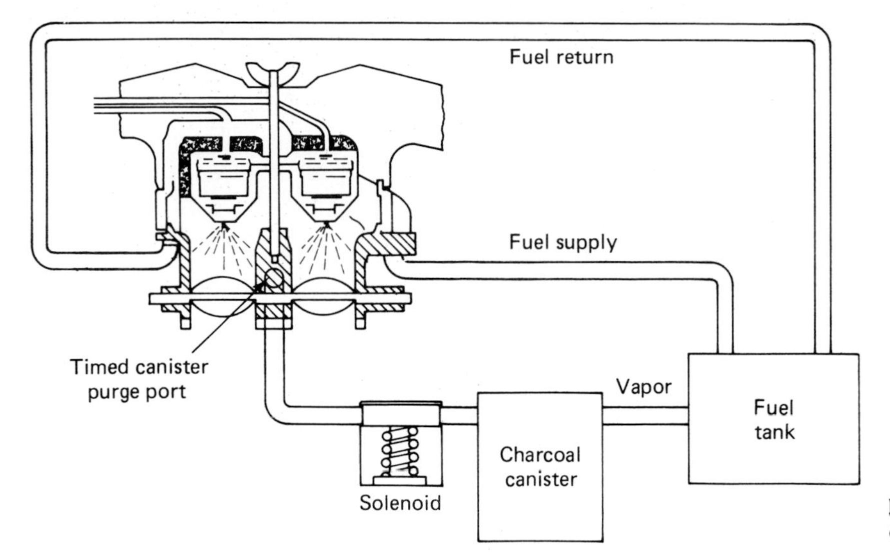

FIG. 9-19 Canister purge. (Courtesy of General Motors Corporation.)

of the coolant is above 80°C. These conditions are best for burning the excess vapors without producing excessively rich mixtures and thereby increasing pollution.

CRUISE-CONTROL FUNCTION

The cruise-control function manages engine speed. Inputs come from an engine speed sensor, the cruise control on–off switch, a set-coast switch, and a resume-acceleration switch. Outputs go to an accelerator linkage control mechanism.

AIR-MANAGEMENT SYSTEM FUNCTION

The computer also controls the flow of air from the air pump. Depending on the input signals, air is directed to the air cleaner, exhaust ports, or catalytic reactor (Fig. 9-20). Two major sets of valves are employed. The first, called the *diverter valve,* sends air to the air cleaner or to a second valve, which is called the *switching valve.* This second valve sends air to the exhaust ports or to the catalytic reactor. The exhaust-air passages also contain the usual check valves to prevent backfiring during deceleration.

The system has three main modes of operation:

1. During normal, or closed-loop, operation, the diverter valve sends air to the switching valve, which, in turn, sends air to the reactor. The catalytic agents in the reactor are formed in two layers or beds (Fig. 9-21). Air injected between the beds helps the platinum and palladium agents oxidize HC and CO.
2. During cold operation and in other open-loop conditions, the reactor is not hot enough to make use of the extra air. Consequently, the diverter valve sends air to the switching valve, which directs the air to the exhaust ports. Performing the function normally assigned to pressurized air from the air pump (until computerized systems), it helps oxidize HC and CO contained in the exhaust gases.
3. When the reactor gets too hot, air added to the exhaust or the reactor can cause damage. Under those conditions, the diverter valve sends air to the air cleaner.

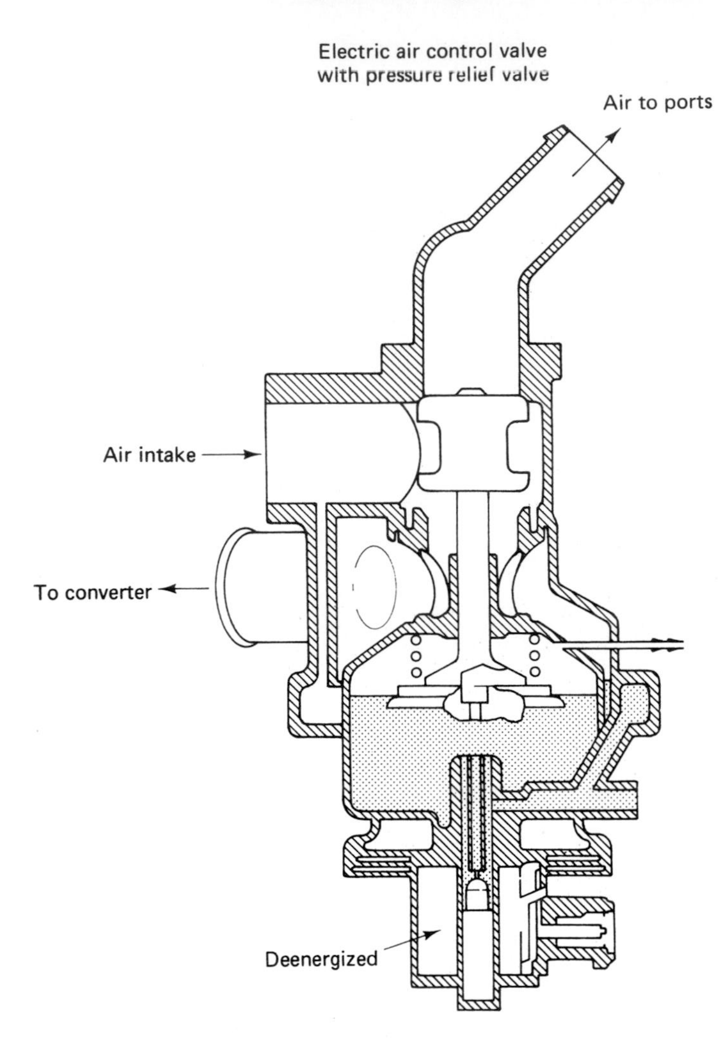

FIG. 9-20 Air management components. (Courtesy of General Motors Corporation.)

CLUTCH FUNCTIONS

Still another application for the on-board computer is controlling an automatic transmission clutch (Fig. 9-22). Normally, there is a certain amount of slippage or wasted energy in a torque convertor or automatic transmission. As the transmission turbine forces fluid through the stator to the pump, the turbine has a tendency to rotate faster than the pump.

The speed difference between the engine and the transmission can be eliminated by a direct connect clutch, pressure plate, spring damper, and control solenoid assembly. The solenoid, which determines the position of the clutch, is energized by the computer in response to inputs from these sources:

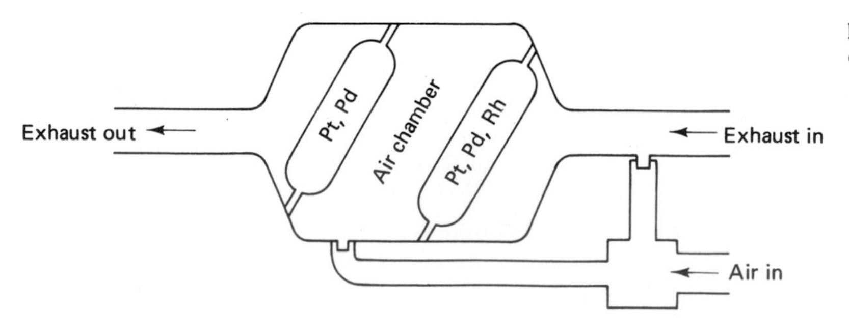

FIG. 9-21 Dual-bed convertor. (Courtesy of General Motors Corporation.)

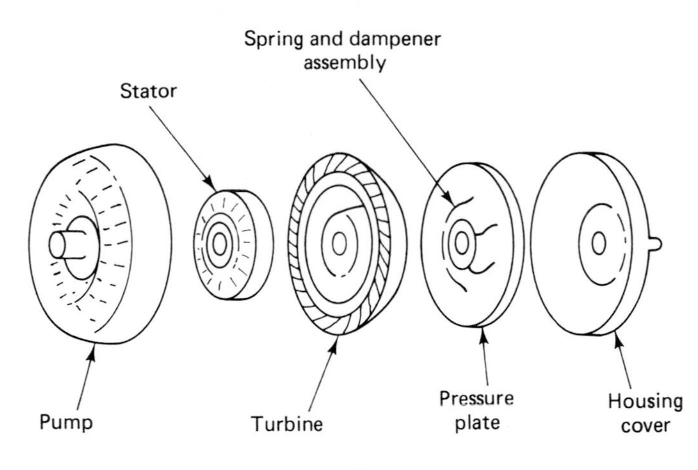

FIG. 9-22 Torque convertor clutch. (Courtesy of General Motors Corporation.)

1. *Brake switch.* If the brake pedal is pressed, the solenoid will be energized regardless of inputs from the other sources. The transmission will operate in a normal manner.
2. *Transmission pressure switch.* This switch tells the ECM what gear the transmission is in. Different vehicles use the clutch feature for different gears. The exact combination is stored as one of the PROM calibration values.
3. *Park-neutral switch.* The control solenoid is disengaged when the transmission is in park or neutral.
4. *Vehicle speed sensor.* The solenoid is energized only after the vehicle reaches a certain preset PROM value (usually 35 to 45 miles per hour).
5. *Coolant temperature sensor.* The solenoid is also engaged only after the engine reaches a certain preset temperature.
6. *Manifold air-pressure sensor.* When the engine is operating above a certain preset load (as indicated by the manifold pressure reading), the clutch is released.

FAIL-SAFE FUNCTION

Because the computer is a vital part of engine operation, it constantly checks itself to make sure that the programs are working properly. On encountering certain problems, a backup injector program is put into action. The backup program, which uses fewer inputs, and consequently is less sensitive to sensor malfunctions, provides three injector pulse durations:

1. 15 milliseconds for cranking
2. 2 milliseconds for idle
3. 5 milliseconds for open-throttle operation

These pulse durations will not give the best operation for all conditions, but will allow the vehicle to be driven to the garage for repairs.

In addition to the fuel-injection backup program, the system also provides an ignition system backup. In the event of a problem, the bypass line noted previously drops to 0 volts. This tells the ignition system to ignore inputs from the computer and to rely instead on signals produced within the ignition system.

FUNCTION SUMMARY Table 9-1 summarizes the engine-control functions just examined. When looking at this table, certain interesting facts emerge:

Functions	Output	Input (standard)														
		Manifold air pressure	Manifold air temperature	Fuel pressure	O_2 exhaust content	Barometric pressure	Engine rpm	Throttle position	Battery voltage	Ignition switch	Engine coolant temperature	Piston position	Engine knock sensor	Transmission position	Air-conditioning compress. switch	Brake switch
Fuel system modes:	Fuel injection															
Normal running	Fuel injection	X	X	X	X											
Altitude compensation	Fuel injection	X				X										
Idle throttle compensation	Fuel injection						X	X								
Acceleration	Fuel injection	X				X										
Deceleration	Fuel injection	X				X										
Clear flood	Fuel injection						X	X								
Battery voltage correction	Fuel injection								X							
Cold engine running	Fuel injection	X									X					
Fail-safe/emergency	Fuel injection															
Ignition system modes:	Primary, ignition coil															
Normal running	Primary, ignition coil	X				X	X		X		X	X	X			
Advance/retard	Primary, ignition coil	X				X	X		X		X	X	X			
Starting	Primary, ignition coil	X				X	X		X		X	X	X			
Battery voltage correction	Primary, ignition coil	X				X	X		X		X	X	X			
Fail-safe/emergency	Primary, ignition coil	X				X	X		X		X	X	X			

Functions	Output	Input (standard)														
		Manifold air pressure	Manifold air temperature	Fuel pressure	O_2 exhaust content	Barometric pressure	Engine rpm	Throttle position	Battery voltage	Ignition switch	Engine coolant temperature	Piston position	Engine knock sensor	Transmission position	Air-conditioning compress. switch	Brake switch
Idle speed control modes:	Throttle plate position motor															
Cold/hot engine	Throttle plate position motor										X					
Low battery voltage	Throttle plate position motor								X							
Transmission position	Throttle plate position motor													X		
Air conditioning	Throttle plate position motor														X	
Deceleration	Throttle plate position motor	X					X									X
Exhaust-gas recirculation (EGR)	EGR solenoid	X														
Cannister purge	Cannister solenoid valve	Closed-loop indicator														
Cruise control	Accelerator linkage motor	(1) Operator set speed, (2) switches, (3) vehicle speed														
Air-management pump modes:	1 Diverter valve 2 Switching valve															
Closed-loop		Closed-loop indicator														
Open-loop		Open-loop indicator														
Hot reactor		Catalytic reactor sensor														

Table 9-1 Continued

Functions	Output	Input (standard)														
		Manifold air pressure	Manifold air temperature	Fuel pressure	O_2 exhaust content	Barometric pressure	Engine rpm	Throttle position	Battery voltage	Ignition switch	Engine coolant temperature	Piston position	Engine knock sensor	Transmission position	Air-conditioning compress. switch	Brake switch
Automatic transmission clutch modes:	Clutch solenoid															
Braking	Clutch solenoid															X
Park/neutral	Clutch solenoid													X		
Normal operation	Clutch solenoid	(1) Transmission fluid pressure, (2) vehicle speed														
Early fuel evaporation	Exhaust heat riser valve solenoid										X					

- The fuel system has the most complicated set of functions.
- Most functions, although they may collect data from many input sources, usually only control one output device.
- Manifold air pressure is the most common input. Anything that affects manifold air pressure will have an effect on a number of functions.
- The cannister purge and air-management pump functions are dependent on the state of the fuel system, whether it is in open- or closed-loop state. Consequently, the proper operation of these two functions depends on the operation of the fuel system functions and I/O devices.

COMPARISONS

Although most systems perform similar functions, as indicated at the beginning of this chapter, implementations are different. The remainder of this chapter examines a few additional representative systems.

BOSCH MOTRONIC

The Motronic system is manufactured by the Robert Bosch Company of Germany for various automobile companies. It provides most of the functions described in the previous system. The fully digital unit manages fuel injection and ignition during normal (closed-loop) running and open-loop operation (acceleration, starting, cold operation, etc.).

Figure 9-23 pictures the components in the system. As shown, the system uses multiple fuel injectors, one located near the intake port of each cylinder.

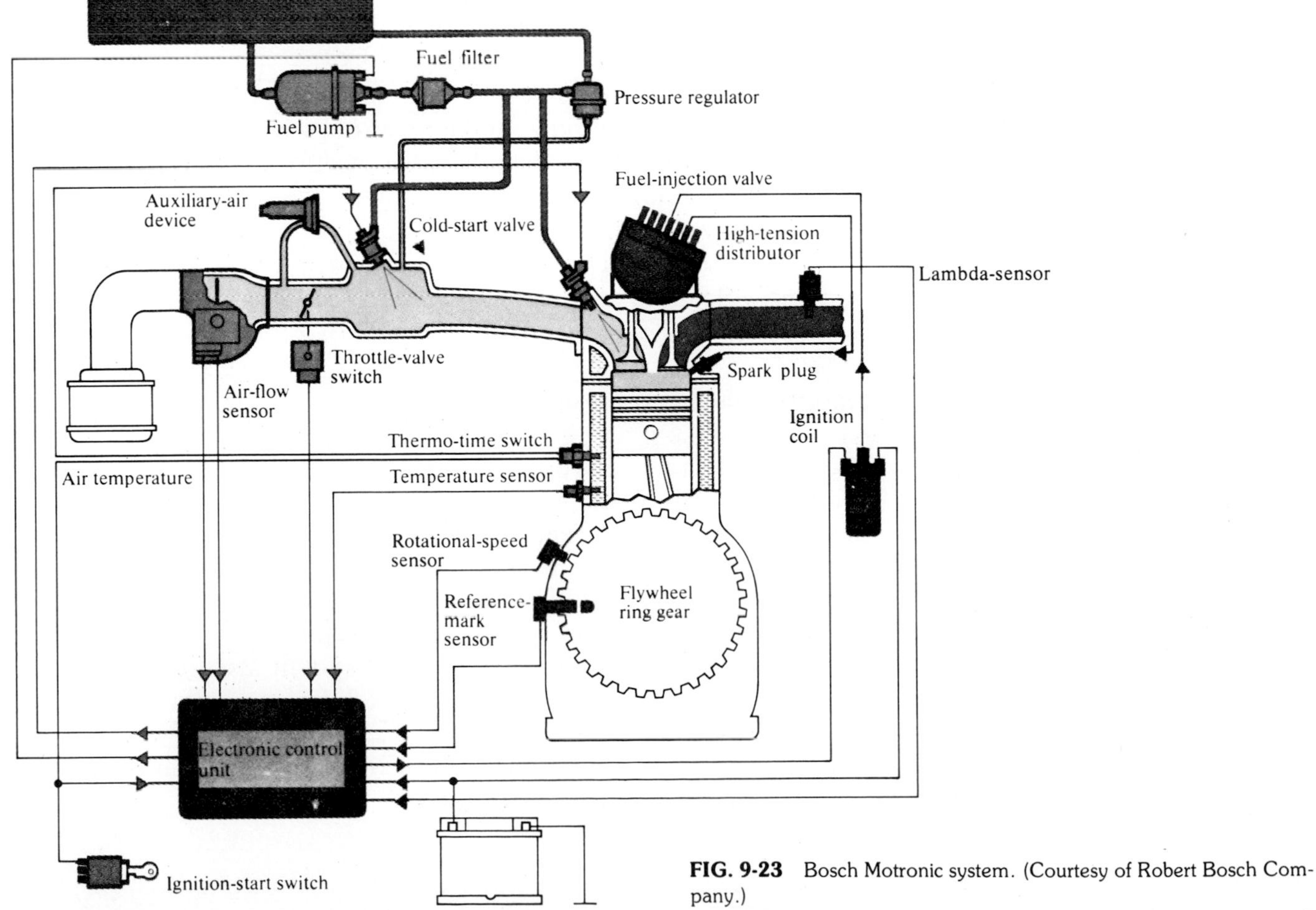

FIG. 9-23 Bosch Motronic system. (Courtesy of Robert Bosch Company.)

Fig. 15/1

1 Vacuum connection
2 Coaxial plug for position indicator on flywheel
3 4-pole plug (sensor signal input)
  1 Coolant temperature sensor
  2 Throttle valve switch
  4 Plug with fixed resistance
4 4-pole plug (supply circuits)
  15 = Circuit 15
  16 = Connection, ignition coil terminal 1
  TD = Transistor-rpm impulse
  31 = Ground

FIG. 9-24 Ignition control module. (Courtesy: Mercedes-Benz)

MERCEDES BENZ

In the Mercedes Benz example, fuel injection and ignition are controlled by different microprocessors.

Ignition-control module

The ignition-control module (ICM) is shown in Fig. 9-24. It receives input information regarding intake manifold vacuum, coolant temperature, engine speed, and throttle-valve position. The microprocessor examines this information and then sends the control pulses to the coil's primary circuit.

Operating modes (functions) are provided for normal running (including acceleration and loading), starting, deceleration, and idle operation. A schematic of the ICM is shown in Fig. 9-25.

Electronic fuel-control unit

The electronic fuel-control unit is shown in Fig. 9-26. Inputs, outputs, and basic operating modes are pictured in Fig. 9-27. As the illustration shows, the same basic fuel system functions are performed as in the representative system examined previously. Closed-loop processing, referred to as the *lambda control,* is performed for normal operation. Open-loop modes are provided for deceleration, full load operation, acceleration, starting and warm-up. The system also goes into open-loop operation if the O_2 sensor is not ready for operation or is defective.

Figure 9-28 shows the location of the components in the system; Fig. 9-29 shows an electrical schematic.

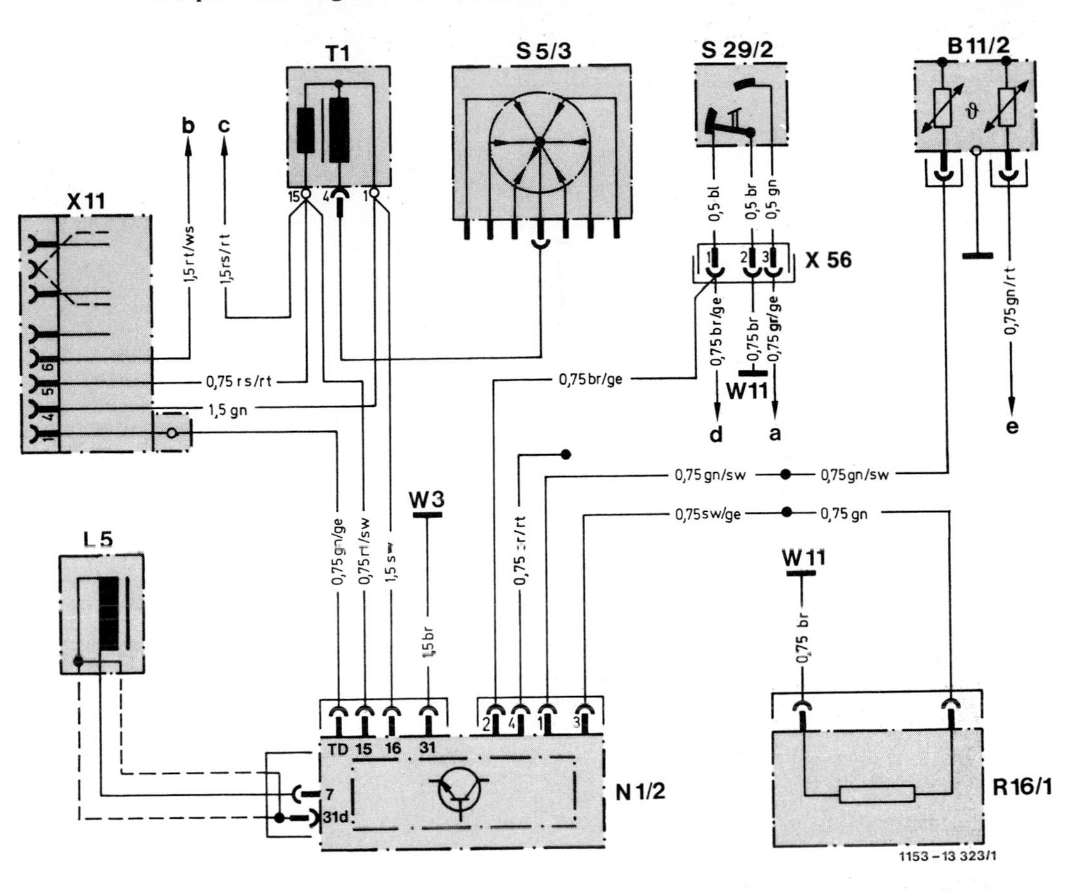

Wiring diagram, electronic ignition system with electronic timing adjustment (EZL)

B 11/2	Coolant temperature sensor
L 5	Position indicator, flywheel/flexplate
N 1/2	Ignition control module
R 16/1	Reference input resistor (EZL)
S 5/3	High voltage distributor
S 29/2	Throttle valve switch
T 1	Ignition coil
W 3	Ground front left wheelhousing (ignition coil)
W 11	Ground, engine (electric wire connection)
X 11	Diagnostic socket (terminal TD)
a	To CIS-E control unit plug (terminal 5)
b	Engine compartment wiring harness plug connection terminal 3 (circuit 30)
c	Engine compartment wiring harness plug connection terminal 1 (circuit 15)
d	CIS-E control unit plug (terminal 13)
e	CIS-E control unit plug (terminal 21)

FIG. 9-25 Wiring diagram, electronic ignition system with electronic timing adjustment (EZL). (Courtesy: Mercedes-Benz)

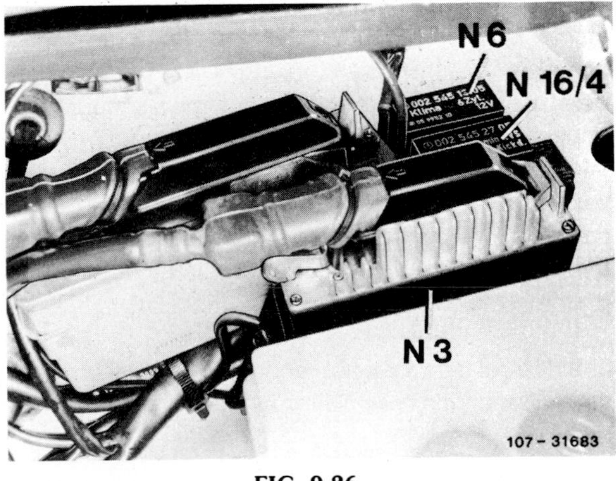

FIG. 9-26

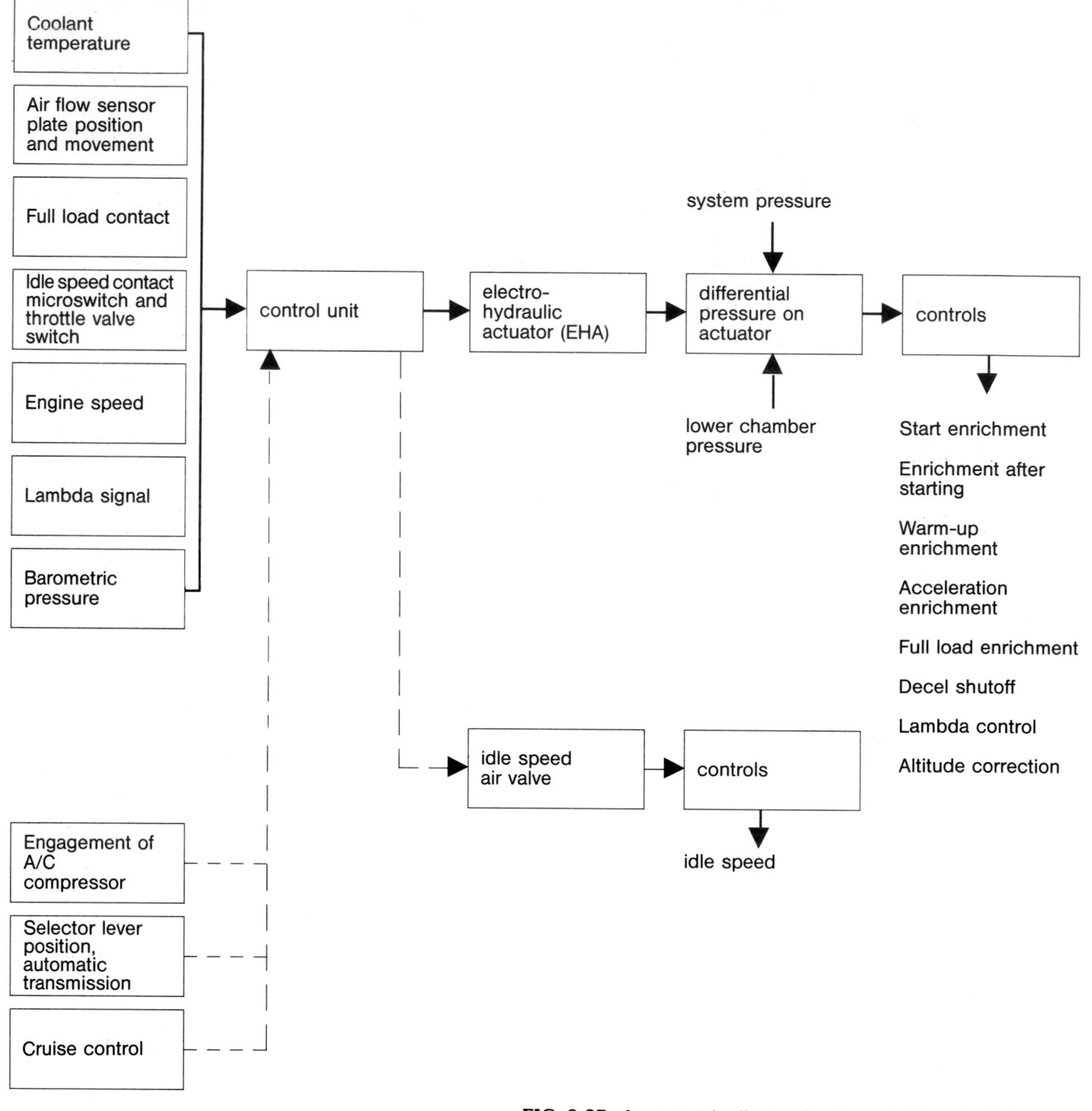

FIG. 9-27 Input signals affecting the electronically controlled mixture correction. (Courtesy: Mercedes-Benz)

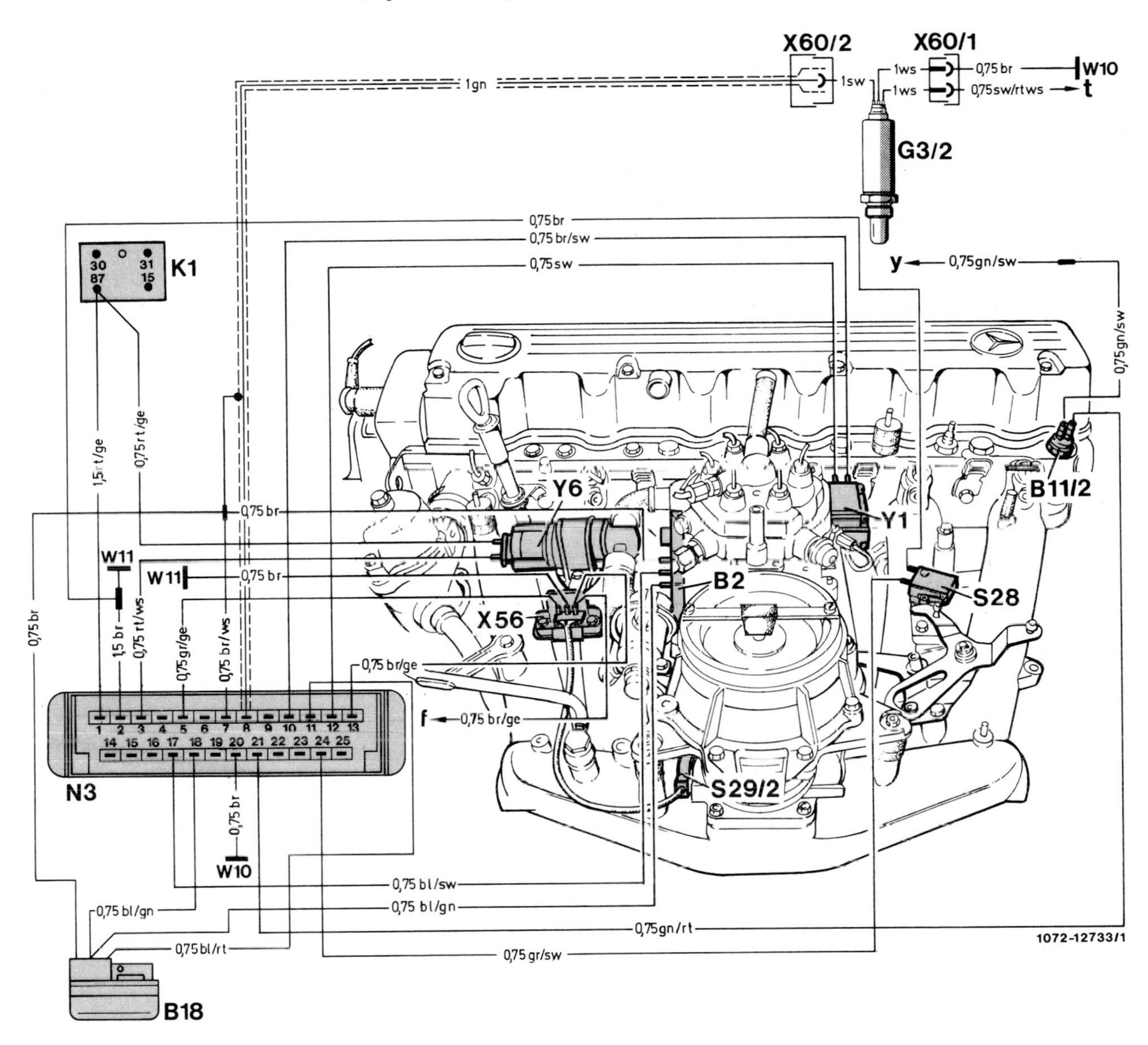

Function diagram, CIS-E gasoline injection system

B	2	Air flow sensor position indicator
B	11/2	Coolant temperature sensor (2-pin)
B	18	Altitude correction capsule
G	3/2	O_2-sensor, heated
K	1	Overvoltage protection relay
N	3	CIS-E control unit (25-pin connector)
S	28	Microswitch, deceleration shut off
S	29/2	Throttle valve switch, idle and full load contacts
W	10	Ground, battery
W	11	Ground, engine (electric wire connection)
X	56	Plug connection, throttle valve switch
X	60/1	Plug connection, O_2-sensor heating coil
X	60/2	Plug connection, O_2-sensor signal
Y	1	Electrohydraulic actuator (EHA)
Y	6	Idle speed air valve

f to ignition control module connector (terminal 2)
t to fuel pump relay connector (terminal 7) circuit 87
y to ignition control module connector (terminal 1)

FIG. 9-28 Function diagram, CIS-E gasoline injection system. (Courtesy: Mercedes-Benz)

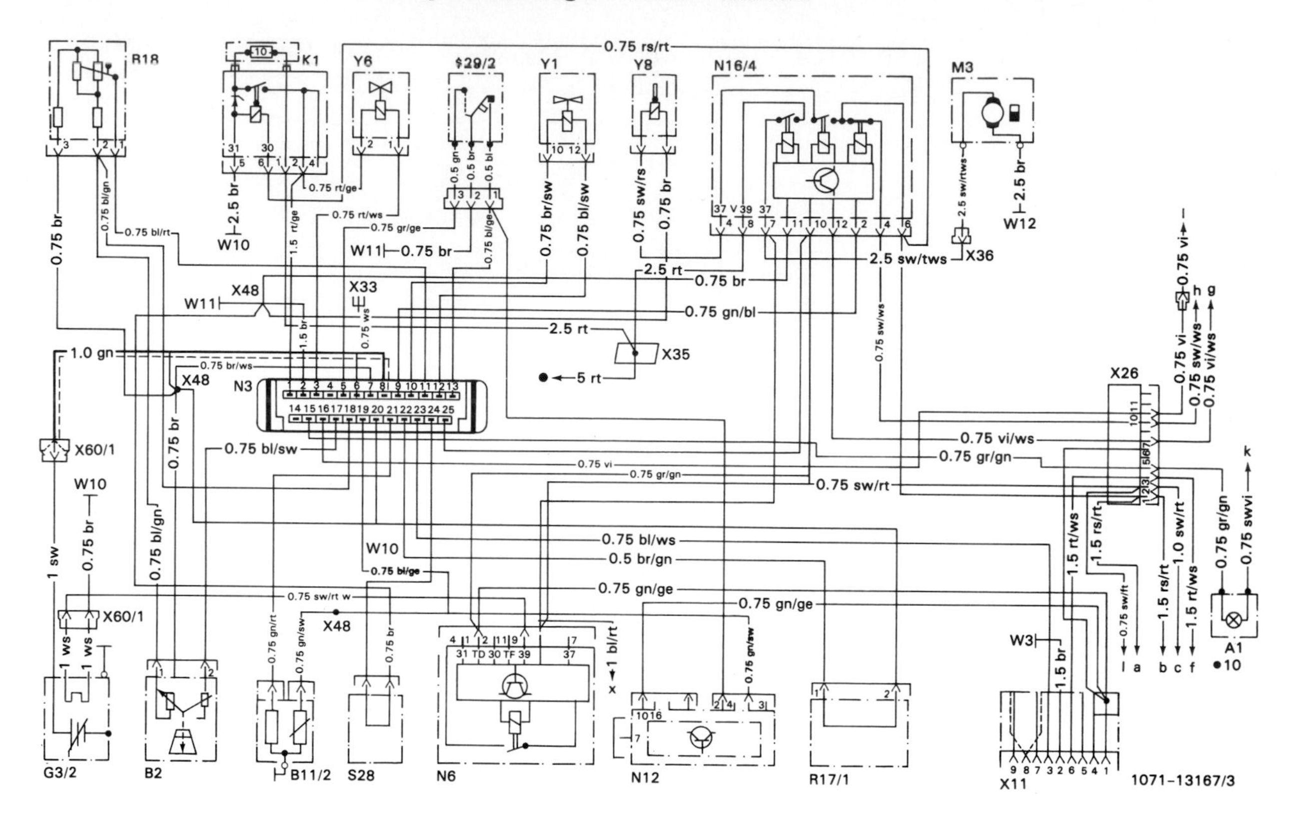

A	1 e 10	Indicator lamp, O_2-sensor
B	2	Air flow sensor position indicator
B	11/2	Coolant temperature sensor
B	18	Altitude correction capsule
G	3/2	O_2-sensor, heated
K	1	Overvoltage protection relay
M	3	Fuel pump
N	1/2	Ignition control module
N	3	CIS-E control unit (25 pin connector)
N	6	A/C compressor cut-out control unit
N	16/3	Fuel pump relay (manual transmission)
N	16/4	Fuel pump relay (automatic transmission)
R	17/1	Bridge, CIS-E
S	28	Micro switch, deceleration shut off
S	29/2	Throttle valve switch, idle and full load contacts
W	3	Ground, left front wheelhousing
W	9	Ground, left front at head lamp unit
W	10	Ground, battery
W	11	Ground, engine (electric wire connection)
W	12	Ground, center console
X	11	Diagnostic socket/wire connector, terminal TD
X	26	Plug connection, engine compartment wiring harness (12 pin)
X	33	Plug connection, CIS-E system to cruise control (1 pin)
X	35	Plug connection, circuit 30/circuit 61 (battery)
X	36	Plug connection, fuel pump wiring harness
X	48	Solder terminal in wiring harness
X	60/1	Plug connection, O_2-sensor heating coil
X	60/2	Plug connection, O_2-sensor signal
Y	1	Electrohydraulic actuator (EHA)
Y	6	Idle speed air valve
Y	8	Cold start valve

a	Ignition coil, terminal 15
b	Fuse 7
c	Fuse 7, circuit 15
e	Wire connector, circuit 30 (fuse and relay box)
f	Fuse 9, circuit 30
g	Plug connection, starter lockout switch (terminal 4)
h	Starter lockout switch (terminal 1)
i	Automatic transmission Plug connection, starter lockout switch (terminal 3), circuit 50 Manual transmission, engine ground (via starter coil)
k	Fuse 6, terminal 15
l	Auxiliary fan relay, circuit 86
x	Pressure switch, A/C compressor

Note: Unmarked ground connections are grounded on engine or chassis. The wire on terminal 6 (circuit 87 K) on the fuel pump relay connector does not apply to vehicles with manual transmission.

FIG. 9-29 Wiring diagram, CIS-E gasoline injection system. (Courtesy: Mercedes-Benz)

10

Input Devices

INTRODUCTION

This chapter describes the input devices that supply information to onboard computers. These devices, which are like the sensory organs in animals, are similar from one manufacturer to another. We focus on representative examples used by Ford, General Motors, and Chrysler Corporation.

BASICS

Before getting into specific units, it will be helpful to go over a few basic points that apply to input in general.

Digital versus analog operation

Sensors convert a physical condition into an electrical signal that represents the condition. Before the signal can be used by the computer, it must be in a digital form. Some devices, such as switches, produce a two-state output that is digital to begin with and does not need to be converted. Most of the sensors discussed in this chapter produce a continuously variable output voltage. These signals are direct analog representations of the condition being sensed and must be changed by a D/A converter into a digital pattern of 1's and 0's.

Reference voltages

Most of the sensors described in this chapter also use reference voltages. A reference voltage is a known voltage signal sent out by the computer to the sensor. The sensor modifies the signal, usually by passing it through a variablc rcsistor network. The resistance in the network depends on the state of the physical condition being measured. After being modified by the sensor, the signal is sent back to the computer. The computer compares the returned signal to the base reference signal.

Transducers

In our context, the term *transducer* refers to devices that convert physical conditions into electrical signals. Some authorities say that a transducer is any device that creates or modifies an electrical signal that is proportional to changes in a physical system. Other authorities say that transducers typically do not modify a known voltage, they create electricity. The piezoelectric crystal used by some sensors to convert vibration into electrical voltage would be regarded by these people as a true transducer.

Feedback relationships

In many instances the information conveyed by input signals to the computer is part of a feedback system. The input information is compared with a reference signal maintained in the computer. The reference signal may be the reference voltage sent out by the computer to the input device. Depending on the variance between the input signal and the reference signal, the control computer will send a compensating signal to an output device. Its job is to change the environment in which the input sensor is located so that the input signal will be brought closer to the reference signal. In effect, the goal of the system is to control incoming data so that it conforms to the view of the world programmed into the computer.

When a control function is operating in a feedback mode, with output affecting input, the system is said to be a closed loop. If for some reason, the system is taken out of a feedback mode, it is said to be in an open loop.

GM PRESSURE SENSORS

GM uses two basic types of air pressure sensors, absolute and differential. (Other manufacturers use similar devices). Figure 10–1 depicts the two sensors.

Absolute units compare manifold or barometric (atmospheric) pressure to a fixed reference pressure. Such devices can be used to determine the altitude by comparing the current atmospheric reading to a sea-level reference pressure. Differential units directly compare manifold and barometric pressures. They are used to calculate manifold pressure (corrected for altitude).

Both types of units contain flexible, metalized quartz plates. The plates bend in response to changing pressure from one or two sources. As the distance between the plates varies, the capacitance of the unit also varies. Consequently, a reference voltage supplied to the assembly increases or decreases depending on the pressures involved.

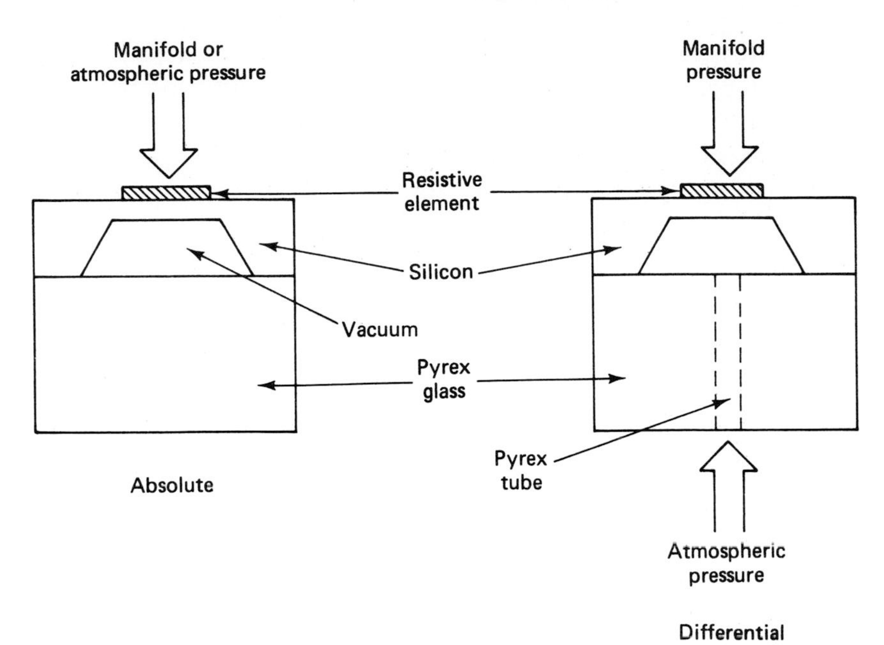

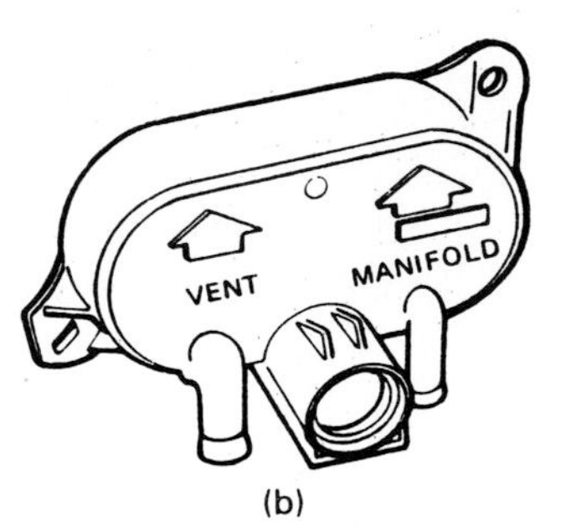

FIG. 10-1 (a) GM pressure sensors (courtesy of General Motors Corporation); (b) Ford pressure sensor (courtesy of Parts and Services Division, Ford Motor Company.)

The primary difference between absolute and differential units is the design of the quartz plate sending unit. In absolute units, the plates surround a sealed reference cavity. The bending action is a product of one variable pressure and one fixed reference pressure. Differential units open up one side of the unit so that atmospheric and manifold pressure work from opposite sides of the same plate assembly. Thus the bending action is the result of two variable-pressure sources.

FORD COOLANT TEMPERATURE SENSOR

The Ford unit consists of a thermostat located inside a brass plug. As shown in Fig. 10-2, the entire assembly is threaded inside a coolant passage in the engine block. As the temperature of the coolant changes,

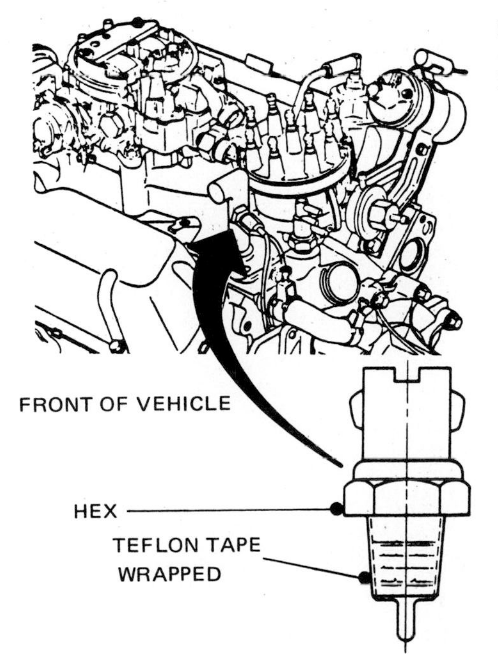

FIG. 10-2 Ford coolant temperature sensor. (Courtesy of Parts and Service Division, Ford Motor Company.)

the electrical resistance of the sensor also changes. Consequently, a reference voltage passing through the sensor is modified in proportion to the temperature changes. The control computer sends the appropriate feedback signals to a coolant valve. Information from this sensor may also be relayed to the instrument system.

FORD THROTTLE-POSITION SENSOR

The Ford throttle-position sensor "tells" the control computer the position of the throttle so that the appropriate feedback signal can be sent. The sensing element is a variable resistor (potentiometer) mounted on the throttle shaft as shown in Fig. 10-3. As the shaft turns, the resistance varies, which alters the reference signal returned to the control computer. (This same sort of potentiometer is used to adjust the volume of a TV or radio.) The entire assembly is attached to a slotted mount on the throttle body. This allows the unit to be moved back and forth for calibration.

GM CRANKSHAFT SENSOR

The GM crankshaft-position sensor provides the GM control computer (called the ECM) with engine rpm signals (sent via the ignition module). The sensor is a Hall-effect switch attached to the engine timing cover, as shown in Fig. 10-4.

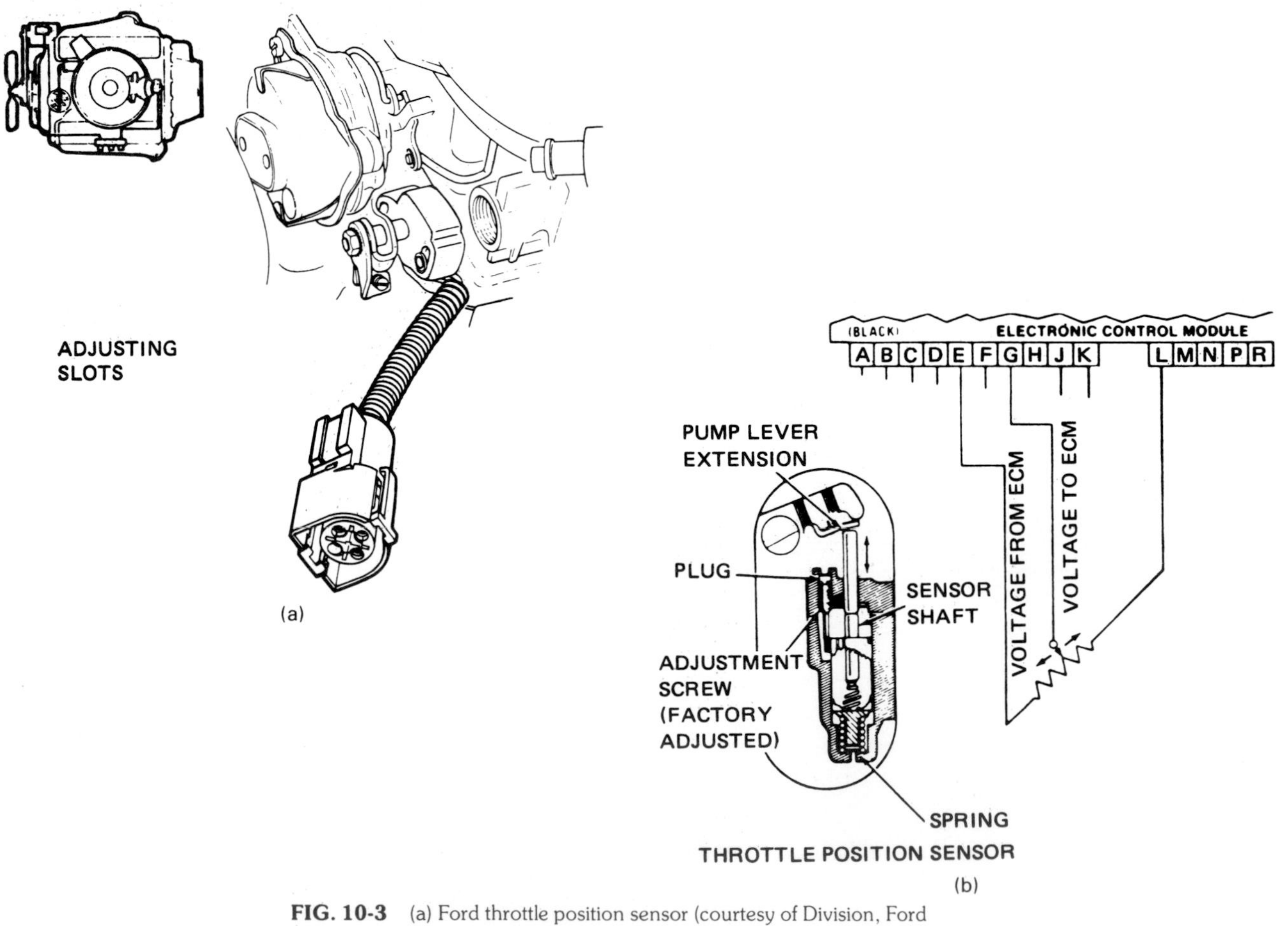

FIG. 10-3 (a) Ford throttle position sensor (courtesy of Division, Ford Motor Company); (b) GM throttle position Chevrolet Motor Division, General Motors Corporation).

In a V-6 engine, three vanes are mounted 120° apart on the rear of the harmonic balancer. A slotted sensor is positioned nearby so that the vanes pass through the slot as the balancer rotates. Each time a vane passes through the slot, a weak voltage signal is produced. As shown in Fig. 10-5, this signal is applied to the base of an NPN transistor. The pulse at the base turns the transistor on and allows a higher voltage signal to pass through the emitter-collector circuit. These higher voltages are used by direct ignition systems (DIS) to determine engine rpm. See Chapter 2 for further information on Hall-effect sensors.

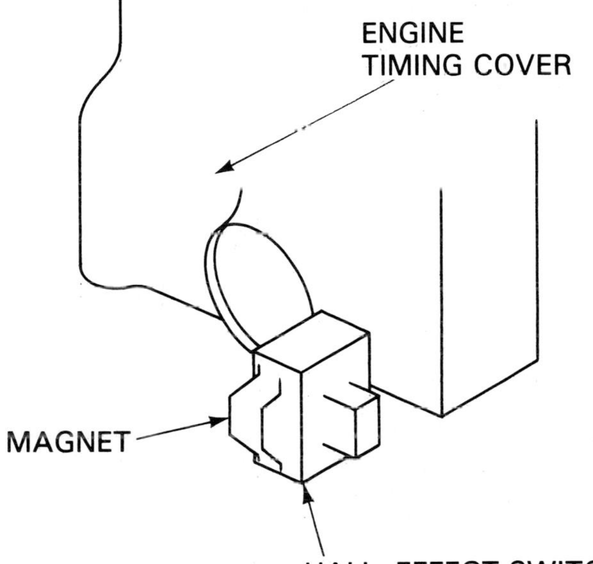

FIG. 10-4 GM crankshaft sensor.

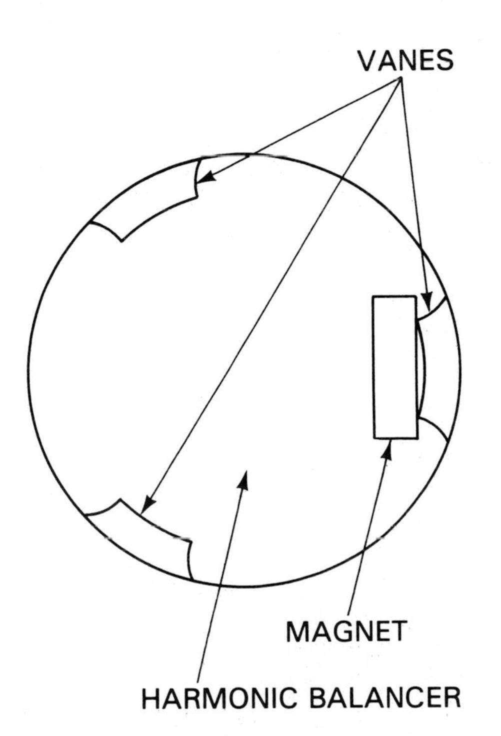

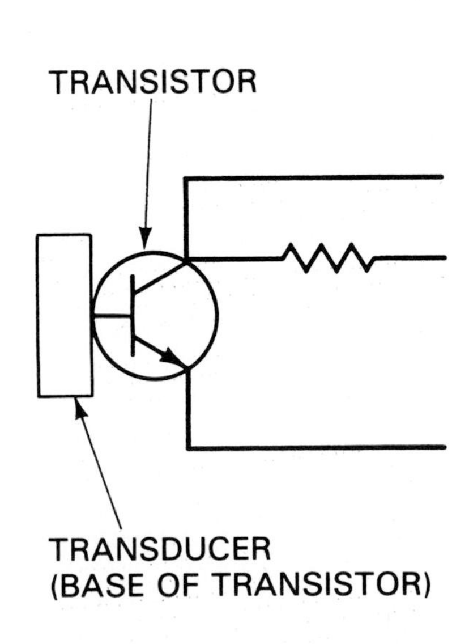

FIG. 10-5 GM crankshaft sensor schematic.

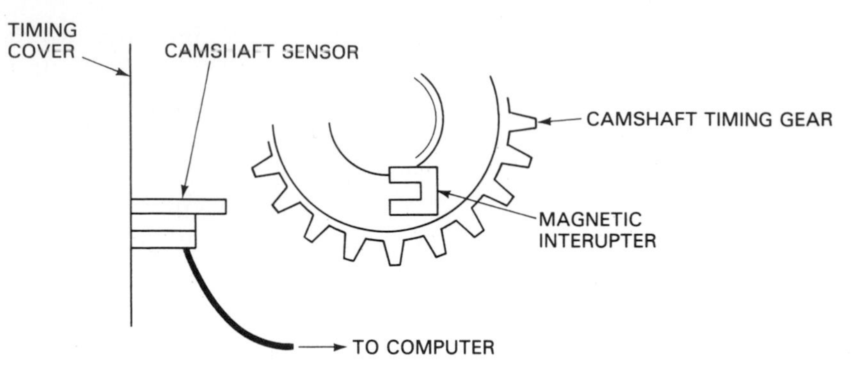

FIG. 10-6 GM camshaft sensor.

GM CAMSHAFT SENSOR

In addition to requiring information about crankshaft position, the GM camshaft sensor also needs to know about camshaft position. The camshaft sensor is a Hall-effect switch located on the front of the engine timing cover, as shown in Fig. 10-6. The switch is used to determine the TDC position of the number 1 cylinder during the compression stroke. Signals from the camshaft sensor are also used to start sequential pulsing of the port fuel injection.

FORD EGR VALVE SENSOR

The Ford EGR valve sensor monitors the position of the pintle valve used in the EGR system. The valve itself is operated by a vacuum diaphragm and spring assembly (Fig. 10-7). Reduced manifold pressure

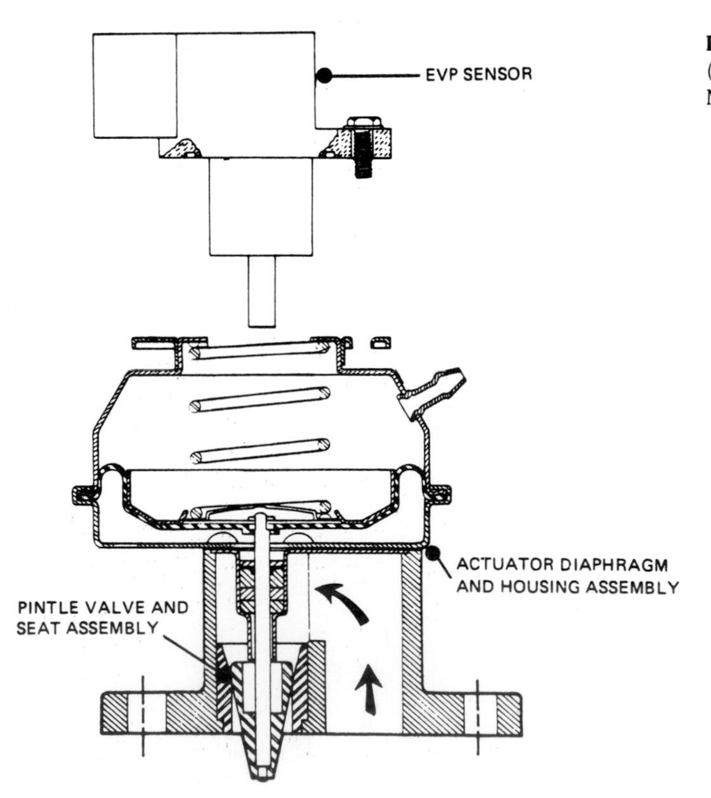

FIG. 10-7 Ford EGR pintle position sensor. (Courtesy of Parts and Service Division, Ford Motor Company.)

causes the diaphragm to open the pintle, thereby allowing exhaust gas to mix with the incoming air and fuel. When manifold pressure increases, the return spring closes the valve.

The sensor, which is mounted on the diaphragm, contains a variable resistor. As the diaphragm moves up and down, the resistor modifies a reference signal from the computer. That way, the computer is able to judge the position of the pintle valve.

The sensor itself does not determine the position of the valve. That control comes from another valve, which, on command from the computer, blocks the vacuum passage to the EGR valve diaphragm. When directed to open, the valve lets reduced pressure from the intake manifold draw in exhaust gas.

GM OXYGEN SENSOR

The O_2 sensor (Fig. 10-8) is actually a small battery. The main components are two platinum plates with a zirconia electrolyte located in between. Positioned in the exhaust manifold, the sensor directs hot exhaust gas to one plate and outside air to the other (Fig. 10-9). Here is how the unit works.

When the zirconia electrolyte is exposed to oxygen, it becomes a carrier of free electrons. The platinum plate nearest the outside air is exposed to more oxygen, so it has more electrons. It becomes negatively charged. The plate nearest the exhaust gas encounters fewer atoms of oxygen, giving it a positive charge. If the two plates are connected into a complete circuit, current will flow.

The voltage or pressure causing the current flow depends on the potential between the two platinum plates. It can vary from 100 to 900 millivolts. The exact reading depends on the O_2 content in the exhaust, which in turn, depends on the air/fuel ratio.

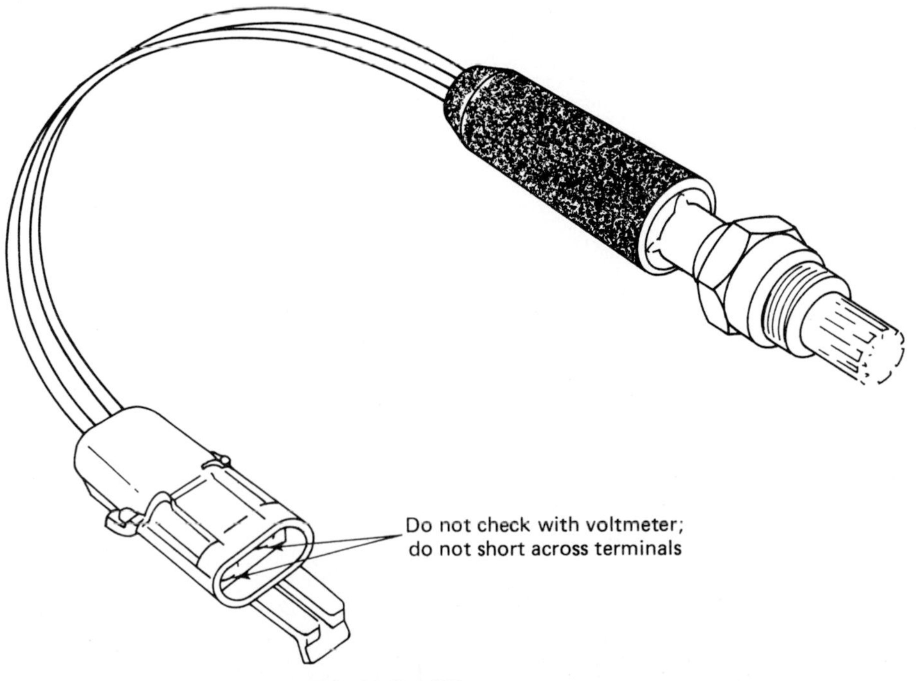
Do not check with voltmeter;
do not short across terminals

FIG. 10-8 GM oxygen sensor.

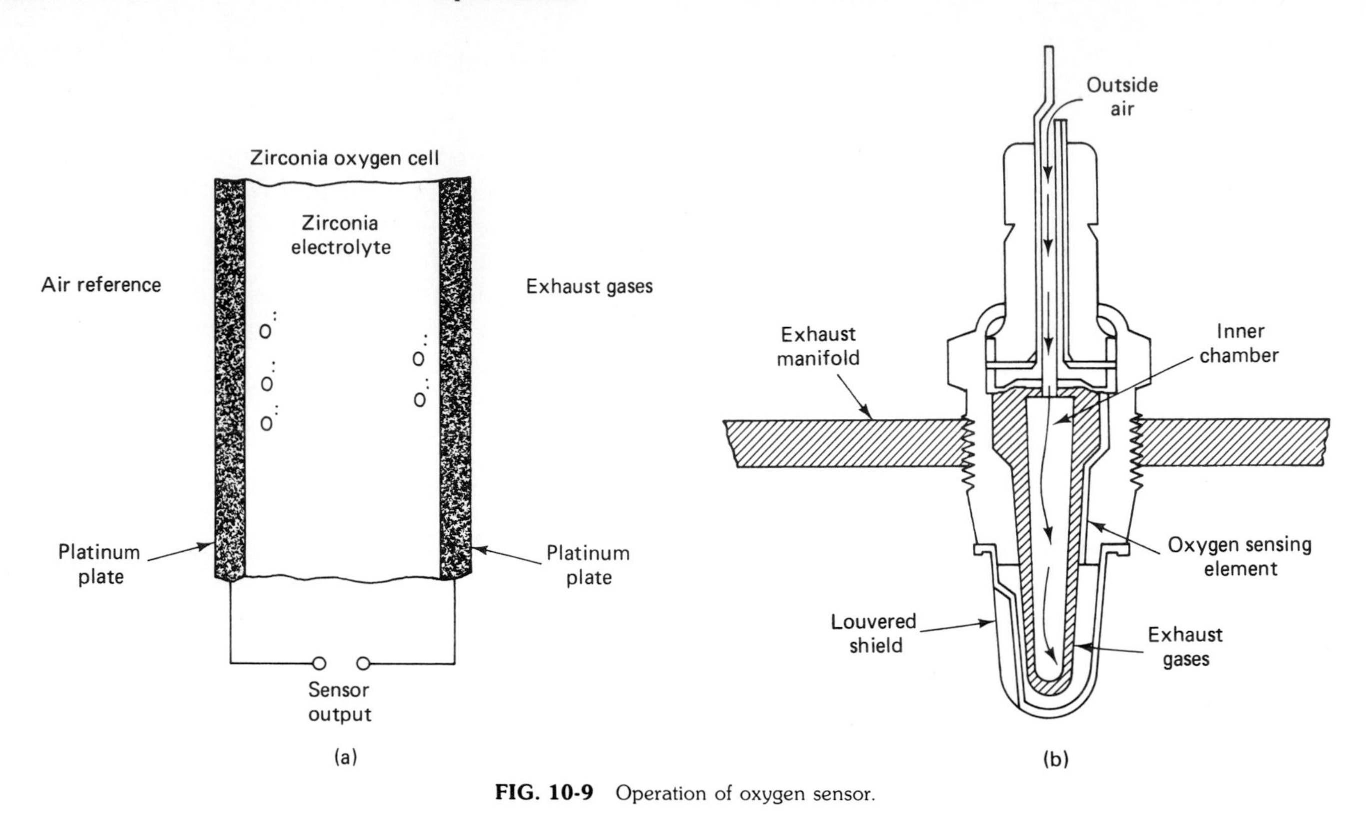

FIG. 10-9 Operation of oxygen sensor.

One essential aspect of the sensor is the speed at which it reacts to changes in the air/fuel ratio. As noted in Fig. 10-10, the voltage curve changes abruptly when the fuel mixture reaches the stoichiometric ratio of 14.7:1. Leaner mixtures produce a voltage level near 900 mV; richer mixtures produce a voltage level near 100 mV. The computer examines the voltage fluctuation, then makes corresponding changes in the injector on-time.

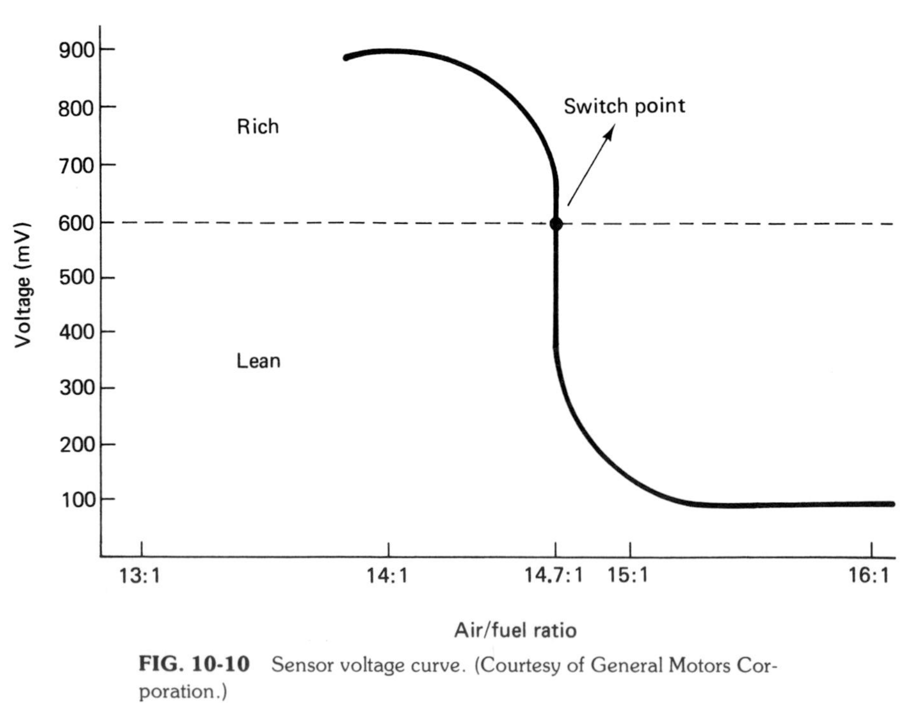

FIG. 10-10 Sensor voltage curve. (Courtesy of General Motors Corporation.)

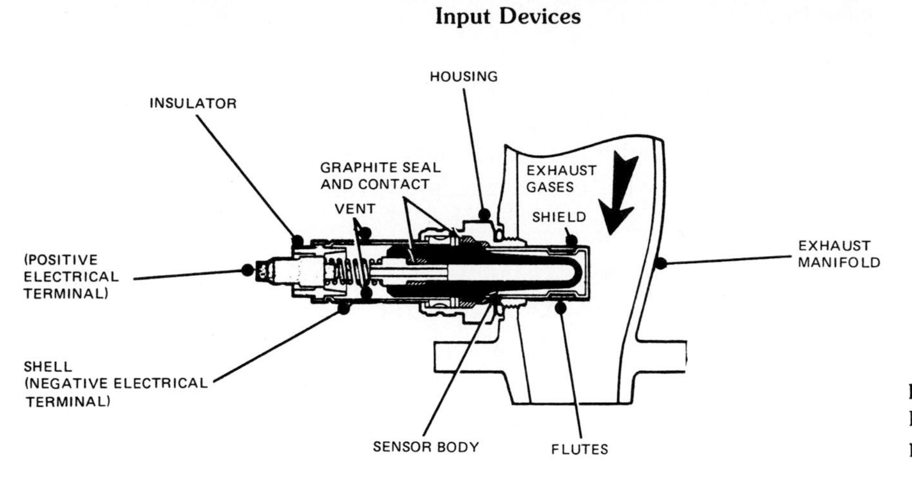

FIG. 10-11 Ford O_2 sensor. (Courtesy of Parts and Service Division, Ford Motor Company.)

However, as we saw in Chapter 9, this closed-loop method of operation works only when the exhaust is above certain temperature levels. At levels below 200°C, the voltage does not switch fast enough to provide useful readings. Figure 10-11 shows a Ford O_2 sensor.

GM AIR-FLOW SENSORS

The GM air-flow sensor determines the amount of air flowing into the air/fuel system. The sensing element is a heated wire placed in the air-flow path. A known reference voltage is sent through the wire. As the air flow increases or decreases, the wire becomes cooler or hotter. This, in turn, affects the reference voltage passing through the wire. The control computer interprets the difference between the actual voltage and the reference voltage as air flow. An air-flow sensor is shown in Fig. 10-12.

ENGINE-KNOCK SENSOR

The engine-knock sensor detects abnormal vibrations caused by engine knocking. The main sensing element may be a piezoelectric crystal transducer, which converts knocking vibrations into electrical signals. The sensing unit is mounted in the engine block, near the cylinders. In one application, a device called the electronic spark module (ESC) re-

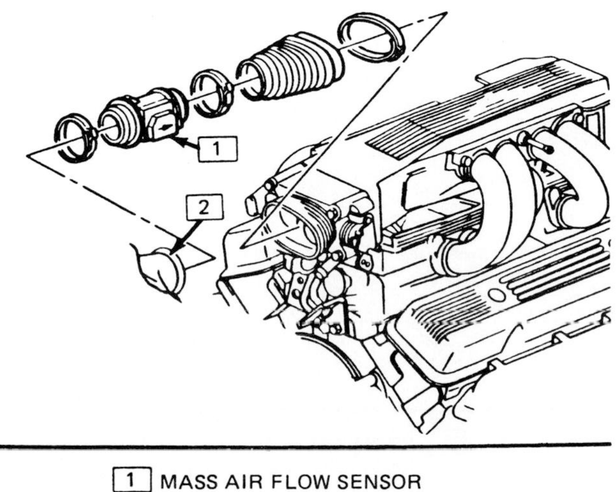

FIG. 10-12 Mass airflow sensor.

ceives a signal from the knock sensor, which it processes and then sends on to the main control computer. The computer, operating in a feedback relationship with the sensor, adjusts spark timing to reduce spark knocking (actually to reduce the signals that indicate spark knocking).

INPUT SWITCHES

Following are a number of switches used to transmit various conditions to the control computer. As noted at the first of the chapter, the output of switch sensors is already a digital value and does not need to be changed by an A/D converter.

- The *park neutral switch* is connected to the transmission selector lever and tells the control computer whether the automatic transmission is in park or neutral.
- The *transmission gear position switches* tell the control computer which gear has been selected in a manual transmission.
- The *fluid pressure switches* tell the computer when critical fluid pressures have been reached (in the cooling system, air conditioning system, brake system, lubricating system, automatic transmission, etc.). The main sensing element in most of these switches is a pressure-sensing element similar to that used in a brake light or oil-pressure light switch. See Chapter 13.

ELECTRICAL CONNECTORS

Snap-apart connectors are used to join the cables leading to the control computer. In many cases, the male side of the cable has a release tab, which, when gently lifted, allows the two sides of the connector to be lifted apart. The connectors are joined by aligning the two halves and then pressing them together. Holding tabs provide the right tension for securing a good electrical connection. Figure 10-13 shows the connections at the rear of a GM control computer (ECM). This type of connector is typical of today's electrical harness.

Most connector terminals are not numbered. However, they are designed to be connected only in the proper way.

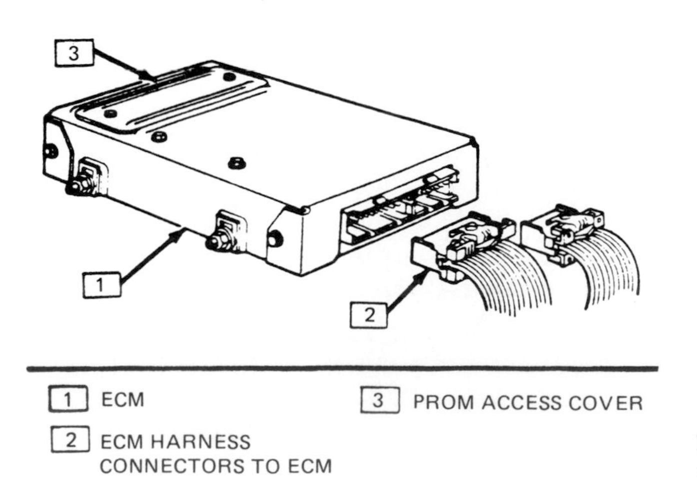

FIG. 10-13 Electrical harness terminal connectors.

11

Output Devices

INTRODUCTION

Just as your hands and feet perform actions in response to signals from your brain, output devices in computer systems perform actions in response to signals from a computer. This chapter deals primarily with the devices used to control engine conditions, such as air/fuel mixture, spark timing, pollution controls, and so on.

GENERAL OBSERVATIONS

Before getting into the output devices used in various systems, we make several observations:

- Output devices convert electrical signals from the computer into physical actions. This is how the computer interacts with the real world. The most common output devices are electric motors (including stepper motors) and solenoids. Refer to the last part of Chapter 2 for an explanation of these devices.
- As noted in the previous chapter on input devices, output devices are often part of feedback-control systems. Responding to information conveyed by an input sensor, the computer sends a control signal to an output device. The output device alters the environment in which the input device is operating. The object of the system is to cause the input signals to match a reference signal contained in the computer's memory. In effect, the purpose of output devices in a closed-loop feedback system is to control the data transmitted by input devices.
- Output devices used in body-control systems are described in Chapters 14 and 15.

- Control system functions (where input and output are viewed together) are described in more detail in Chapter 9.
- This chapter deals primarily with output that results in a physical action. Engine-control computers can also send pure information to body computers and to off-board diagnostic computers. Interfaces to body computers are discussed in Chapters 14 and 15. Interfaces to diagnostic computers are discussed in Chapter 19.

FUEL SYSTEM

Fuel system control is a major function of automotive-control computers. Responding to input from various sensors, the computer sends out commands to make the fuel mixture leaner or richer. Output devices respond by changing the amount of air and fuel mixed together. The type of output device employed depends on the type of fuel delivery system used.

Carburetor controls

Figure 11-1 shows solenoid-operated mixture controls in a carburetor. The solenoid, which is energized by a signal from the computer, controls the position of the paddle and thus the amount of air and fuel mixed together.

Figure 11-2 shows a stepper motor control system. Four separate windings are energized sequentially by the computer to move a control rod in or out 120 positions, or "steps." The total travel is 0.400 inch.

Throttle body injection systems

Throttle body injection systems replace the main body of traditional carburetor systems with one or two fuel injectors. The throttle body itself, containing the throttle plates, idle ports and mounting flanges, is retained.

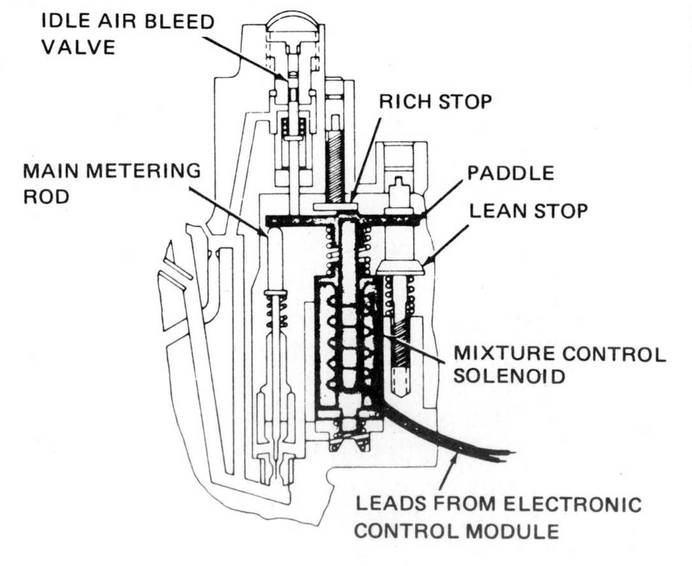

FIG. 11-1 GM mixture control solenoid (E2ME, E4ME). (Courtesy of Chevrolet Motor Division, General Motors Corporation.)

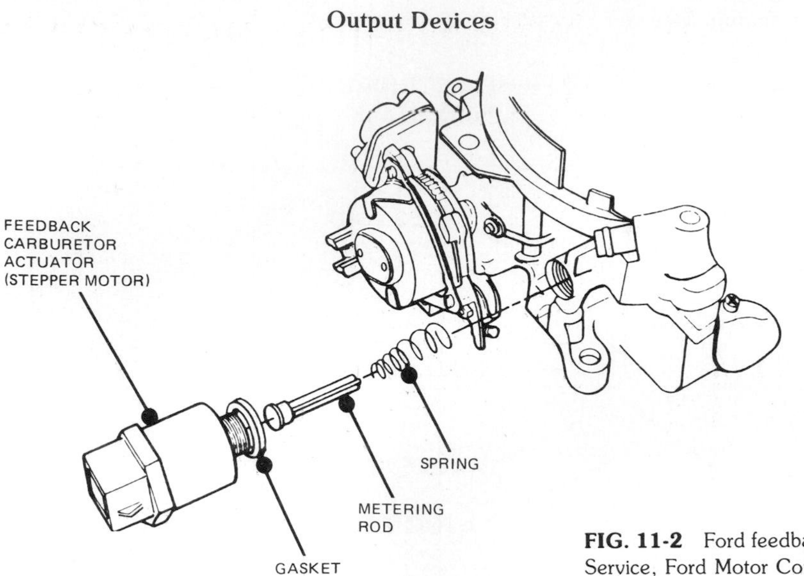

FIG. 11-2 Ford feedback carburetor actuator. (Courtesy of Parts and Service, Ford Motor Company.)

The air flowing through the throttle bore is controlled, as in carburetor systems, by the throttle plates, whose position is controlled by the accelerator, which is controlled by the driver's foot. The driver's foot is the primary output device of a feedback program running in the driver's head.

The amount of fuel delivered is controlled by solenoid-operated injectors positioned above the throttle plates. The solenoids are controlled by feedback programs running inside the control computer (which is ultimately controlled by the feedback program running in the driver's head).

Figure 11-3 shows the fuel flow in a throttle-body-based injection system. Figure 11-4 pictures the injector used in a throttle-body system.

Ported-fuel injection systems

The primary difference between the output devices in throttle-body and ported-fuel injection systems is the number and location of the fuel injectors. As the name implies, a ported system has a fuel injector for

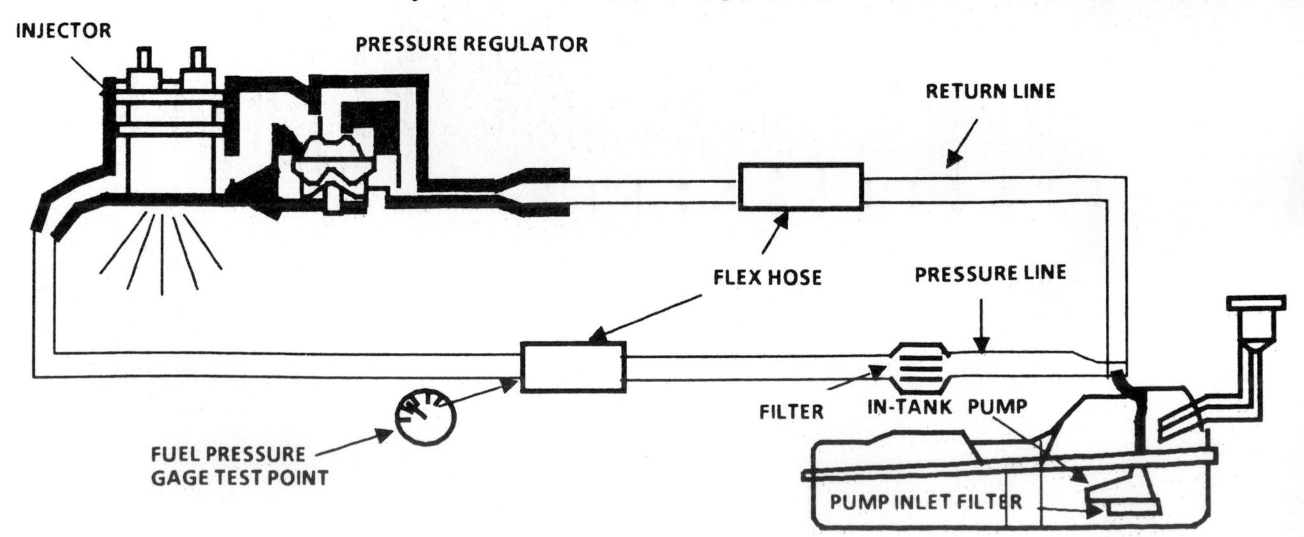

FIG. 11-3 Fuel circuit, throttle body injection (TBI). (Courtesy Pontiac Division, General Motors Corporation.)

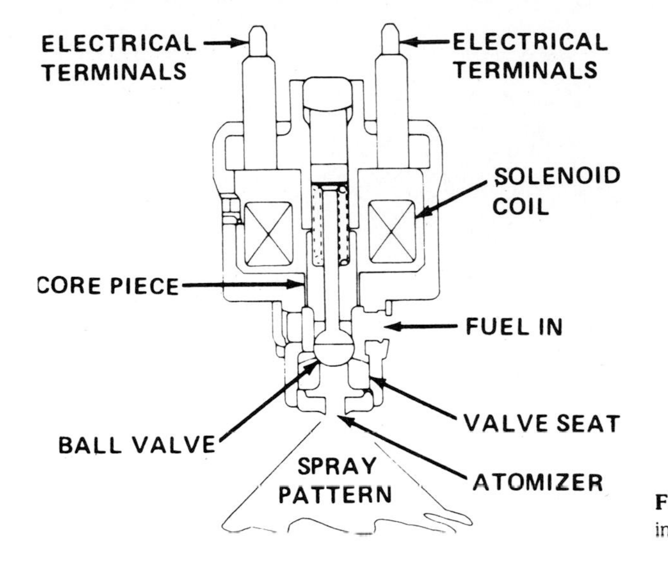

FIG. 11-4 Cross section of throttle body fuel injector.

each intake port, near the intake valve (not in the cylinder, as in diesel injection systems). Ported systems typically have a throttle body mounted, as usual, atop the intake manifold. Throttle plates in the throttle body provide the means of controlling air flow.

As in a throttle-body injection system, the control computer manages the duration and timing of the injector pulses. The longer the injectors stay open, the more fuel is delivered and the richer the mixture.

Some systems operate injectors by groups. For example, in a vee-type engine, all the injectors in one bank might be operated, followed by all the injectors on the other bank. Other systems, sometimes called SFI (sequential fuel injection), operate only the injector whose cylinder is ready for combustion.

Figure 11-5 pictures a fuel injector used in a ported-fuel injection

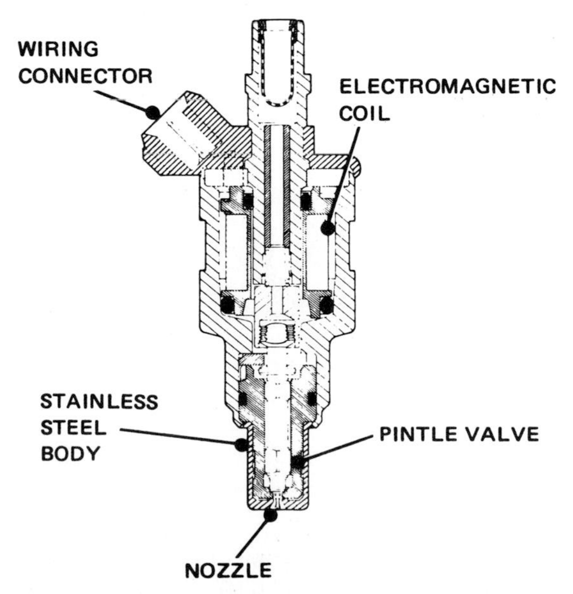

FIG. 11-5 Ford EFI fuel injector. (Courtesy of Parts and Service Division, Ford Motor Company.)

system. This injector has a stainless steel ball-and-seat valve and director plate for fuel control. The computer opens the valve, holds it open, and then closes it. Compared to earlier injectors, these units provide faster response time, improved fuel atomization, better spray control, and lower operating voltage (an important consideration when cranking in cold weather).

IGNITION SYSTEMS

All ignition systems today are managed by on-board computers. The main object of these systems is to control the moment of ignition. As we know from previous chapters, that is when a high voltage surge of electricity is induced in the ignition's coil's secondary circuit. Since the secondary surge is caused by an interruption in primary circuit, the output of the control computer must be directed to the current flow through the primary circuit.

In precomputerized systems, the device used to control the flow of current through the primary circuit was a mechanically operated switch consisting of the points. They were located inside the distributor, along with the rotor.

Today, the major output device is a transistor. Although implemented in different ways in different systems, the principle is the same in most implementations. The control computer, operating in response to input signals, controls the flow of current to the base of a transistor. The base, in turn, acts as a switch for current flowing through the emitter-collector circuit. This circuit controls the current flow through the primary windings of the coil. Figure 11-6 shows first- and second-generation distributors produced by Ford.

Given the developments just described, the job of the ignition distributor has been reduced to nothing more than routing high-voltage surges to the spark plugs. In some engines, even this function has been eliminated. Multiple ignition coils have replaced the distributor.

GM has been one of the main developers of these kinds of systems, which they refer to in different divisions as the distributorless ignition system (DIS), computer-controlled coil ignition (C3I), and integrated direct ignition (IDI).

The advantages claimed for such systems are

- Fewer moving parts
- More compact mounting
- Remote mounting capability
- Eliminating of mechanical timing adjustment
- Elimination of possibility misadjustment or tampering
- Use of reliable electronic components

Looking at a Buick V-6 engine as an example, three coils are used, one for every two cylinders. Each coil has two secondary towers to supply secondary voltage to two spark plugs simultaneously. The Buick

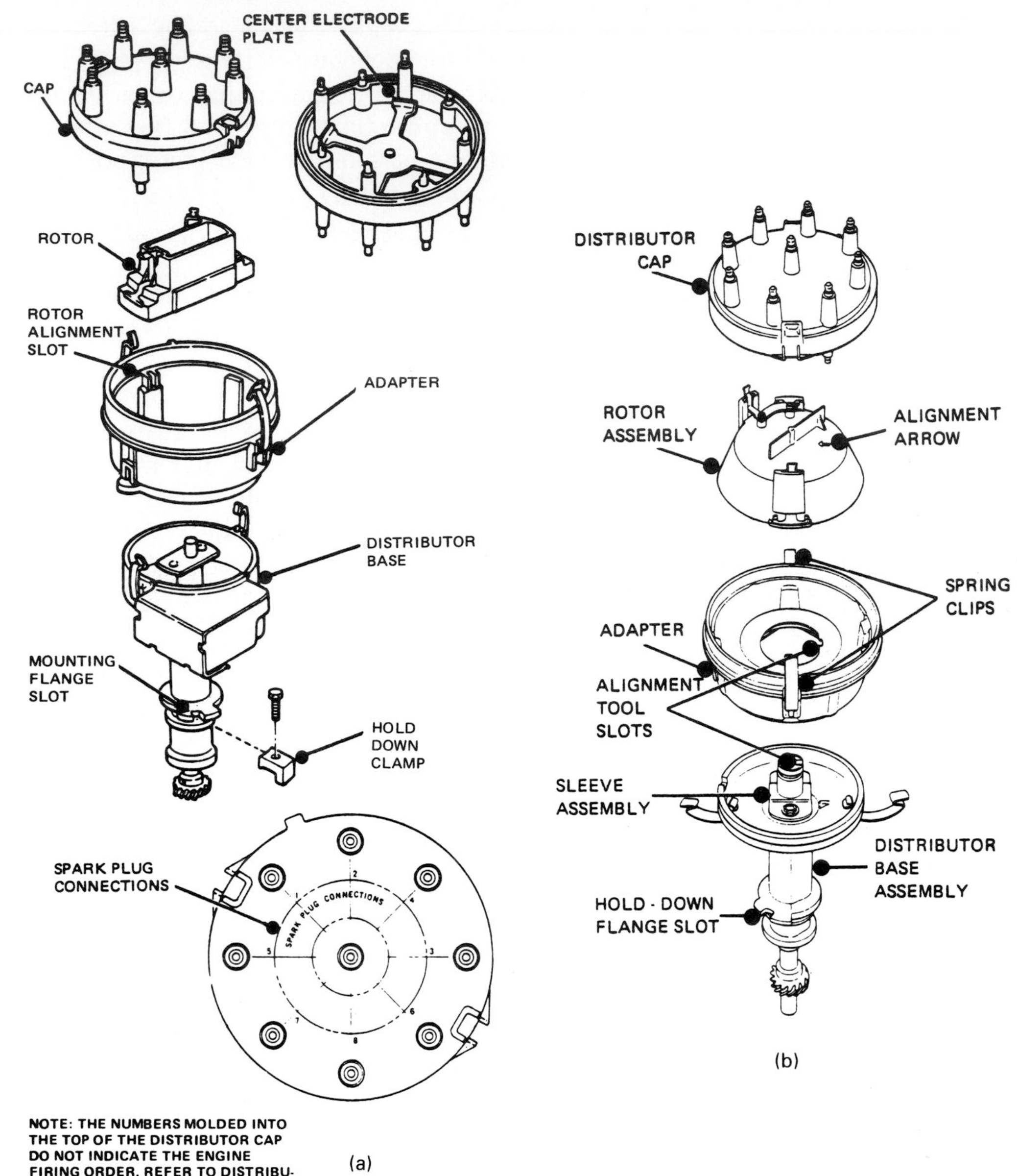

FIG. 11-6 (a) First-generation Ford electronic distributor; (b) second-generation Ford electronic distributor. (Courtesy of Parts and Service Division, Ford Motor Company.)

system is available in separate versions called type I and type II. Figure 11-7 shows the differences.

These systems use what is referred to as a "waste spark" method of spark distribution. Each cylinder is paired with the cylinder opposite it (1–4, 3–6, 2–5). The spark occurs at the same time in the cylinder coming up on the compression stroke and in the cylinder coming up on the exhaust stroke. Very little energy is required to fire the spark plug on the exhaust stroke. The remaining energy is used to fire the cyclinder on the compression stroke. The same is true when the cylinders reverse roles.

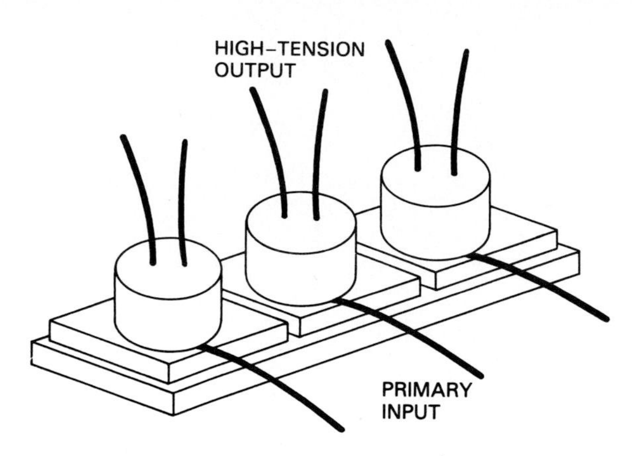

FIG. 11-7 Ignition system identification.

In the Buick system, a separate electronic module controls the operation of the coils below 400 rpm. Over 400 rpm, the system is managed by the main engine-control computer (called the ECM in GM systems). Figure 11-8 is a schematic of the GM system.

AIR-MANAGEMENT SYSTEM

The air-management system is used to provide extra air to the exhaust manifold or catalytic reactor. The added oxygen, delivered on command from the control computer, helps oxidize hydrocarbons and carbon monoxide. The main mechanical components of the system include a vane-type pump driven by a belt from the crankshaft and a series of

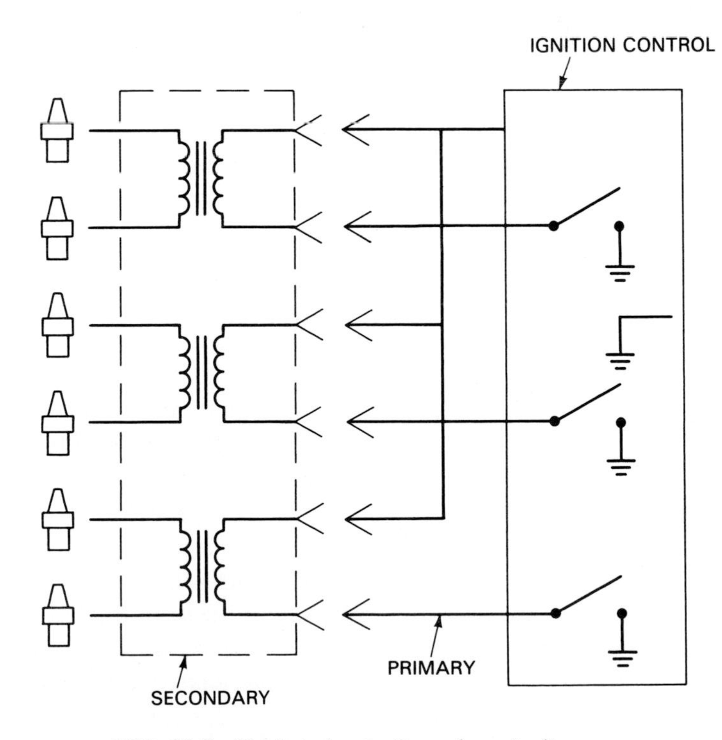

FIG. 11-8 Distributorless ignition schematic diagram.

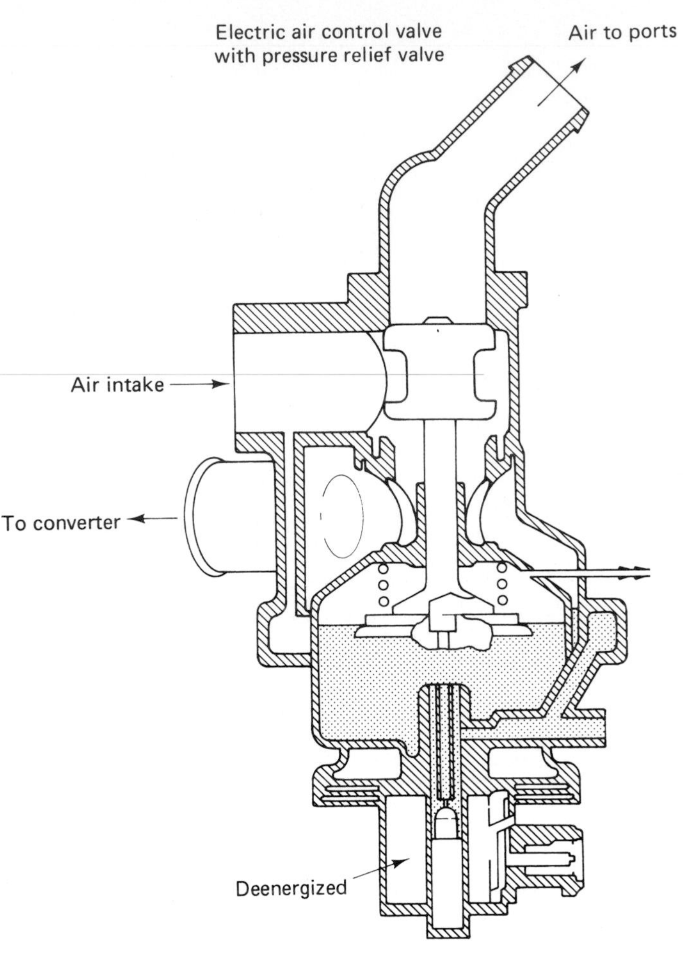

FIG. 11-9 Air management components.

solenoid operated valves. The valves are the primary output devices, operated on command from the computer. Figure 11-9 shows some basic air-management components. Figures 11-10 through 11-14 picture a Ford ''Thermactor'' system.

IDLE-SPEED CONTROL

The idle-speed control system controls idle speed by adjusting the position of the throttle plates. In the system shown in Fig. 11-15, a small electric motor mounted on the side of the throttle body is the main output element. On command from the computer, the motor operates a gear-and-shaft assembly, which determines the closing stop position of the throttle plates.

Ford uses a different output element. Instead of a motor, a solenoid-operated throttle kicker is employed. When energized by the control computer, the solenoid moves the kicker, which controls the position of a vacuum-actuator diaphragm. The diaphragm extends the throttle stop to increase engine idle speed.

IDLE-AIR CONTROL

The purpose of an idle-air control valve is to control engine speed when the engine is running at idle speeds. Typical computer-controlled sys-

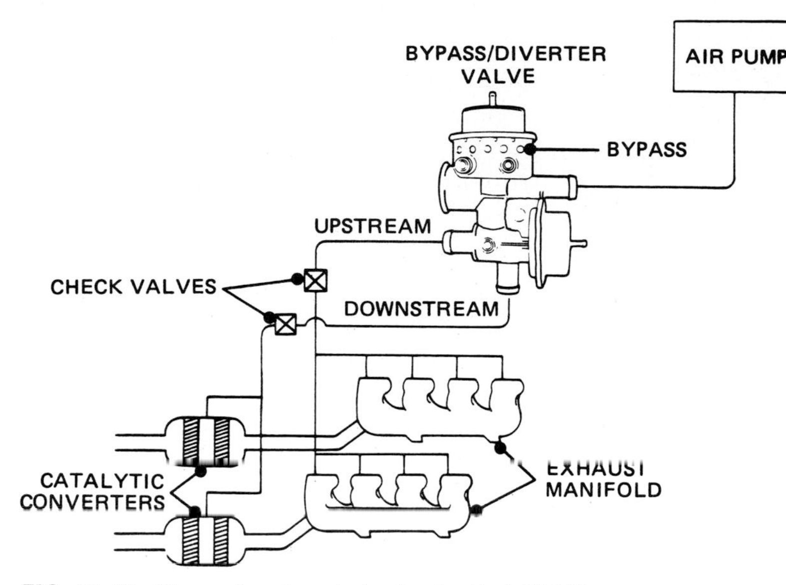

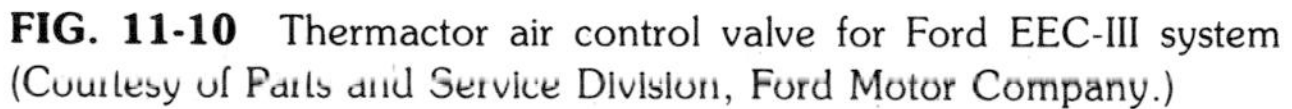

FIG. 11-10 Thermactor air control valve for Ford EEC-III system (Courtesy of Parts and Service Division, Ford Motor Company.)

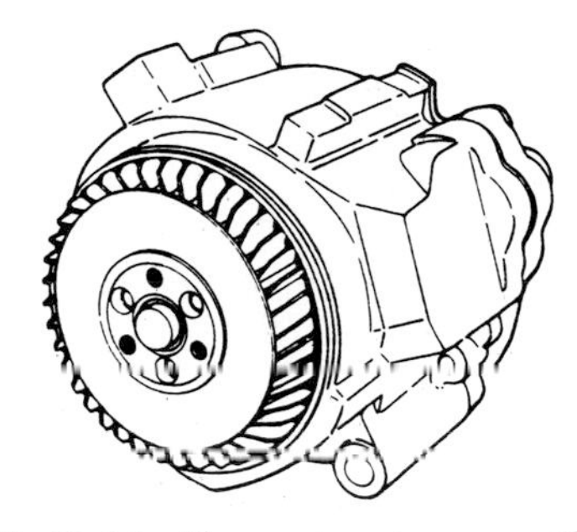

FIG. 11-11 Thermactor air pump (Courtesy of Parts and Service Division, Ford Motor Company.)

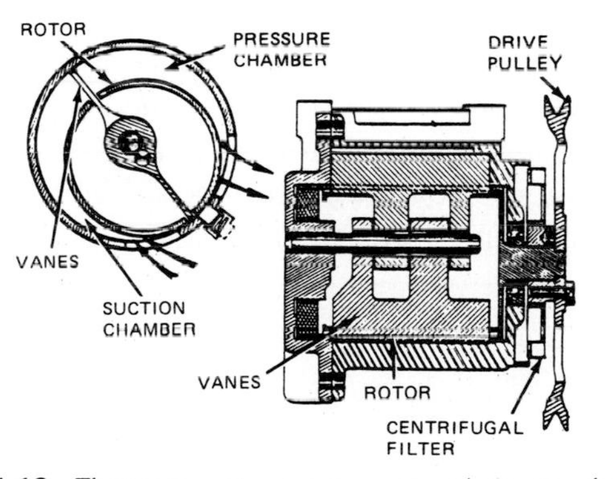

FIG. 11-12 Thermactor air pump pressure/vacuum chambers. (Courtesy of Parts and Service Division, Ford Motor Company.)

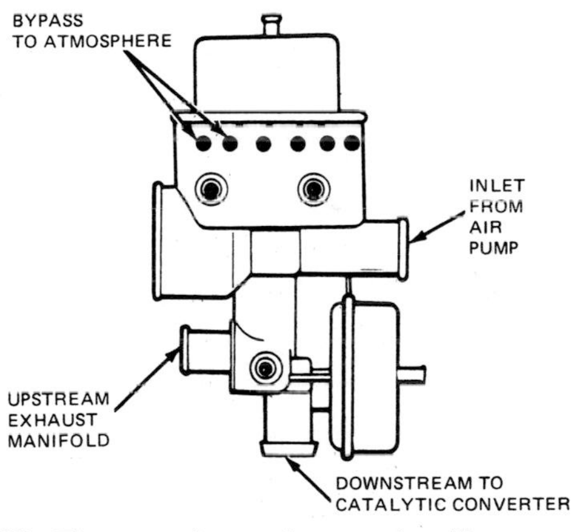

FIG. 11-13 Thermactor bypass diverter valve. (Courtesy of Parts and Service Division, Ford Motor Company.)

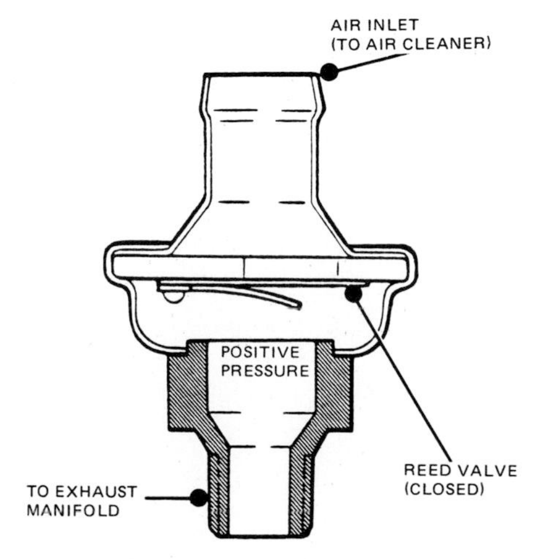

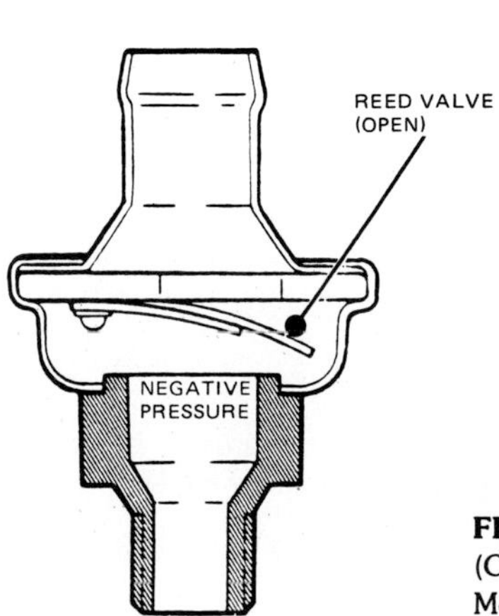

FIG. 11-14 Thermactor air check valves. (Courtesy of Parts and Service Division, Ford Motor Company.)

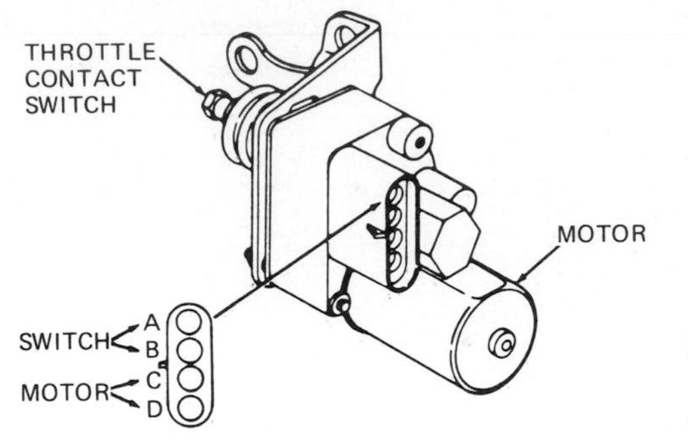

FIG. 11-15 GM idle-speed-control motor. (Courtesy of Chevrolet Motor Division, General Motors Corporation.)

tems are shown in Fig. 11-16 and 11-17. The main output element is a stepper motor, which operates a cone-shaped pintle valve. The assembly, called the IAC valve in these examples, controls bypass air flow around the throttle valve. When the pintle is moved in, air flow around the throttle plate is decreased; when it is moved out, air flow is increased. Early versions of four-barrel carburetors used a manual screw to control bypass air flow.

TORQUE CONVERTER CLUTCH

Several manufacturers use a solenoid-operated clutch mounted in the converter area of automatic transmissions. Examples are pictured in Figs. 11-18 and 11-19. Under certain conditions, the control computer energizes the solenoid, causing the clutch to couple the engine directly to the transmission. In this mode, transmission slippage is reduced. When operating conditions indicate that power should be transmitted in the normal, fluid coupling manner, the solenoid and clutch are disengaged.

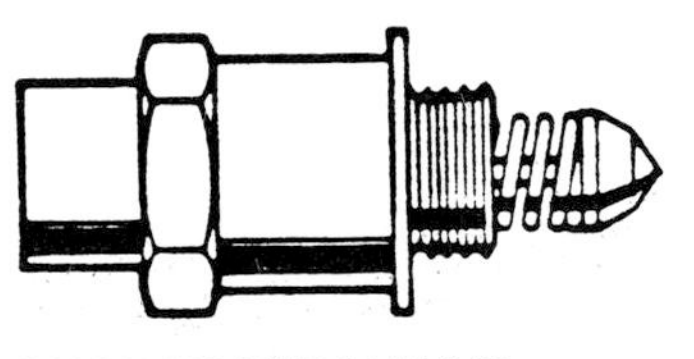

FIG. 11-16 Idle-air control (IAC) valve. (Courtesy Chevrolet Division, General Motors Corporation.)

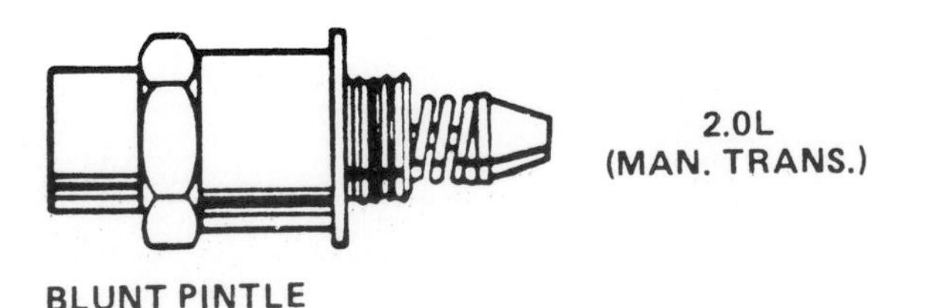

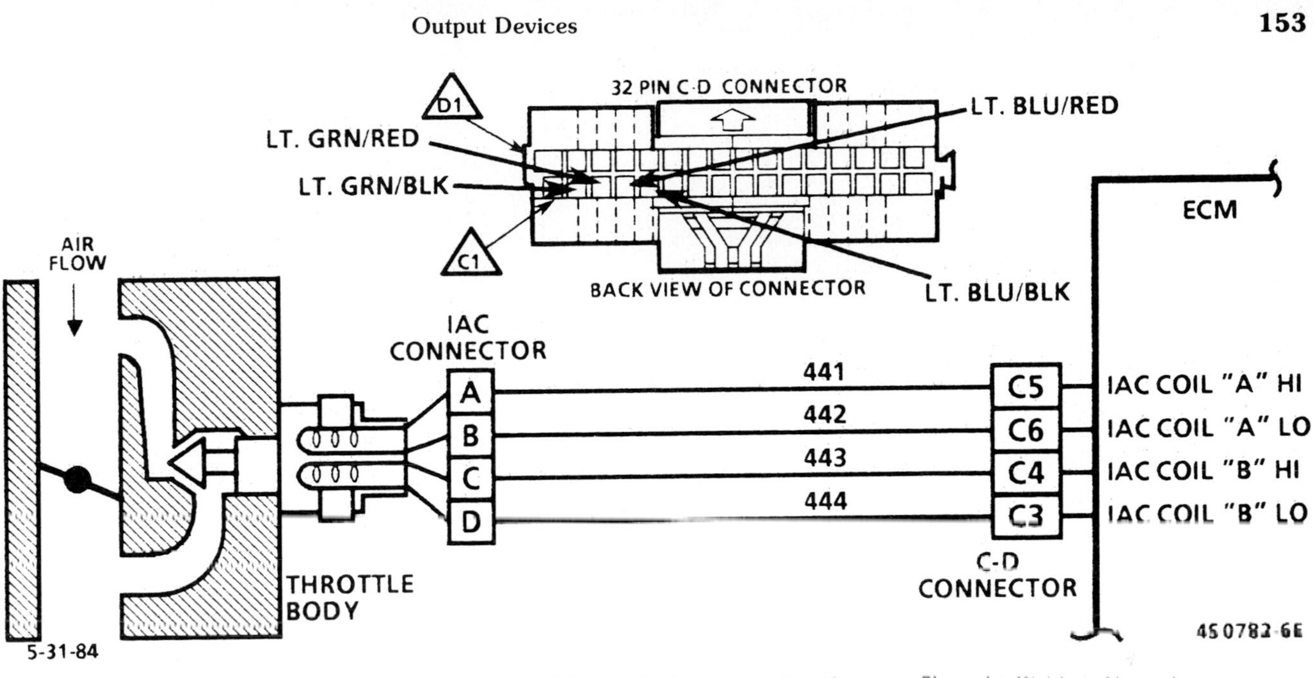

FIG. 11-17 IAC schematic. (Courtesy Chevrolet Division, General Motors Corporation.)

CONTROL VALVES

As shown in Fig. 11-20, the computer uses solenoid-operated valves to control the ported vacuum to the EGR. Consequently, there is no direct output link between the EGR valve and the computer. The EGR valve responds to the vacuum source the same as it did before computer controls. It is the control valve that is operated by the computer.

When the solenoids are energized, the control valves are closed, thereby shutting off the vacuum source to the EGR valve. The EGR valve cannot operate at this time. When the solenoids are de-energized, the control valves are opened. Manifold vacuum then determines whether the EGR valve will be open or closed. See Figs. 11-21 through 11-24 for additional details.

EARLY FUEL EVAPORATION (EFE) CONTROL VALVE

The EFE control valve is used to supply extra heat to the incoming air and fuel when the engine is cold. The extra heat helps fuel evaporate, thereby promoting more complete combustion, which improves performance and reduces emission.

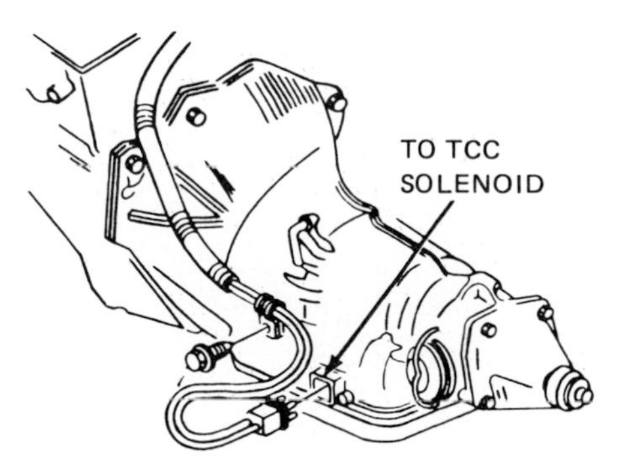

FIG. 11-18 Transmission clutch control solenoid. (Courtesy of Chevrolet Motor Division, General Motors Corporation.)

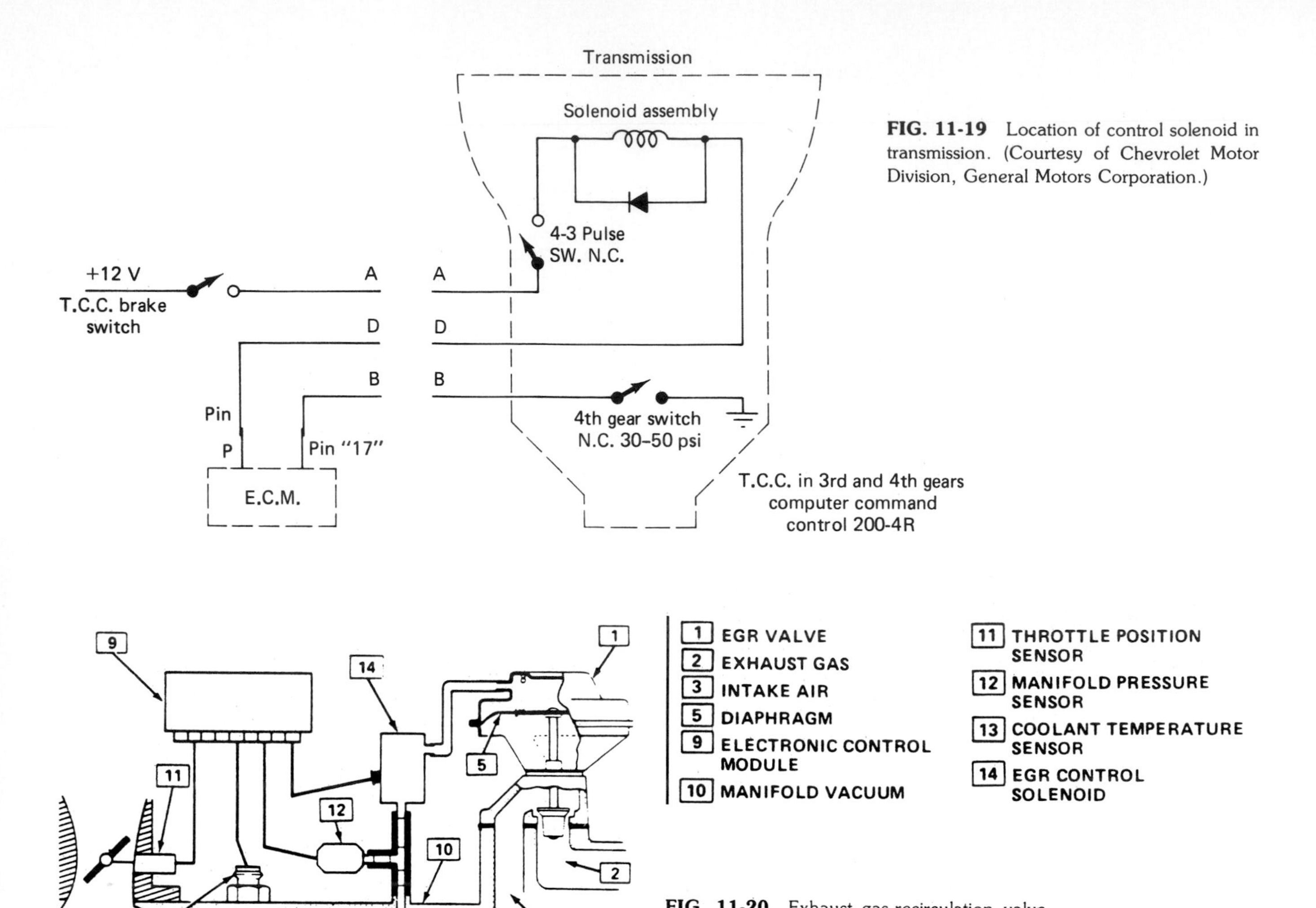

FIG. 11-19 Location of control solenoid in transmission. (Courtesy of Chevrolet Motor Division, General Motors Corporation.)

FIG. 11-20 Exhaust gas-recirculation valve, EGR system. (Courtesy Chevrolet Division, General Motors Corporation.)

FIG. 11-21 Ford exhaust gas-recirculation system. (Courtesy of Parts and Service Division, Ford Motor Company.)

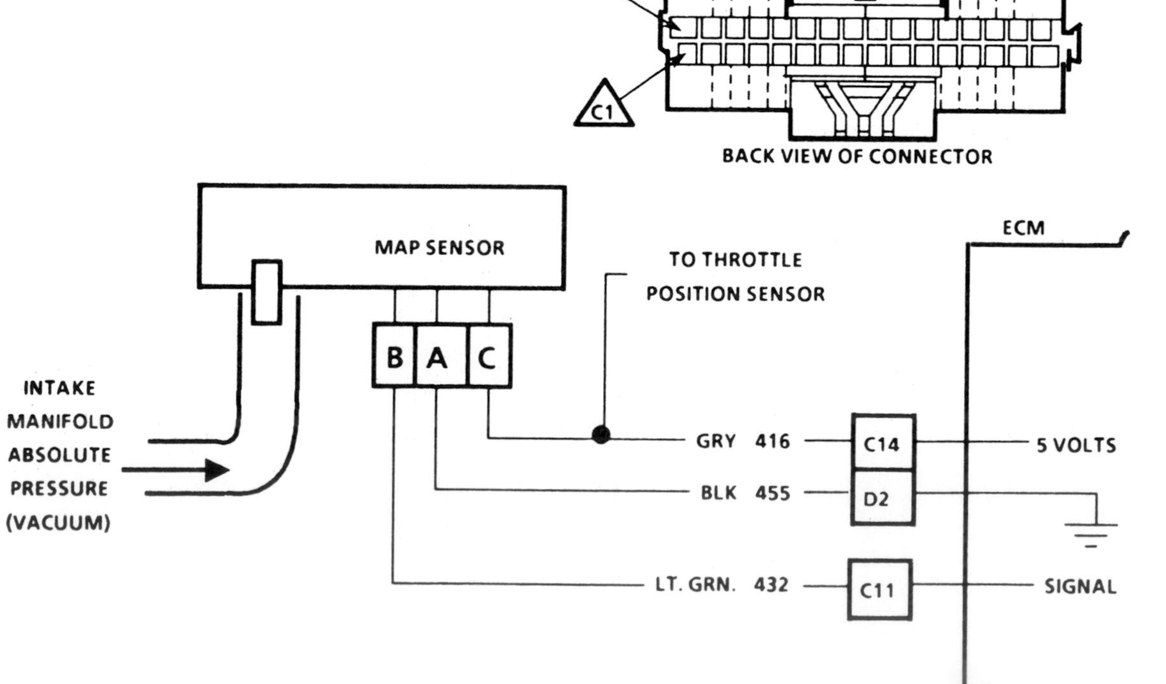

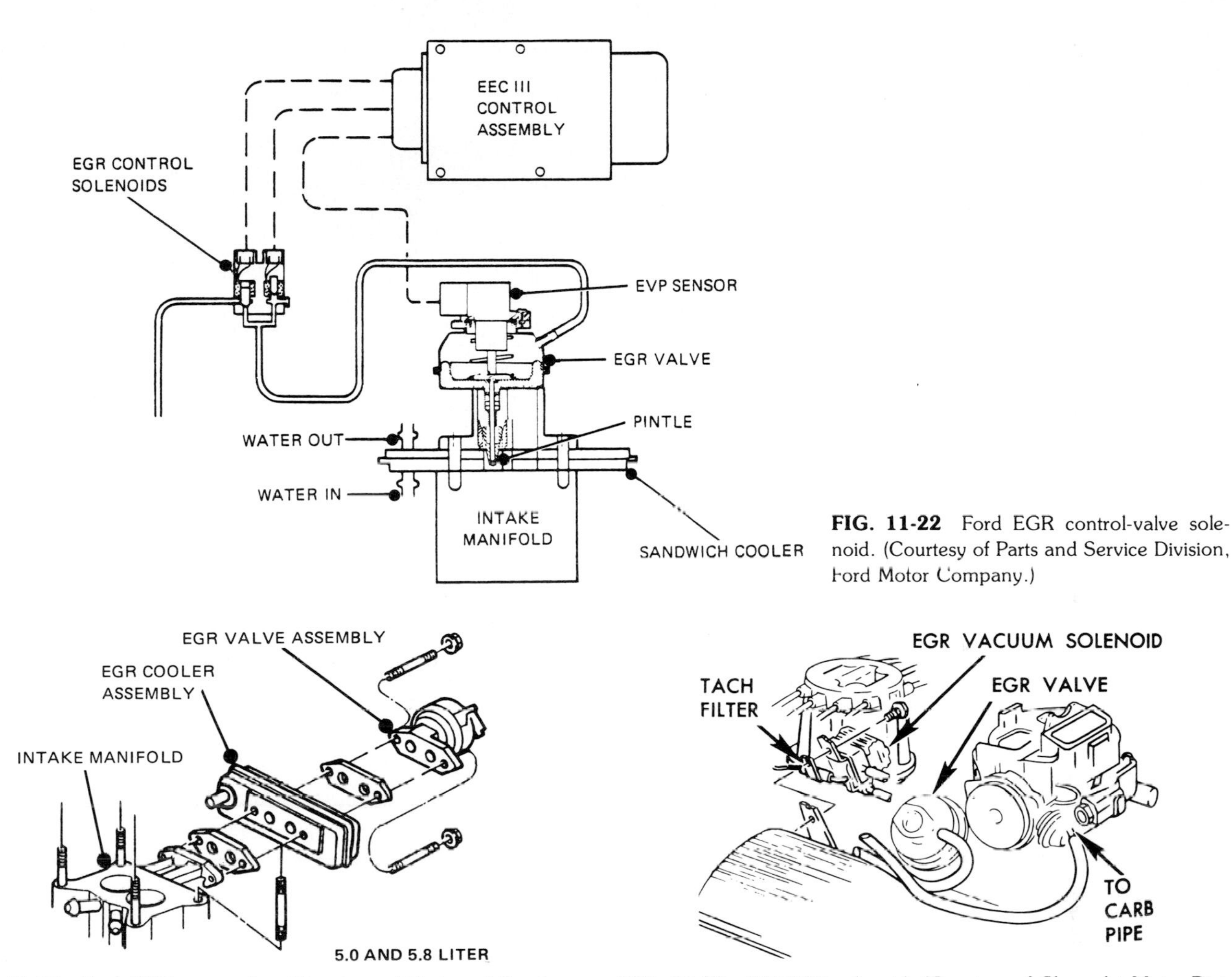

FIG. 11-22 Ford EGR control-valve solenoid. (Courtesy of Parts and Service Division, Ford Motor Company.)

FIG. 11-23 Ford EGR gas cooler. (Courtesy of Parts and Service Division, Ford Motor Company.)

FIG. 11-24 GM EGR solenoid. (Courtesy of Chevrolet Motor Division.)

One computer-controlled EFE system uses a solenoid-controlled vacuum actuator motor that operates an exhaust heat valve. The valve is located between the exhaust manifold and the exhaust pipe. By regulating the vacuum level of the motor, the computer controls the position of the valve and the heat applied to the incoming air. This EFE system is pictured in Fig. 11-25.

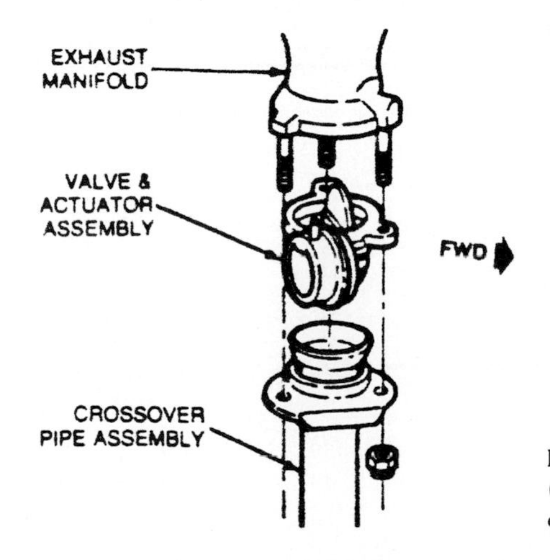

FIG. 11-25 Early fuel-evaporation valve. (Courtesy of Chevrolet Motor Division, General Motors Corporation.)

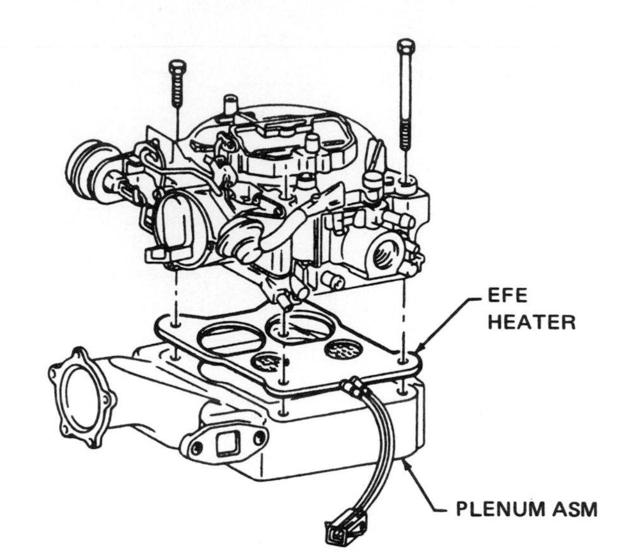

FIG. 11-26 Early fuel-evaportation heater. (Courtesy of Chevrolet Motor Division, General Motors Corporation.)

Another EFE system is shown in Fig. 11-26. It is used on smaller, carburetor-based engines. The heating element is a ceramic heater grid located under the primary bore of the carburetor. When the ignition switch is turned on and the coolant temperature is low, the computer operates a relay connected to the heater grid. The relay closes the circuit to the heater, which then heats the incoming air. After the temperature reaches a certain point, the computer de-energizes the relay, which shuts off the heater circuit.

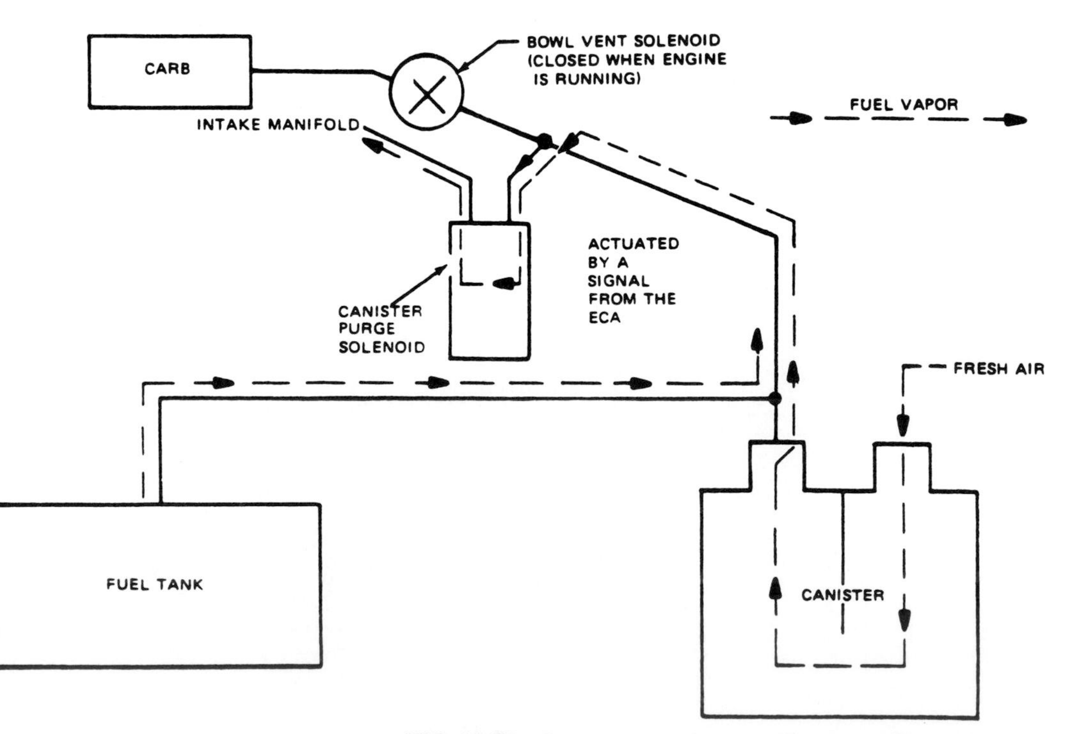

FIG. 11-27 Canister purge schematic. (Courtesy of Parts and Service Division, Ford Motor Company.)

CANISTER PURGE VALVE A solenoid-operated valve is used in many systems to control the flow of excess fuel vapors from a charcoal storage canister to the intake manifold. Figures 11-27 and 11-28 show a canister purge system.

THROTTLE LINKAGE MOTOR Some systems use a computer-controlled motor to adjust the position of the throttle during certain operations (such as cruise control). The cruise-control function is discussed more completely in Chapter 12.

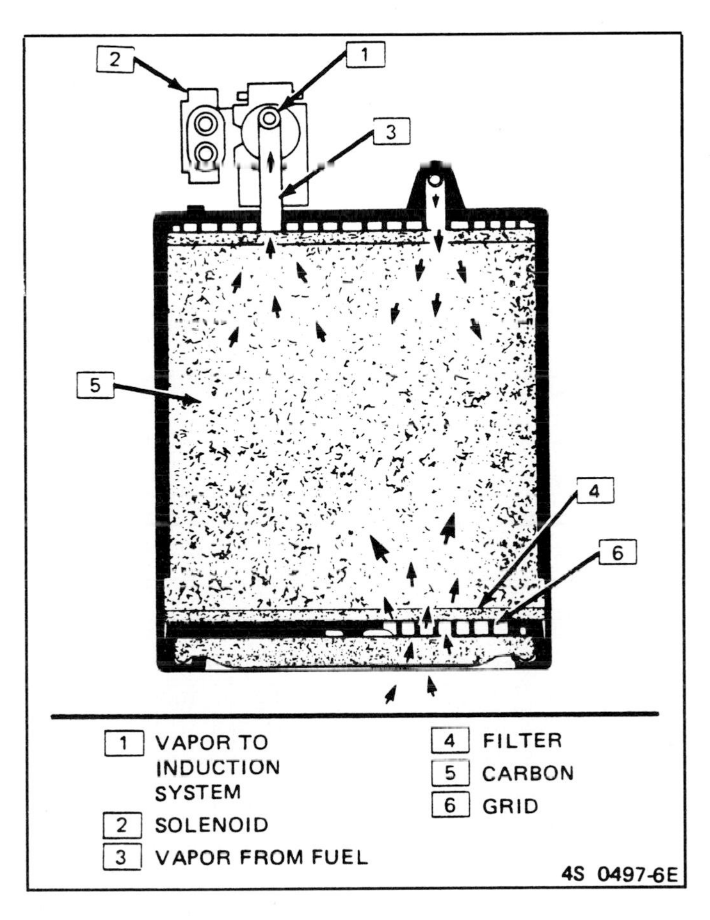

FIG. 11-28 Purge canister. (Courtesy Chevrolet Division, General Motors Corporation.)

12

Overview of Body Computer Systems

GENERAL The next chapters explain body computer systems. These systems can be defined as performing any function not directly related to engine control.

Initially, the only body computer function was managing a digital dashboard instrument display. Digital instrumentation appeared not long after the introduction of computerized engine-control systems. Unlike engine systems, which were developed for political, environmental and economic reasons, instrument systems were probably developed totally for financial reasons.

Now, body computer systems do much more than manage instrument displays. Body computer functions may be involved in:

- Radio operation
- Digital instrumentation
- Cruise control
- Air conditioning
- Navigation
- Automated braking control
- Active suspension system

Some vehicles have none of these body computer systems. Others, typically expensive luxury cars, may have all the functions.

COMMUNICATIONS One key to understanding the body computer systems introduced in the next chapters is communications. That is because in most vehicles, body computer functions are performed by multiple computers that must "talk" to each other.

Typically, communication between different computers is conducted over serial lines. (See Chapter 4 for an explanation of serial communications.) Certain rules or protocols are observed by sending and receiving units. For example, if one computer needs information from another computer on the network, it must first take control of the line and then establish contact with the other computer. The other computer must reply (or respond to the "handshake"). After communication has been completed, the line must be relinquished.

When studying body computer systems, it may be helpful to view the entire system as being like one large computer with multiple processing points. In some vehicles, the network may only include one processor; in others it may include many. In some systems, the body computer network is joined with the engine-control system. In other systems, the on-board computers may be connected by a serial communications link to a diagnostic computer located in the shop. That computer may itself be connected to a master computer operated by the manufacturer.

Figure 12-1 shows some of the communication possibilities.

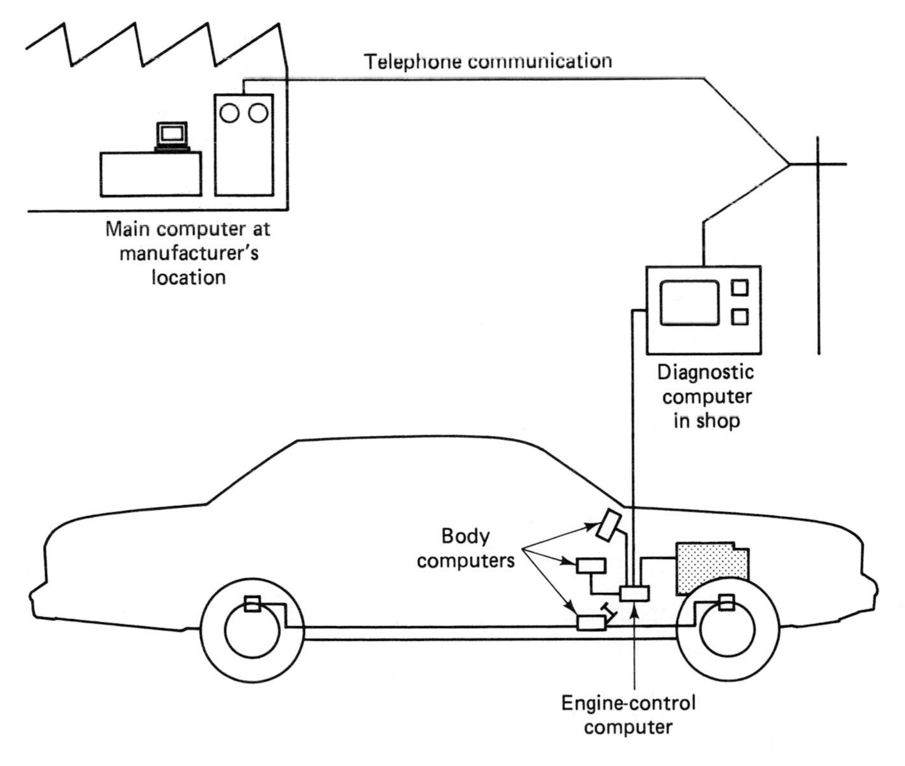

FIG. 12-1 Communications between on-board and remote computer systems.

13

Precomputerized Instrument Systems

GENERAL

Modern automobiles are equipped, for the most part, with electronically operated instruments. Many of these instruments represent data in a digital manner and are managed by microprocessors.

The next chapter examines computerized instrumentation functions. The present chapter reviews some of the older-style dash-mounted gauges and instruments. Not only will this be helpful if you ever have to repair an old unit, being introduced to basic instrumentation principles will help you understand computerized functions. Also, it is important to remember that some of these traditional instrument components, especially the sensors used to detect input data, are still being used.

TRADITIONAL GAUGES

Two basic types of gauges are able to display information about a variety of vehicle conditions. Look behind the dial face of a temperature, oil-pressure or fuel gauge in many pre-1980 cars and you will likely find one or the other of these gauges.

Thermostatic gauges

A thermostatic gauge (sometimes called a thermoelectric gauge) is shown in Fig. 13-1. The main components in the gauge itself are a set of heating coils, a bimetal blade, a pointer, and a linkage connecting the blade to the pointer. The main component in the sending unit is a variable resistor.

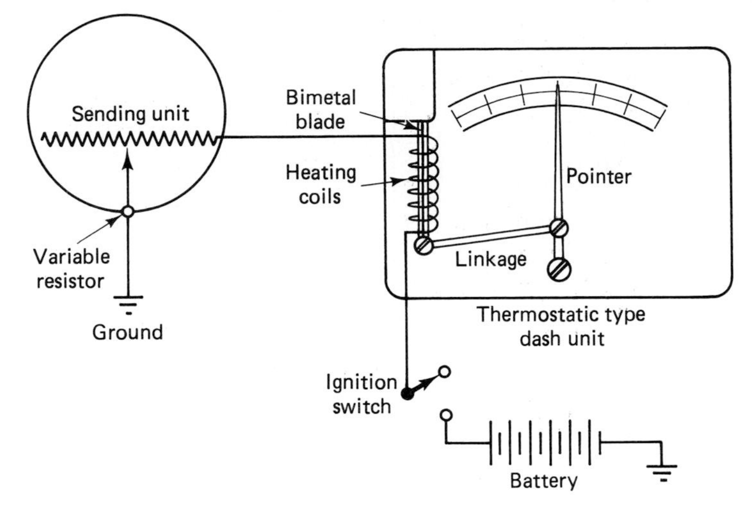

FIG. 13-1 Thermostatic dash-gauge unit.

Current passing through a variable resistor in the sending unit goes through the heating coils. Heat from the coils causes the bimetal blade to bend. The bending movement of the blade is transferred by the linkage to the pointer. Movement of the pointer against a dial face represents the physical condition being measured.

Several principles are involved in the operation of this analog (as opposed to digital) system.

In the sending unit, changes in a physical condition (such as the level of fuel in a tank) work through a linkage to change the resistance of the variable resistor. Changes in resistance affect the electricity flowing from the sending unit to the thermostatic gauge. (Variable resistors are used in many devices to provide an analog representation of a physical action. For example, the volume control on most radios and TV's uses a variable resistor.)

In the thermostatic gauge, two types of changes combine to cause movement of the pointer across the dial face:

- Changes in the amount of current flowing through the heating coil affect the molecular activity within the coil, causing changes in its temperature.
- Changes in the temperature of the heating coil cause the bimetal blade to bend one way or the other because the two metals used to make the blade expand or contract at different rates as the temperature increases or decreases. Since the two metals bend at different rates, the blade cannot expand or contract in a uniform manner; it must bend.

Balancing-coil gauges

The *balancing-coil gauge,* sometimes called the electromagnetic gauge, is similar to the thermostatic gauge. However, instead of using heat to move the pointer, the balancing-coil gauge uses magnetic forces.

The main elements of this gauge are two small electromagnets with a needle pointer balanced between them. One of the magnets is

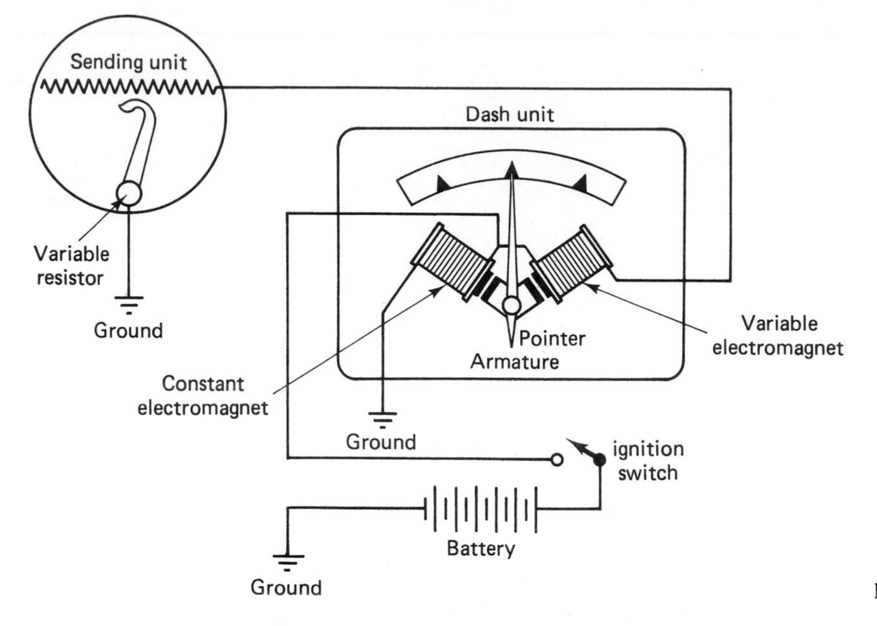

FIG. 13-2 Balancing-coil type.

grounded within the meter and, as a result, maintains a constant magnetic strength. The other is grounded through the variable resistance of the sending unit. Consequently, as the resistance of the sending unit varies, the strength of the magnet grounded through the sending unit varies. The needle pointer is attracted to the magnet with the strongest magnetic force, which means that any change in the resistance of the sending unit will cause the pointer to move. Figure 13-2 shows the balancing coil circuitry.

Instrument voltage limiter

Both the thermostatic and balancing coil gauges are affected by variations in the voltage of the automobile's electrical system. To minimize the inaccuracy that will result from this unavoidable condition, many manufacturers include a voltage limiter (regulator) in instrument circuits.

The main components in the limiter, as shown in Fig. 13-3, are a heating coil, a bimetal spring, and a switch connected to the spring.

Current for dash-mounted instruments flows through the coil, causing it to heat up. As the coil heats up, the bimetal spring bends. When the current is great enough, the coil heats up enough and the spring bends enough to open the switch connected to the spring. This

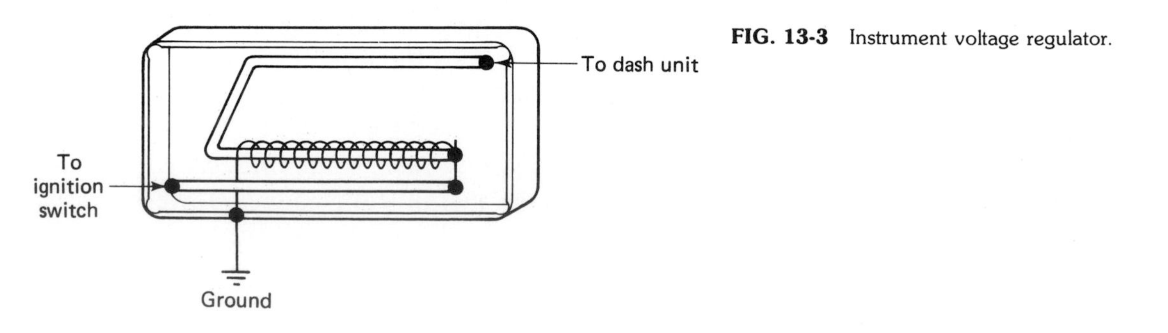

FIG. 13-3 Instrument voltage regulator.

breaks the circuit and stops current flow. The coil then cools off and the spring bends in the other direction. When the spring bends back far enough, the switch is closed again, allowing current flow to resume.

The process is repeated whenever the current flow exceeds the predetermined level for which the coil and bimetal spring have been calibrated.

TRADITIONAL SENDING UNITS

Three types of sending units are typically associated with the gauges just described:

- Temperature-sending unit
- Oil-pressure-sending unit
- Fuel level–sending unit

As the previous discussion has indicated, each of these sending units uses a variable resistor to convert the physical condition being measured into a variable, analog voltage signal. The output from any of the sending units can be interpreted by either a thermostatic or balancing coil gauge.

Temperature-sending unit

A temperature-sending unit, like the one diagrammed in Fig. 13-4, is typically screwed into the engine block, with the tip projecting into a coolant liquid passage. Current flows from the temperature gauge:

1. Into the terminal at the top of the sending unit.
2. Through the variable resistor at the base of the unit.
3. Through the threaded connector to the engine block (ground).

The amount of current flowing through the gauge (and hence, the temperature registered) depends on the resistance to flow offered by the variable resistor. The resistance is, in turn, affected by the temperature of the coolant liquid.

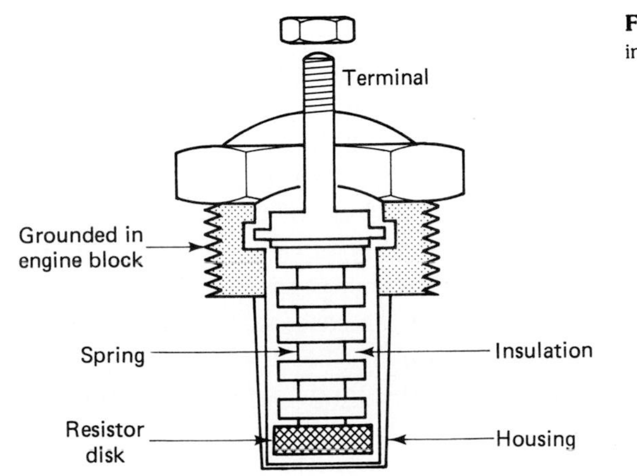

FIG. 13-4 Temperature gauge block sending unit.

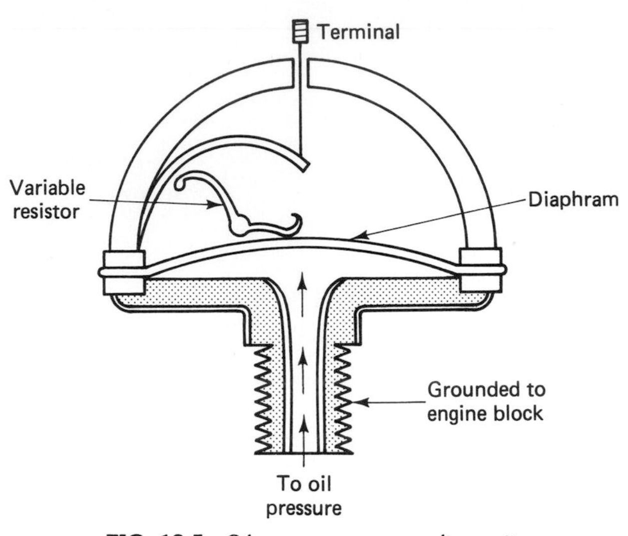

FIG. 13-5 Oil pressure gauge sending unit.

As the temperature rises, the resistance decreases and the current flow increases. As the temperature drops, the resistance increases and the current flow decreases.

Oil-pressure-sending unit

The oil-pressure-sending unit, like the one shown in Fig. 13-5, is screwed into the engine block with the tip projecting into a small oil-flow passage. The pressure exerted by the oil moves a flexible diaphragm within the oil-pressure-sending unit. As the diaphragm flexes up and down, the movement is transferred to a contact arm, which slides along a resistor. The position of the arm determines the resistance, which controls the amount of current flowing through the oil pressure gauge. As we have seen before, variations in current are transformed by the gauge into readings on a meter dial face.

Note: Relatively few automobiles within the past years, even prior to computerized instrumentation, used oil-pressure gauges. Most have relied on dash-mounted indicator lights to indicate whether oil pressure is within acceptable levels. These and other so-called idiot lights are discussed later.

Fuel level-sending unit

A fuel level-sender, like the one pictured in Fig. 13-6, is equipped with a float-arm unit. The unit is constructed from materials impervious to any corrosive effects of the fuel in which it is contained.

As the fuel level rises and falls, the float rides up and down. This movement is transferred by the arm to a contact, which slides along a resistor. As in the case of the oil-pressure-sending unit, the position of the contact determines the electrical resistance, which affects the current going to the gauge, which determines the position of the fuel-gauge pointer.

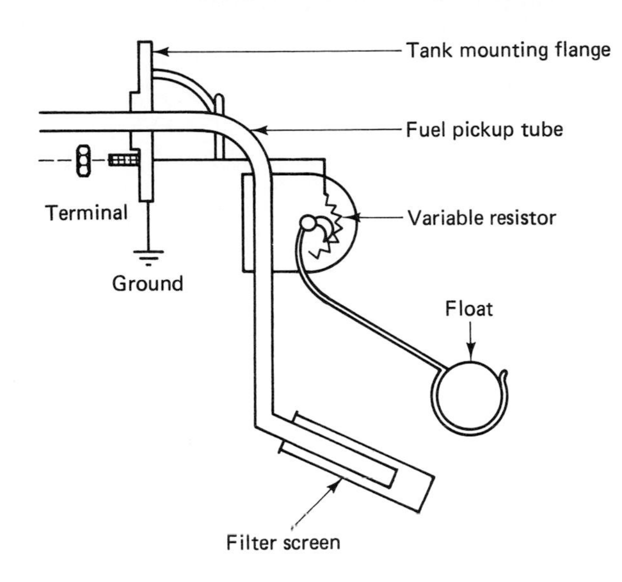

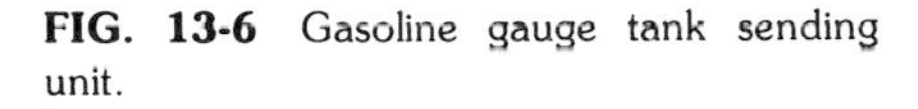
FIG. 13-6 Gasoline gauge tank sending unit.

OTHER TRADITIONAL INSTRUMENTS

Other traditional instruments include:

- Ammeters
- Indicator lights
- Speedometers
- Odometers

Ammeters

An ammeter gauge provides information about the vehicle's charging system in particular and the entire electrical system in general. Typically, an ammeter dial face is divided into two parts, one showing charging rate and the other showing discharging rate. When output from the charger exceeds demand from the electrical system, the indicator points to the charging side of the dial. When demand exceeds charger output (when the battery supplies the load), the indicator points to the discharge side of the dial.

Two types of ammeters have been typically used in automotive applications:

- Moving-vane ammeters
- Loop ammeters

MOVING-VANE AMMETERS

A moving-vane-type ammeter is diagrammed in Fig. 13-7. The main elements and their relationships include: a pointer needle attached to a armature. This pivots between two poles of a horseshoe magnet. A conductor located beneath the armature is connected to leads going to and from the battery.

When no current is flowing to or from the battery, the needle and armature are balanced between the poles of the horseshoe magnet. At that time, the pointer indicates no charge and no discharge.

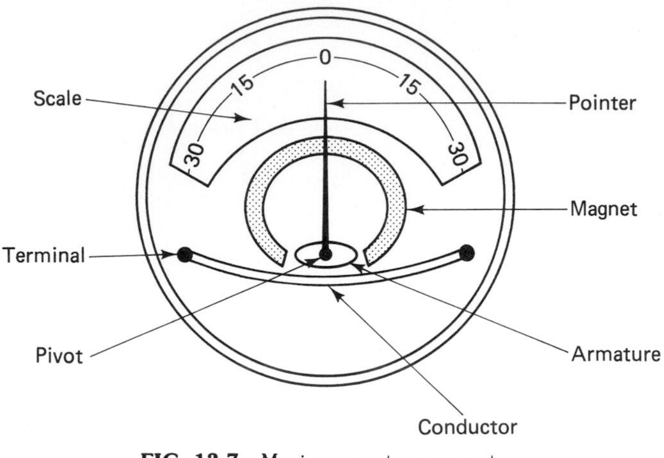

FIG. 13-7 Moving vane-type ammeter.

When the electrical system is closed, current flows through the conductor. As noted in the chapter on magnetism, lines of force then surround the wire. The lines of force counteract the magnetic field from the permanent magnet, causing the needle-armature assembly to move off center.

The direction in which the needle points (charging or discharging) depends on direction taken by the lines of magnetic force circling the conductor. The direction of the circling force lines, in turn, is determined by the direction that current flows through the conductor.

When the battery is supplying the vehicle's electrical needs and is discharging, the current flows in one direction. The needle then points toward the discharge side of the scale. If the charger is supplying the vehicle's electrical needs, current flows in the other direction and the needle points to the charging side of the scale.

LOOP-TYPE AMMETER

Figure 13-8 shows a loop-type ammeter. The main elements and their relationships include: A pointer needle attached to a small permanent magnet and a conductor passing through a loop located behind the magnet and connected to leads going to and from the battery.

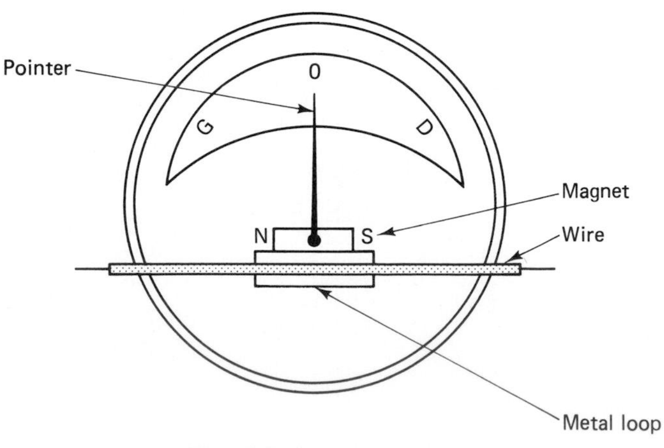

FIG. 13-8 Loop-type ammeter.

The basic operating principles are the same as the moving-vane ammeter. This time, however, the magnetic lines of force from the conductor react directly with the lines of force from the small magnet, causing the attached needle to move in one direction or the other.

Dash indicator lights

Many gauge-type instruments found in earlier vehicles were replaced by indicator lights mounted on the dash. Some people argue this was done because a colored light attracts the operator's attention to a critical condition faster than a gauge. Other people argue that indicator lights were used because they are cheaper than gauges.

Indicator lights are a kind of digital signal, either on or off. Like all digitized data, the light presents a simplified, categorized representation of the world. When the light is on, something is one way; when the light is off, it is another way. Unlike the data presented by gauges, there are no in-between states.

Indicator lights themselves are small bulbs covered by colored lenses or, in some cases, by silhouetted openings in the shape of the condition being represented. The sending units for indicator lights are typically on–off switches that are opened or closed by changes in the condition being monitored.

Temperature light

Figure 13-9 shows the components in a temperature light system. The sending unit is screwed into the block with the tip projecting into the coolant liquid jacket. The sensing element is a bimetal spring. When

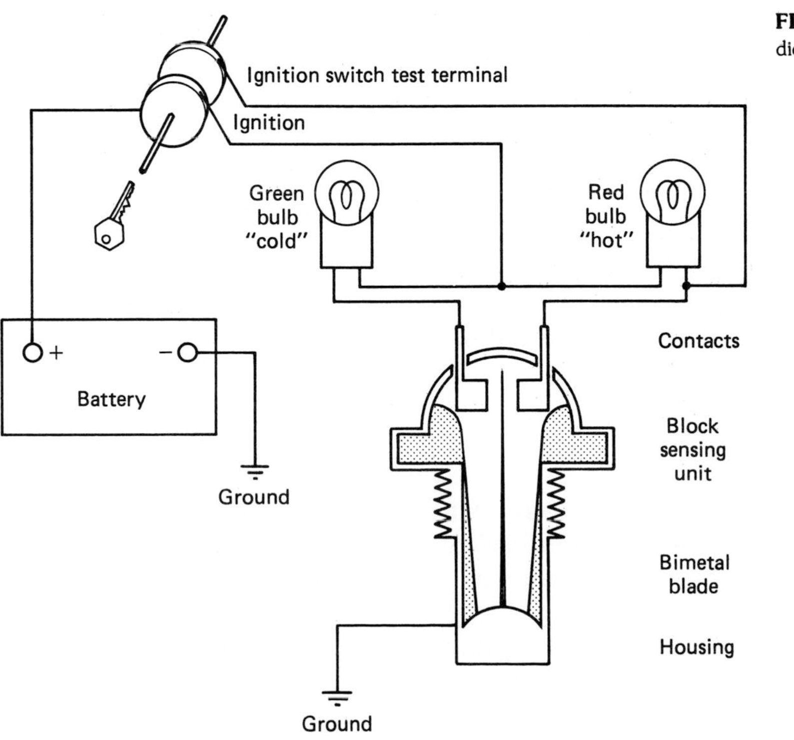

FIG. 13-9 Cold-normal-hot temperature indicator light circuit.

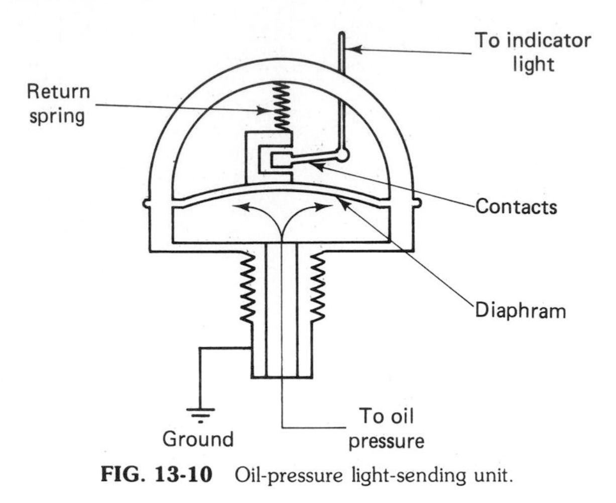

FIG. 13-10 Oil-pressure light-sending unit.

coolant becomes too hot, the calibrated spring bends to the right (in this illustration) and closes the circuit going to the "hot" indicator bulb.

Until the coolant reaches the predefined normal temperature level, the blade stays to the left, closing the contact to the "cold" bulb. (By coming on when the ignition switch is turned on, the cold light also provides an indication that the system is functioning.) At normal operating temperatures, the blade remains in the middle, not touching either contact.

Oil-pressure light

Figure 13-10 shows a diagram for the sending unit of an oil-pressure light. The unit is screwed into the block, with the tip projecting into an oil passage. The sensing unit is a diaphragm, which is flexed back and forth in response to changes in oil pressure.

When the engine is first cranked or whenever oil pressure is low, the diaphragm allows a set of contacts to close the circuit leading to the oil-pressure light. The light comes on. As the oil pressure builds up, the diaphragm pushes the contacts apart to open the circuit. The light goes off then.

Other indicator lights

Among other things, lights are also used for:

- The turn-signal indicator
- Low-beam, high-beam headlight indicator
- Charging-system indicator
- Cruise-control-position indicator

None of these indicator-light systems are very complicated. Consult the manufacturer's shop manuals for additional details.

Speedometers

The main elements in a traditional speedometer-odometer mechanism are diagrammed in Fig. 13-11. The speedometer cable is connected to the output shaft of the transmission. Rotation of the output shaft is transferred along the cable to a permanent magnet located within speedometer assembly. Surrounding the magnet is a metal "speed" drum. The speedometer pointer hand is attached to the drum. As noted in the illustration, there is no solid connection between the pointer and the speedometer cable.

As the magnet is turned by the speedometer cable, the magnetic field produced by the magnet also turns. This produces eddy currents, which cause the drum to follow the movement of the magnet. The shaft leading from the drum transfers the motion to the pointer hand.

Sudden, erratic movement of the speedometer pointer is damped by a hairspring connected to the pointer shaft. The spring resists drum rotation. The speedometer face plate is calibrated to account for the vehicle's gearing, tire size, and so on.

Odometer

As shown in Fig. 13-11, the odometer is geared directly to the speedometer cable output shaft. Typically, the odometer consists of six wheels, each numbered 0 through 9 around the outside rim. The numbers are viewed through a slot in the speedometer face and can display mileage from 00000.0 to 99999.9 miles (although some manufacturers optomistically use seven wheels, giving the capability to display one tenth less than 1 million miles traveled).

The wheels are geared together so that one revolution of a wheel on the right moves the next-left wheel one position; a complete revolution of that wheel moves the next-left wheel one position, and so on, to the last wheel on the left.

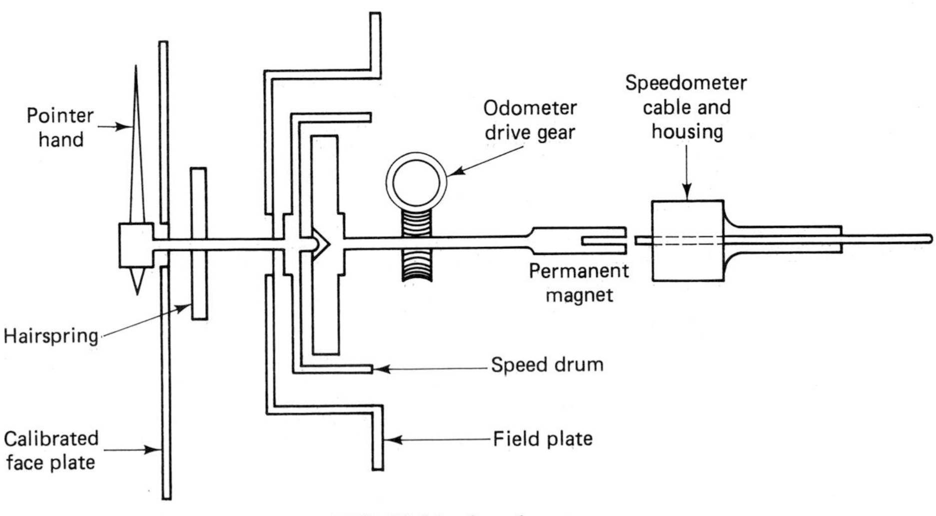

FIG. 13-11 Speedometer.

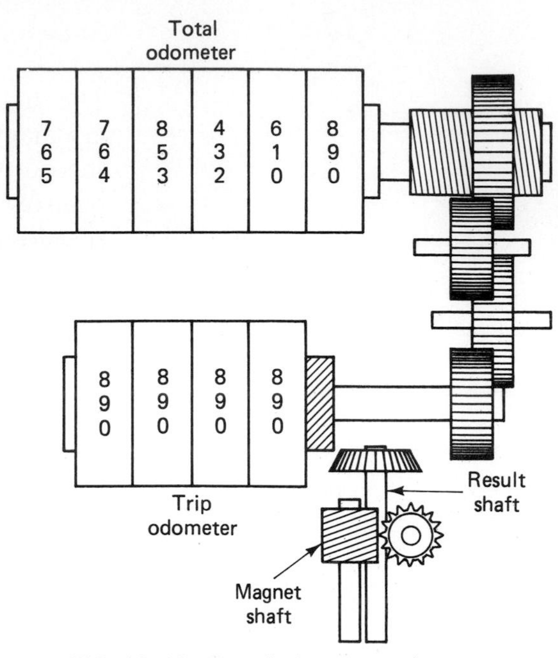

FIG. 13-12 Typical odometer mechanism.

Some cars also have a second set of odometer wheels located near the first. This set usually contains only four individual wheels and is typically called a trip odometer. The wheels of the trip odometer can be reset to 0's at any time by the operator.

As in the case of the speedometer, the odometer must be calibrated to take into account the vehicle's gearing, tire size, and the like. The operation of a typical odometer is shown in Fig. 13-12.

14

Computerized Instrumentation Functions

In the past, analog instruments provided information about vehicle speed, mileage traveled, fuel level, and, optionally, engine rpm. So-called idiot lights let the driver know if there were problems with oil pressure, engine temperature, and the state of the charging system. Beyond that, the only information likely to be conveyed was the current setting of the radio and possibly the air conditioner.

Today's electronic instrumentation systems display all this information on a variety of digital and analog style displays. In addition, there may be built-in maps and position locators, mileage and fuel-consumption calculators, diagnostic displays, and virtually anything else that the consumer is willing to pay for.

This chapter examines some of the fundamentals of electronic, computerized instrumentation.

THE INSTRUMENTATION COMPUTER

The heart of an electronic/digital instrument system is one or more microprocessor units (MPUs). The operation of the instrument computer is similar to the operation of the engine-control computer.

Like its control counterpart, the instrument MPU receives signals from input sensors, then transmits processed data to output devices. However, in a computerized instrument system, the primary output devices are various kinds of visual displays (or, sometimes, audible-tone generators, such as speech synthesizers). The instrument computer provides operating information for the driver; the engine computer provides control information for the engine.

Depending on the manufacturer and the application, a single MPU may be responsible for one instrument function or many. If the processor is devoted to one function, it is likely to be called an MPU. If it performs multiple functions, it may be called the central processing unit (CPU).

Some manufacturers refer collectively to the instrument computer in which the CPU is contained as well as the instruments themselves as the digital cluster. A digital cluster and the display functions it provides are pictured in Fig. 14-1.

Also, in the written materials produced by some manufacturers, the CPU contained within the instrument computer and the computer itself are often treated as the same thing. Although this book occasionally follows the same practice, you should be aware that the computer has more components than just the CPU.

For the most part, this section will treat all instrument functions as being performed by a single CPU contained in an instrumentation computer. These functions could be performed by other computers, including the on-board engine control computer.

INPUT CONSIDERATIONS

An instrumentation CPU may receive information regarding a number of engine conditions. Some of these conditions include:

- Battery voltage
- Coolant temperature
- Oil pressure
- Rpm
- Vehicle speed
- Fuel level

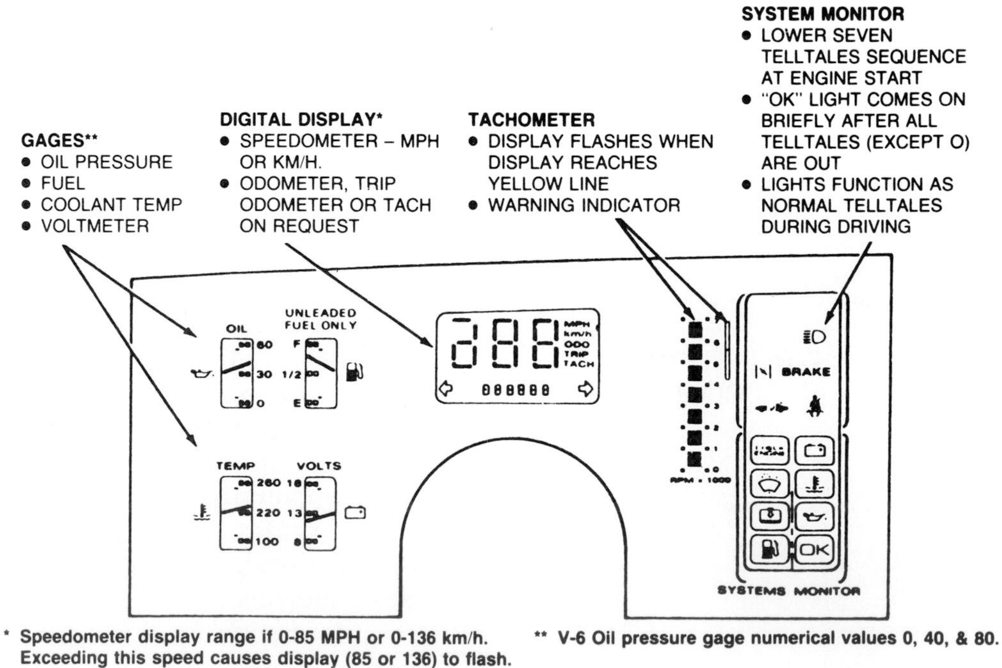

FIG. 14-1 Instrument cluster. (Courtesy of Chevrolet Motor Division, General Motors Corporation.)

Sampling periods

The instrument computer samples (checks incoming data) from the input sensors on a regular basis. Conditions that change rapidly over short periods of time are sampled more frequently than conditions which change less often. For instance, rpm is likely to be checked more often than engine temperature.

A/D conversion

As shown in Fig. 14-2, some input data must be changed through A/D conversion. These data include:

- Fuel level
- Oil pressure
- Coolant temperature
- Battery voltage

Data from these sensors is in the form of continuously variable voltage signals. The A/D converter changes this analog data into digital, numeric values that the computer can then process.

Note. As shown in Fig. 14-2, vehicle speed and rpm are not processed by the A/D converter because this data arrives at the computer in a digital form. Rpm is represented by a series of binary pulses produced by a filter device. Vehicle speed is converted to digital form by a buffer unit. Both of these processes are discussed under their respective instrumentation functions.

Multiplexing and de-multiplexing

Since the CPU can process only one input at a time, some capability needs to be provided for selecting the function to be performed. As

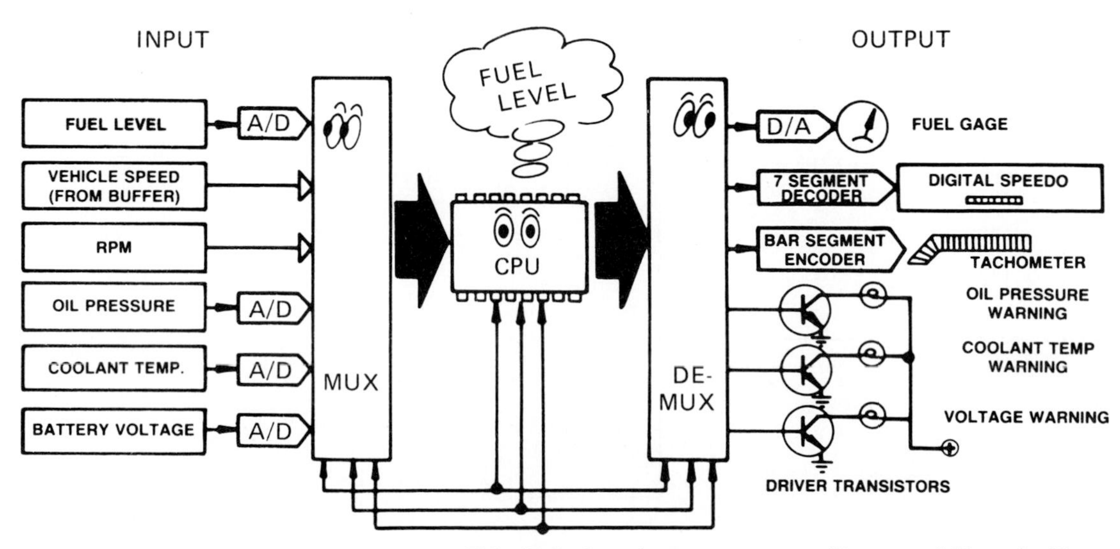

FIG. 14-2 Input/output processing. (Courtesy of Chevrolet Motor Division, General Motors Corporation.)

noted in Fig. 14-2, this operation is called *multiplexing (MUX). Demultiplexing (DE-MUX)* is the name given to the output selection operation.

BASIC OUTPUT FUNCTIONS

After being processed by the computer, output data goes to a variety of instruments, including the:

- Fuel gauge
- Digital speedometer
- Tachometer
- Oil-pressure warning light
- Coolant temperature warning light
- Voltage warning light

The next sections describe these basic instrumentation functions.

ELECTRONIC SPEEDOMETER

Figure 14-3 shows the basic operations in an electronic speedometer.

1. Signals relating to vehicle speed are generated by a permanent-magnet (PM) generator, usually driven from the transmission.
2. Signals from the PM generator go to a buffer, which converts the signals into a digital form usable by the processing unit.
3. From the processing unit, signals go to a display driver, which processes the output signals for swing needle or digital display.
4. Signals also go from the processing unit to an odometer driver, which operates the odometer.

PM generator

Figure 14-4 shows a PM generator.

This small generator takes the place of a speedometer cable. Gear-driven from the transaxle or output shaft of the transmission, the PM generator produces a sine-shaped signal whose frequency and voltage level are directly proportional to the speed of the vehicle.

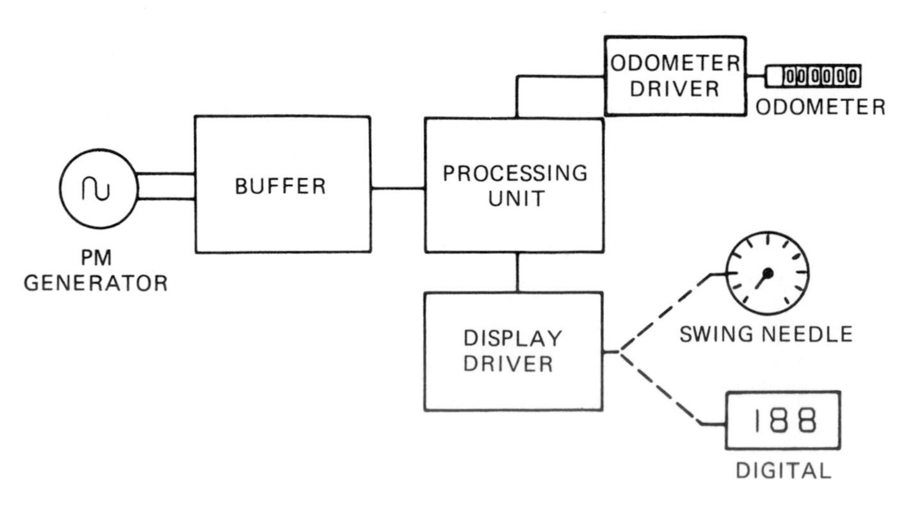

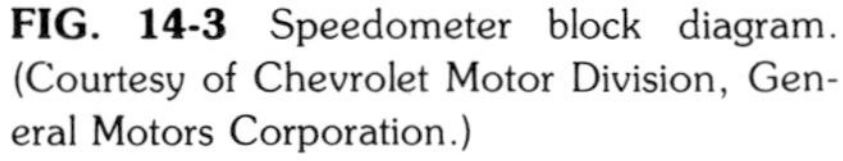
FIG. 14-3 Speedometer block diagram. (Courtesy of Chevrolet Motor Division, General Motors Corporation.)

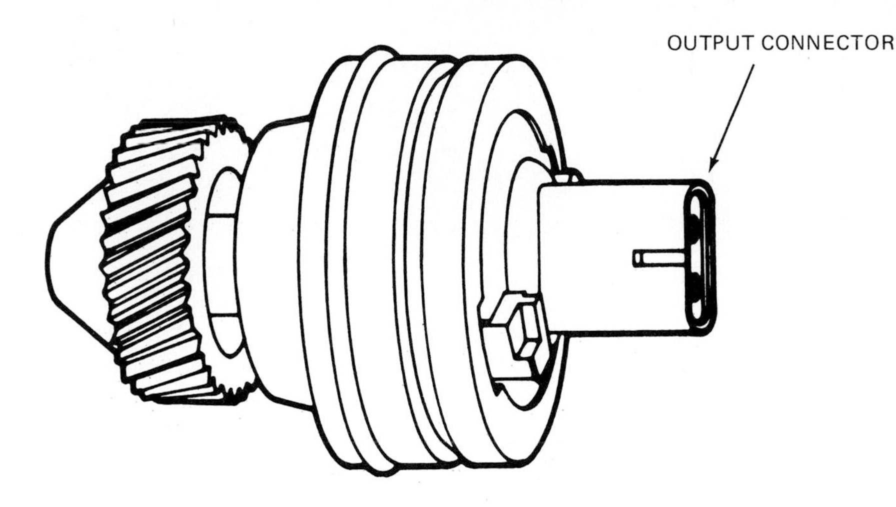

FIG. 14-4 PM generator vehicle-speed sensor. (Courtesy of Chevrolet Motor Division, General Motors Corporation.)

Note. Frequency is a measure of the time between peaks in the sine-shaped voltage curve. The unit used to express frequency is the hertz. One hertz is 1 cycle per second. Two hertz is two cycles per second. Sixty hertz, the frequency of ac power delivered by the utilities in the United States, is 60 cycles per second.

The frequency of the signal in one system is 1.112 hertz per mile per hour. Therefore, if a vehicle traveled at 1 mile per hour, the PM generator would produce signal peaks 1.112 times per second. If the vehicle traveled at 60 miles per hour, signal peaks would occur at a rate of 66.72 hertz, or 66.72 times per second. (For the purposes of odometer operation, the pulse rate translates into 4004 pulses per mile traveled, something less than one pulse per foot.)

The voltage in this same system ranges from 5 volts ac at 2 miles per hour to 80 volts at 125 miles per hour.

Buffer amplifier

The sine-shaped voltage curve produced by the PM generator is a continuously variable analog signal and, as such, cannot be used by a digital computer. The buffer produces a digital, or on–off, pulse pattern which the computer can interpret. Figures 14-5 and 14-6 picture a buffer amplifier circuit.

The buffer continuously monitors the sine wave coming from the PM generator. Each time a positive wave peak passes, an electronic switch in the buffer control circuit is turned on. As the wave peak

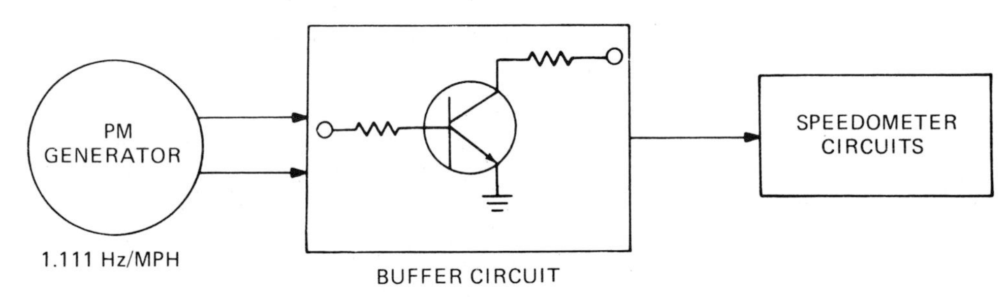

FIG. 14-5 Buffer amplifier. (Courtesy of Chevrolet Motor Division, General Motors Corporation.)

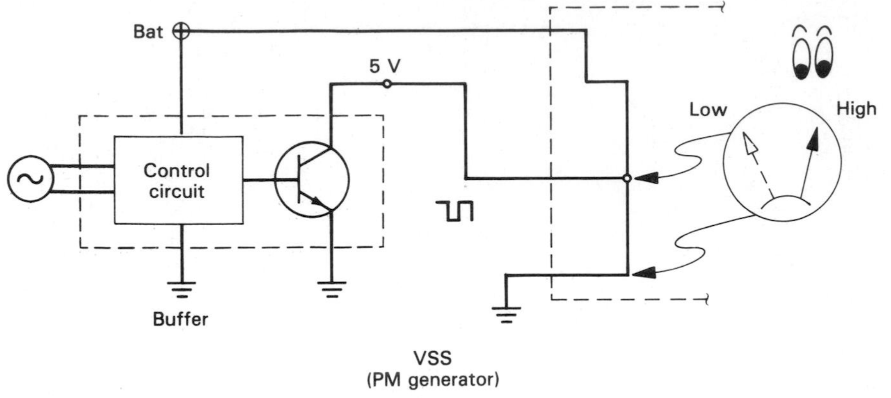

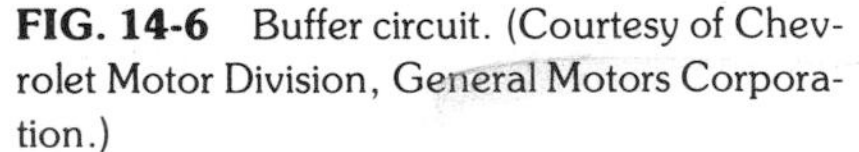
FIG. 14-6 Buffer circuit. (Courtesy of Chevrolet Motor Division, General Motors Corporation.)

passes, the switch is turned off. (In computer terms, this is called toggling.)

The control-circuit switch is connected to the base of a transistor. The instrument computer sends a +5 V current (called a sensor voltage) to this transistor. When the control circuit switch is on, the transistor is also switched on and current flows through the transistor emitter and collector. (Because the collector is open unless the switch is on, this is called an open-collector transistor.) When the control-circuit switch is off, no current flows through the transistor.

The result is a square-shaped voltage pattern whose frequency (time between square-shaped peaks) is proportional to vehicle speed. The instrument computer uses this signal to determine vehicle speed.

In some systems, the square-shaped vehicle speed pulse is also used by the main engine-control computer as well as an MPU located in the cruise control system. The signal going to these computers passes through a divider circuit in the buffer which divides the basic pulse rate by two.

Display drivers

After being processed by the instrument computer, the digitized speedometer reading is sent to a display driver. The driver processes the signal for either a conventional-looking swing-type speedometer needle or a numeric digital display.

Quartz electric display

The swing-type speedometer is sometimes referred to as a quartz electric display. The speedometer driver output goes to a collection of air-core coils surrounding the needle pivot point. The electromagnetic pattern of these coils causes the needle to swing to the appropriate speedometer reading. Figure 14-7 shows a quartz electric system.

Digital display

In a digital display, each bit of output data is used to drive a segment of the display driver circuit. Figure 14-8 shows a digital system. Note

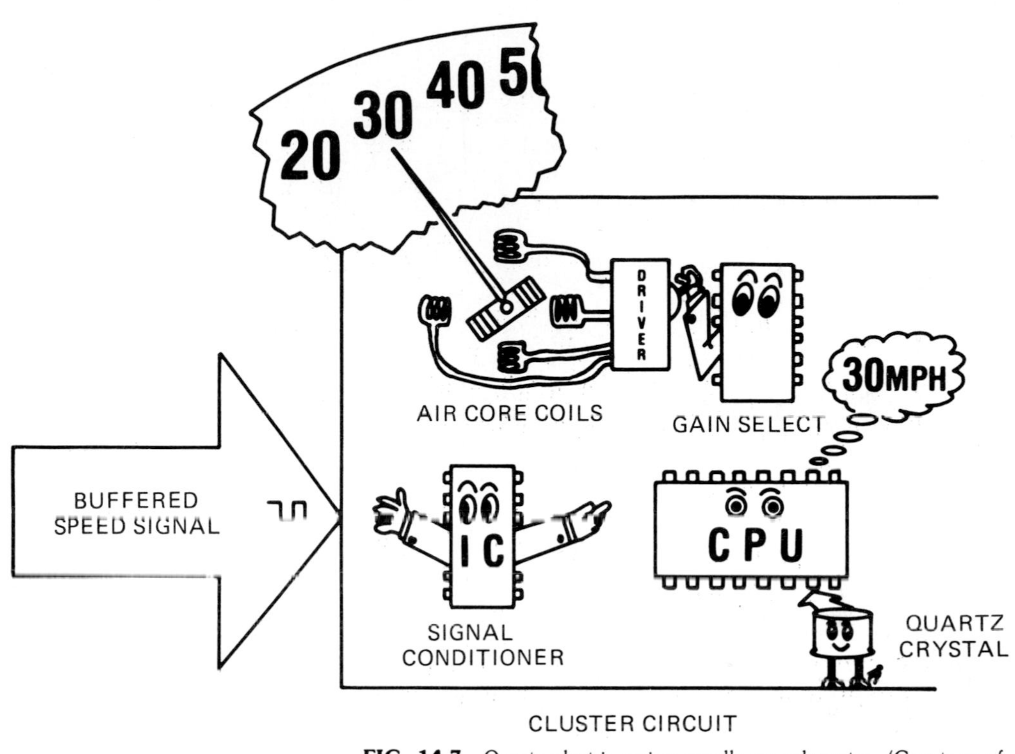

FIG. 14-7 Quartz electric swing needle speedometer. (Courtesy of Chevrolet Motor Division, General Motors Corporation.)

that the vehicle operator can switch the display in this system from English to metric units. Various types of digital display media are discussed later in this chapter.

Photoelectric vehicle speed sensor

Some electronic speedometers use a photoelectric speed sensor, rather than a PM generator. Such a system is pictured in Fig. 14-9.

The main components include:

- A more or less conventional speedometer cable.

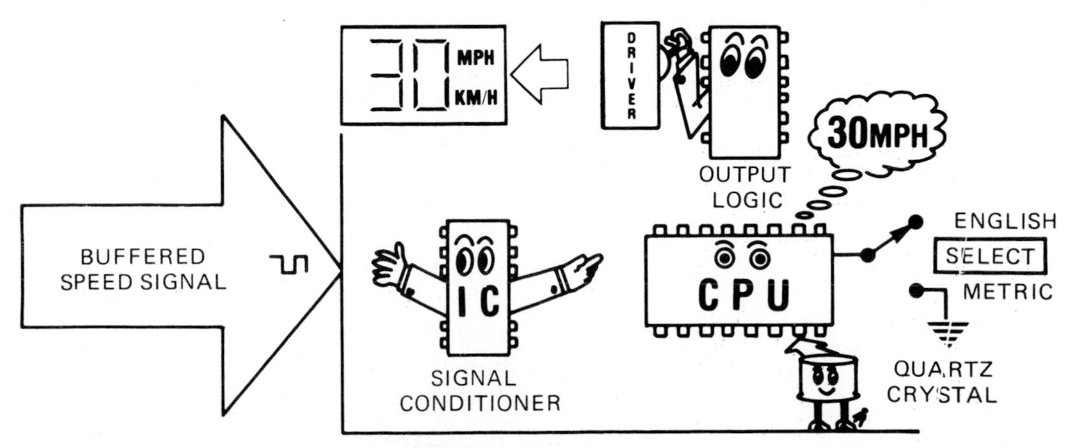

FIG. 14-8 Digital speedometer. (Courtesy of Chevrolet Motor Division, General Motors Corporation.)

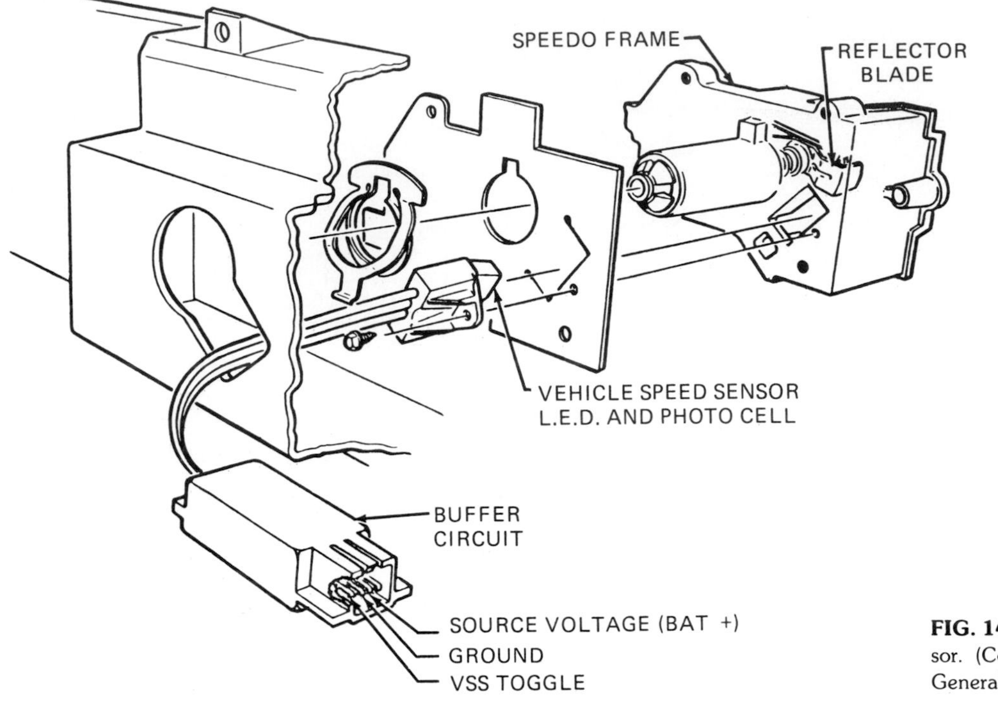

FIG. 14-9 Photo-electric vehicle speed sensor. (Courtesy of Chevrolet Motor Division, General Motors Corporation.)

- A reflector blade attached to the end of the cable.
- A speed sensor unit consisting of an LED light source and a photocell sensor.

The speedometer cable causes the reflector blade to rotate. Light from the LED shines on the blade, then bounces back to the speed sensor. The photocell in the sensor generates a slight pulse of electricity each time it is struck by reflected light from the LED.

This pulsating signal is processed by a buffer. A reference voltage from the buffer, as described in the preceding section, is used by both the instrument and engine-control computers to determine vehicle speed.

ELECTRONIC ODOMETER

Two types of electronic odometers are commonly used—a stepper-motor odometer and a digital odometer.

Stepper-motor odometer

The *stepper-motor odometer* uses the digital signal from instrumentation circuits to drive a stepper motor. The stepper motor operates the wheels of a conventional-appearing odometer as shown in Fig. 14-10.

A stepper motor is different from a conventional electric motor. Instead of being operated by continuous current, the stepper motor is driven by square-shaped voltage pulses. Each time a pulse occurs, the motor makes one precise movement or step.

Output from the buffer amplifier is divided by two, prior to being sent to the stepper motor. As noted earlier, the buffer produces precisely 4004 pulses per mile traveled. Therefore, half of that figure, or 2002 pulses, arrive at the stepper motor for each mile traveled.

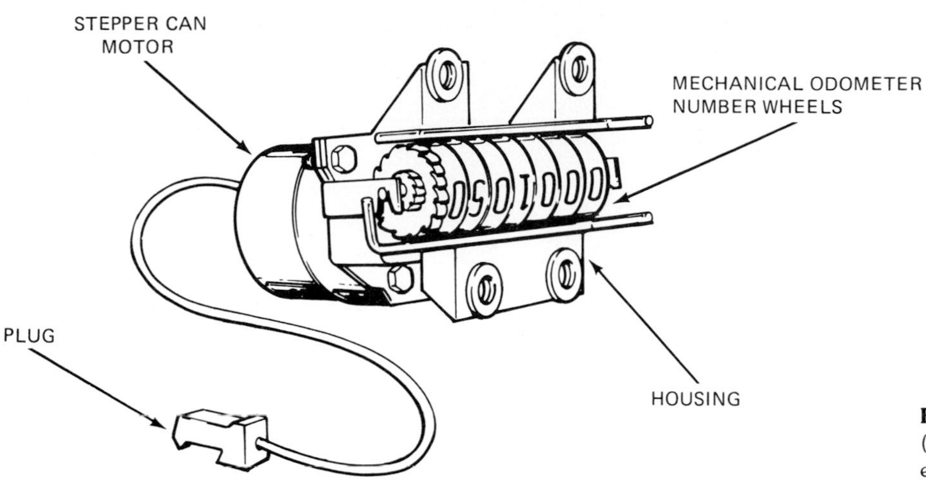

FIG. 14-10 Stepper-motor odometer. (Courtesy of Chevrolet Motor Division, General Motors Corporation.)

Digital display odometer

The *digital odometer* uses the instrumentation signal to operate a digital display driver. Mileage values appear on a digital display unit.

As the mileage display is being updated, so is a special type of RAM memory called nonvolatile RAM (NVRAM). It maintains a record of miles traveled. NVRAM retains data when power is turned off. Otherwise, odometer readings would be lost every time power to the computer was turned off.

Because the NVRAM chip and the data it contains are very important, be very careful when servicing an instrument system with a digital odometer. Read the manufacturer's shop manual before doing any work.

Figure 14-11 pictures some of the relationships in a digital odometer display system.

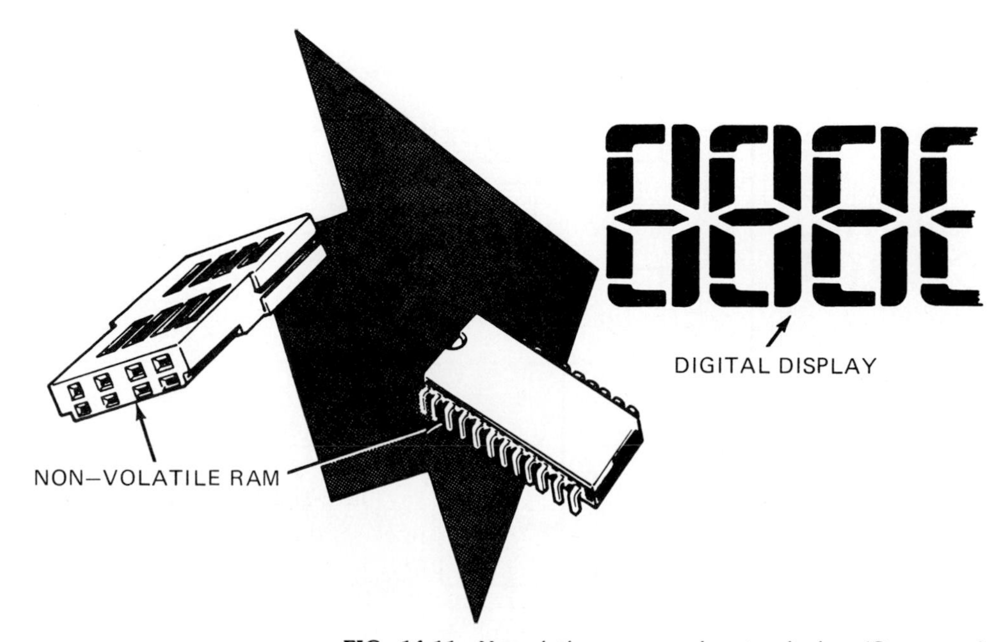

FIG. 14-11 Nonvolatile memory odometer display. (Courtesy of Chevrolet Motor Division, General Motors Corporation.)

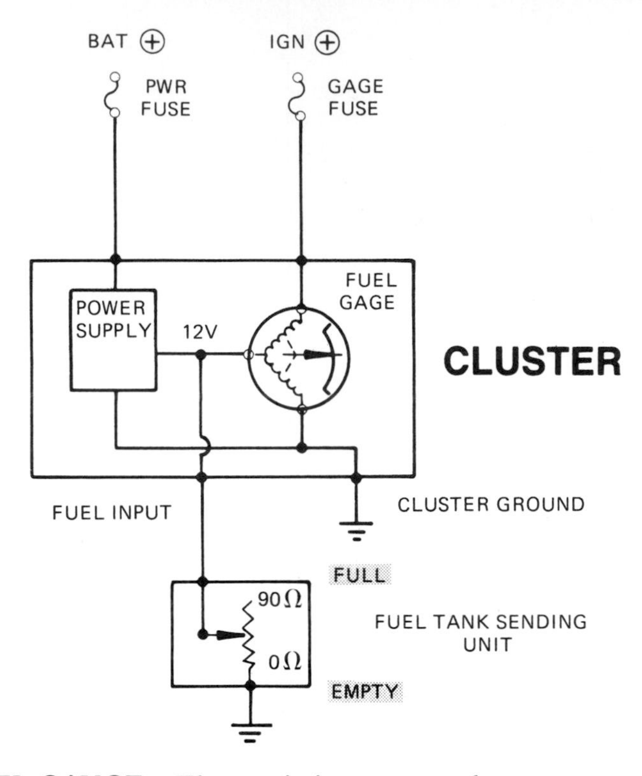

FIG. 14-12 Analog fuel measurement. (Courtesy of Chevrolet Motor Division, General Motors Corporation.)

FUEL GAUGE

Electronic instrument clusters may use either an analog or digital fuel gauge display.

Figure 14-12 pictures an analog system. It uses a variable resistor sending unit and a balancing-coil type display. In these respects, it is very similar to the traditional fuel gauges described earlier in this chapter.

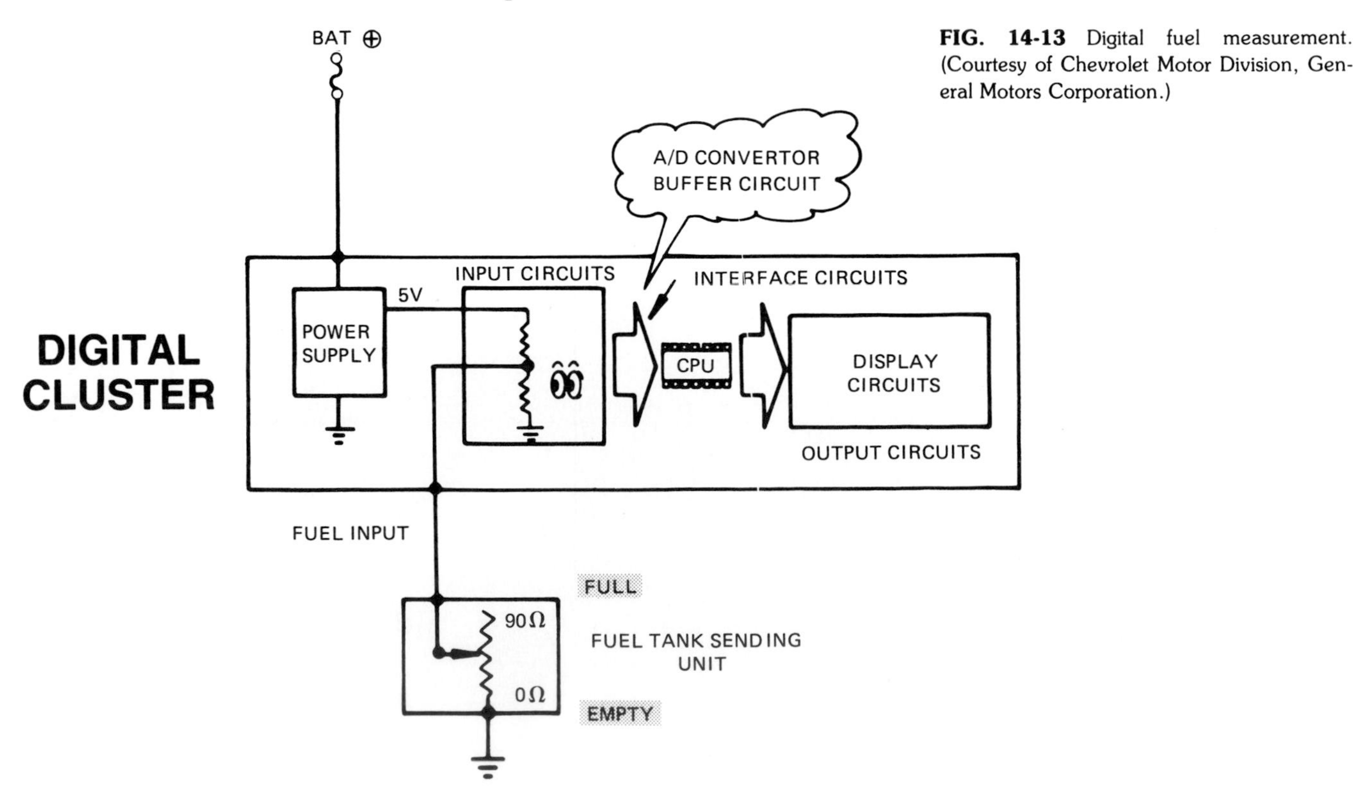

FIG. 14-13 Digital fuel measurement. (Courtesy of Chevrolet Motor Division, General Motors Corporation.)

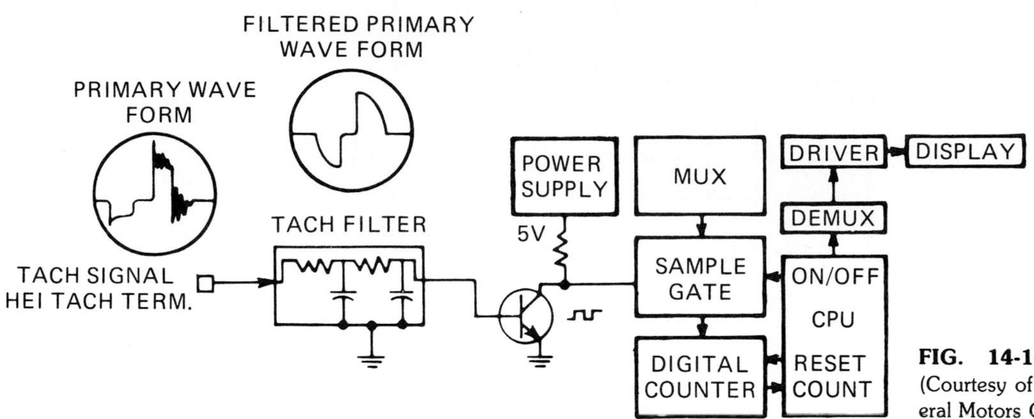

FIG. 14-14 Digital tachometer circuit. (Courtesy of Chevrolet Motor Division, General Motors Corporation.)

Figure 14-13 pictures a digital fuel display. It also uses a variable-resistor sending unit. However, in the digital display, the analog signal produced by the sending unit is changed into digital values by an A/D converter. This digitized signal is processed by the CPU and then sent to a display driver.

DIGITAL TACHOMETER

Figure 14-14 shows the circuitry used by one manufacturer to operate a digital tachometer. The signal originates from a solid state ignition system (the HEI, or high-energy ignition). The signal is "cleaned up" by a filter so that the positive pulses can operate the base of an open-collector transistor switch. As each positive pulse switches the transistor on, a square pulse is allowed to pass in the 5-V sensor voltage circuit from the instrument computer. The computer uses these pulses to drive either a swing-type or digital display.

DIGITAL OIL-PRESSURE INSTRUMENTATION

Figure 14-15 pictures a digital oil-pressure measuring system. Oil pressure in the sending unit pushes against a diaphragm, which pushes against a piezoelectric transducer, which produces a variable analog voltage signal. This voltage signal is converted by an A/D converter into a digital value. The digital value is processed by the CPU. The output from the CPU can be used to operate a digital oil-pressure display, and/or it can be used to operate a transistor switch which controls a warning light if the oil pressure drops too low.

As suggested by the illustration, oil-pressure readings for a given RPM are compared to reference values stored in memory to determine if the current reading is satisfactory.

DIGITAL COOLANT-TEMPERATURE MEASUREMENT

Figure 14-16 shows a coolant temperature measurement system. An analog signal from a thermistor sensor is converted by an A/D converter to digital form, then compared by the computer to high-temperature limits stored in memory. The output from the CPU can be used to operate a digital temperature display, and/or it can be used to operate a transistor switch which controls a warning light if the temperature limit is exceeded.

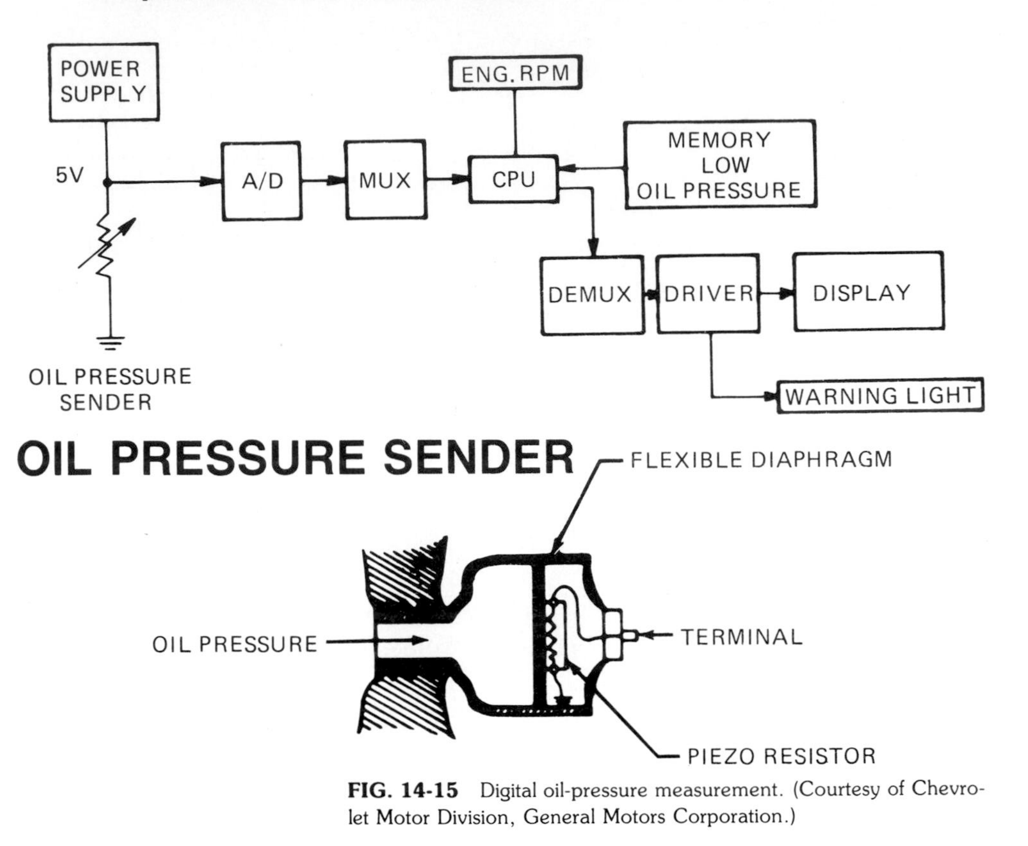

FIG. 14-15 Digital oil-pressure measurement. (Courtesy of Chevrolet Motor Division, General Motors Corporation.)

DIGITAL VOLTAGE MEASUREMENT

Figure 14-17 diagrams a typical digital voltage measuring system. Voltage variations are digitized and then compared by the computer to high and low limits stored in memory. Values that lie outside these ranges cause the computer to turn on a dash warning light. The same digitized and processed voltage readings can also be used to drive digital or swing-needle displays.

TYPES OF DISPLAY DEVICES

The remainder of this chapter examines some of the main types of visual media used in automobiles to display electronically processed data.

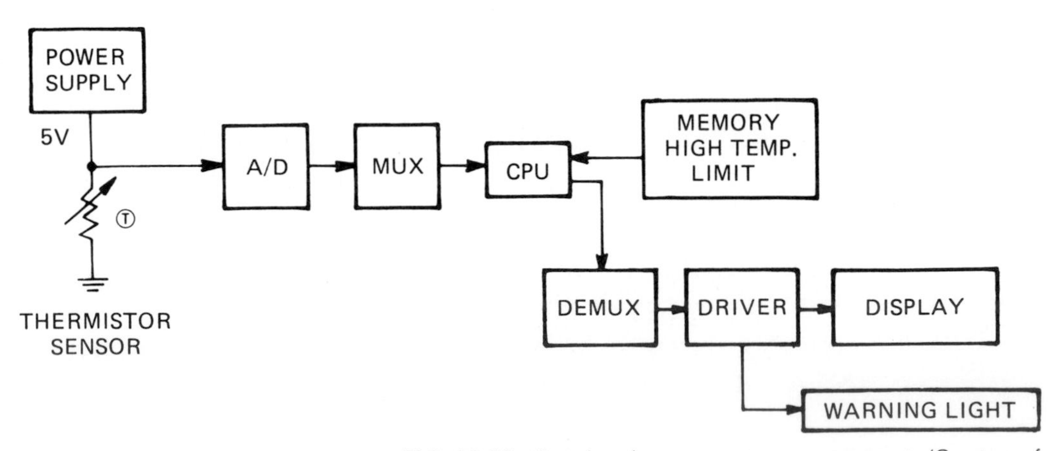

FIG. 14-16 Digital coolant temperature measurement. (Courtesy of Chevrolet Motor Division, General Motors Corporation.)

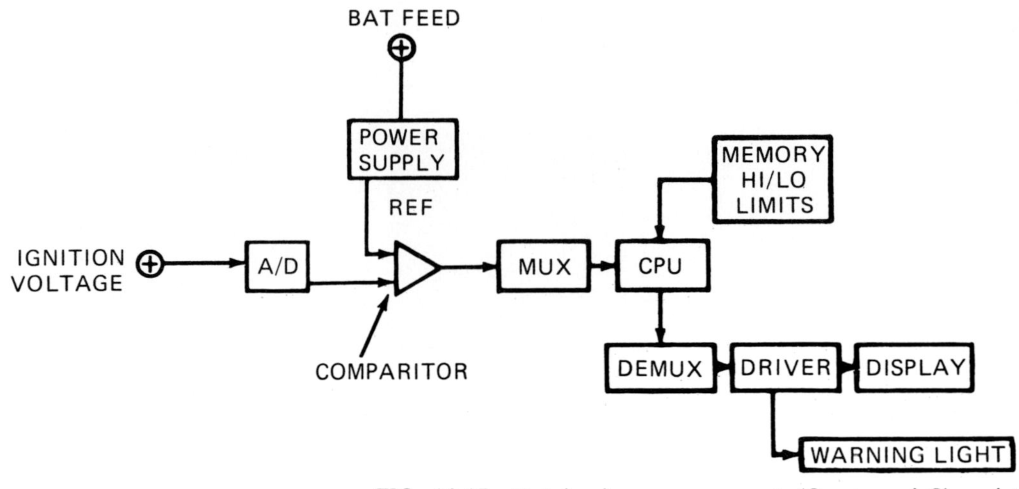

FIG. 14-17 Digital voltage measurement. (Courtesy of Chevrolet Motor Division, General Motors Corporation.)

Light-emitting diode (LED)

An LED is a special type of diode similar in some respects to the diodes introduced in Chapter 4. However, unlike a switch-type diode, an LED produces light energy whenever current passes the junction between the two semiconductor layers. The color of the light emitted depends on the material used in the LED. Yellow-producing LEDs are used in many types of displays.

Figure 14-18 shows an LED. These diodes can be arranged in any pattern: as a bar graph, dot matrix–type display, and the like.

Probably the most popular arrangement is the seven-segment pattern. Depending on the segments which are illuminated, any number as well as a variety of alphabetic characters can be represented. Figure 14-19 pictures a seven-segment arrangement of LEDs.

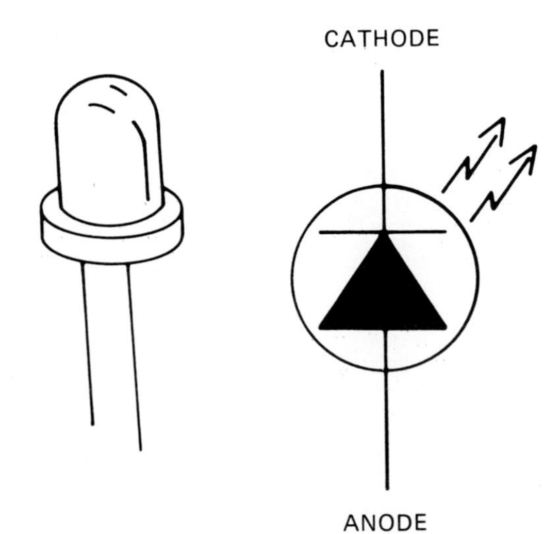

FIG. 14-18 Light emitting diode (LED). (Courtesy of Chevrolet Motor Division, General Motors Corporation.)

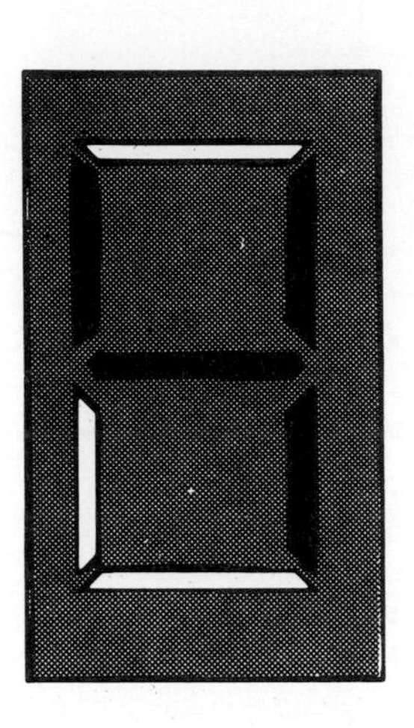

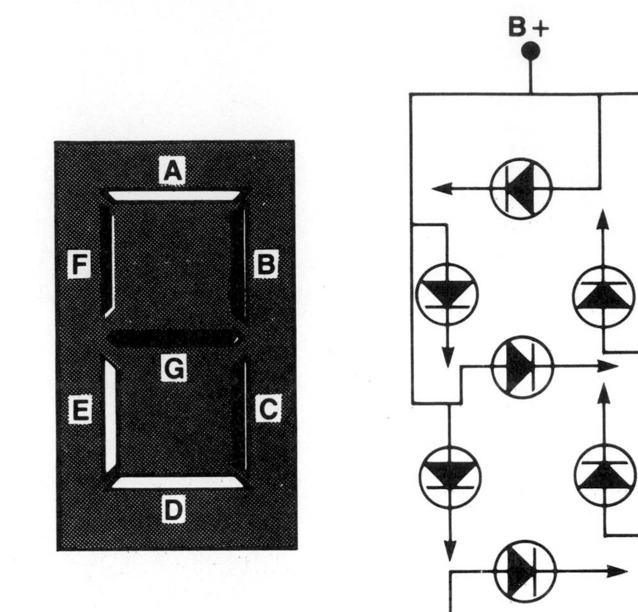

FIG. 14-19 LED's arranged for digital display. (Courtesy of Chevrolet Motor Division, General Motors Corporation.)

Liquid crystal display (LCD)

LCDs have become the display medium of choice for many microprocessor-controlled devices. Almost all lower-priced digital wristwatches use LCD displays; so do most of the briefcase-size, portable computers.

Aside from relatively low cost, the main reason for using LCDs in these applications is low power consumption. An LCD does not require power to generate light; it changes the light-transmitting qualities of the display material. In low-power devices, such as wristwatches and briefcase computers, no built-in light source is provided. Available light illuminates the display.

In automotive applications, a built-in light source is usually employed (otherwise the display would not be available at night). Therefore, the benefit of low power consumption may not be realized. However, LCDs can be fashioned into a variety of informative and/or exotic displays, which may help to explain their use in a number of digital instrument clusters. The heart of an LCD is a special liquid called "twisted nematic fluid." It can change the polarity of light passing through the fluid (like 3D movie glasses).

In automotive applications, the nematic fluid is sandwiched between two polarized glass plates which are coated with electrically conductive film. The nematic fluid elements are randomly oriented. A halogen light bulb is located behind the sandwich.

Square-shaped voltage pulse patterns are passed through the conductive film on both sides of the fluid. When the peaks are directly opposite one another (in phase), there will be no voltage potential across the fluid. The nematic fluid elements remain randomly oriented and no light will pass. However, when the peaks are not opposite one

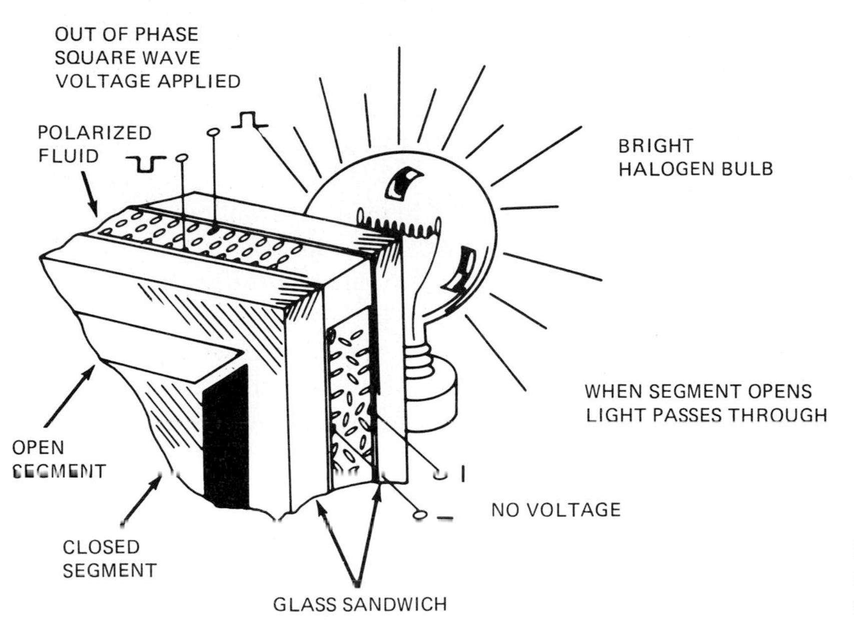

FIG. 14-20 Liquid crystal display, basic sketch. (Courtesy of Chevrolet Motor Division, General Motors Corporation.)

another (e.g., when they are out of phase) a voltage potential is developed between the two conductive films. Current passes through the fluid, causing the elements to line up. Light from the halogen bulb can then pass through the sandwich. The pattern established by the out-of-phase pulses will appear as a pattern on the display.

In some cases, a colored filter may be placed over the display to enhance the image. Refer to the appropriate shop manual when replacing the halogen bulb. Do not replace halogen bulbs with lower-powered bulbs. They may not be bright enough to shine through the dense glass/fluid sandwich. Figure 14-20 pictures the components of an LCD.

Vacuum-tube fluorescent display (VTF or VFD)

The VTF generates light something like a TV (cathode-ray tube, or CRT). As noted in Fig. 14-21, the main elements in a VTF are the:

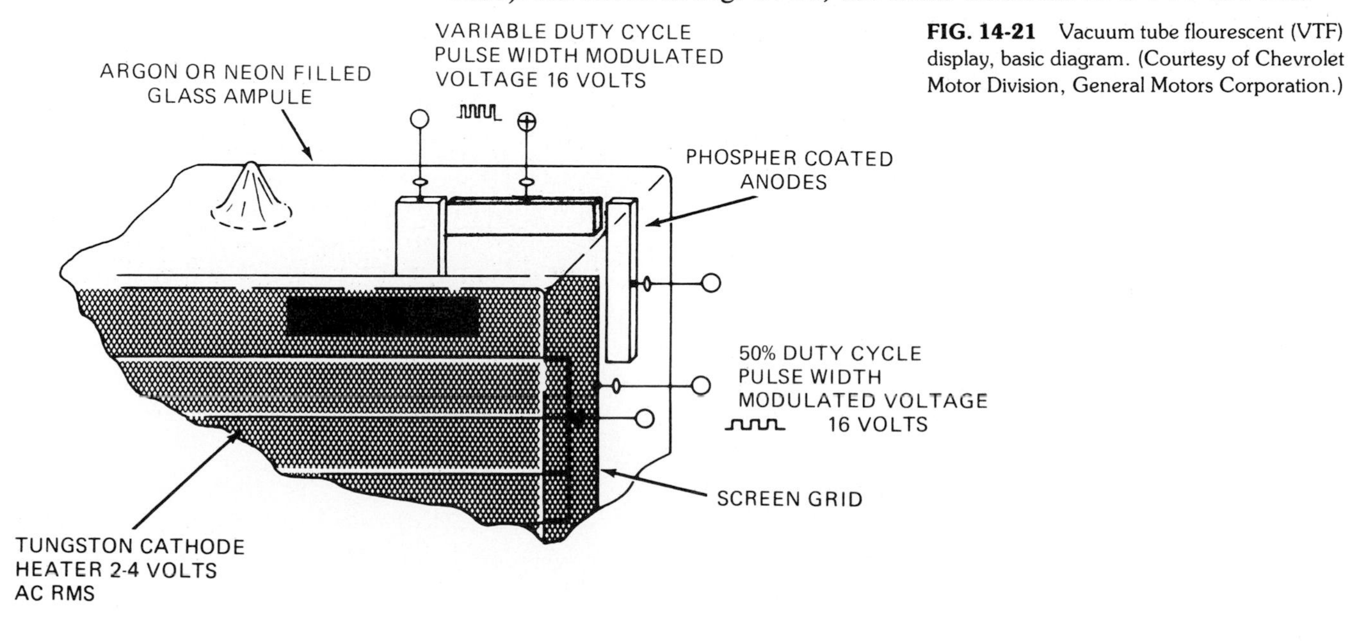

FIG. 14-21 Vacuum tube flourescent (VTF) display, basic diagram. (Courtesy of Chevrolet Motor Division, General Motors Corporation.)

Ampule: contains the following elements:

- *Cathode filaments:* emit electrons
- *Grid:* selectively passes electrons
- *Anode:* attracts electrons and glows when struck by them

The ampule, like the CRT in a TV, is a hollow glass chamber with a significant amount of the air removed (evacuated) and replaced by argon or neon gas. Inside the ampule are the cathode filaments, the grid, and the anode.

Cathode filaments are fine wires arranged horizontally, several millimeters apart, across the rear of the ampule. The filaments are coated with tungsten. As a result of heat caused by a 2- to 4-volt current passed through the filaments, tungsten electrons are driven out of the coating. They form a cloud at the rear of the ampule, something like a cloud of steam from a pot of boiling water.

The grid is a fine wire mesh placed very close to the anodes. A square wave, 16-volt current at a 50% duty cycle is passed through the grid. The term *duty cycle,* in this case, relates to the shape of the square waves. The shape of the 50% duty cycle wave is shown in the illustration. As a result of the current, electrons from the cathode cloud are attracted to the grid.

The anode, at the front of the ampule tube, is the light-producing element. The anode is divided into segments, each segment coated with phosphorus. When struck by electrons, the phosphorus will release energy in the form of light.

A particular segment of the anode is illuminated whenever it receives a square-wave, 16-volt current from the display driver. The segment is illuminated because the current has a higher duty cycle than the square waves placed on the grid. This higher duty cycle wave, as pictured in the illustration, has a greater attraction for electrons than the 50% duty cycle wave in the grid. As a result, electrons are accelerated through the grid, to the particular segment of the anode.

15

Automated Braking Systems

INTRODUCTION

Just as other components have been improved over the years, so have automotive brakes. Old mechanical brakes have been replaced with four-wheel, hydraulically operated drum brakes. Drum brakes have been replaced in many applications by disc brakes. In addition, most modern vehicles have a power assist to make braking easier.

This chapter examines antilock braking systems (ABS), the latest improvement in brakes. We do not review the fundamentals of automotive braking systems. That is best done in a text specifically devoted to the subject.

ORIGINAL HUMAN-FEEDBACK SYSTEM

A good driver knows that a vehicle can be steered only while the wheels are turning. A good driver also instinctively knows that a vehicle stops faster if the wheels are never allowed to go into a skid. Sliding tires offer much less traction than tires in rotating contact with the road. So, for many years, good drivers have pumped the brakes during emergency stopping maneuvers. This procedure reduces speed rapidly by keeping the wheels from locking up and throwing the vehicle into an uncontrollable skid.

Viewed as a feedback-control system, visual and "seat-of-the-pants" input received by the driver is compared to reference goals contained in the driver's head. If the comparison indicates an emergency, output signals are sent to the driver's foot. It operates the brake pedal, which slows the vehicle, which alters the environment, which alters the input signals received by the driver. These signals are compared with secondary reference goals and the appropriate output signals sent back to the driver's foot. For example, if the brakes are being pumped too hard or fast and a wheel starts to lock up, the output is moderated. Conversely, if the vehicle is not stopping fast enough, braking effort is increased. The object is to maintain a fine balance between adequate control and maximum deceleration.

As you might imagine, in a critical situation where events take place in split seconds, even a very skilled driver will have trouble implementing such a program. That is why many manufacturers have offloaded this control function from the human driver to a digital computer. Although not as versatile as its human creator, this inorganic computer does not panic in emergencies and is very fast. A typical system can pump each brake up to 15 times a second. It can also control each front brake separately and can control the rear brakes as a pair as soon as any single brake begins to lock up.

ABS FUNDAMENTALS

Using an ABS

Using an ABS feels somewhat the same as using a conventional braking system. A rise in brake pedal height and pulsations in the brake pedal may be the only indication that the system has been engaged. The force required to initiate the antilock function depends on the road surface. When the surface is slippery, the tires slide easier and the ABS is activated sooner. The converse is true of less slippery surfaces.

Secondary feedback control

The ABS is a secondary feedback system. The driver still supplies the primary feedback control, applying the brakes when a need is perceived and then releasing the brakes when the situation changes. The force of the driver's foot on the brake pedal still determines the maximum force that may be applied to the brakes. The ABS determines *whether* the force will be applied at any given moment.

An ABS does not pulse the brakes by applying more force but by periodically interrupting the force already provided. The duration and frequency of these interruptions is determined by the wheel speed. When a wheel begins to lock up, hydraulic pressure to that particular brake is diverted. When the wheel begins to turn normally, pressure is switched back to the brake. As indicated, the process may be repeated many times a second.

Deciding when a wheel is about to lock up is typically determined by comparing speeds of all the wheels. If a wheel slows up significantly relative to the other wheels, it is judged to be locking up and the appropriate action is taken by the ABS.

Basic components

The ABS function is implemented by certain basic types of components that are added to an otherwise conventional braking system. As shown in Fig. 15-1, these components can include:

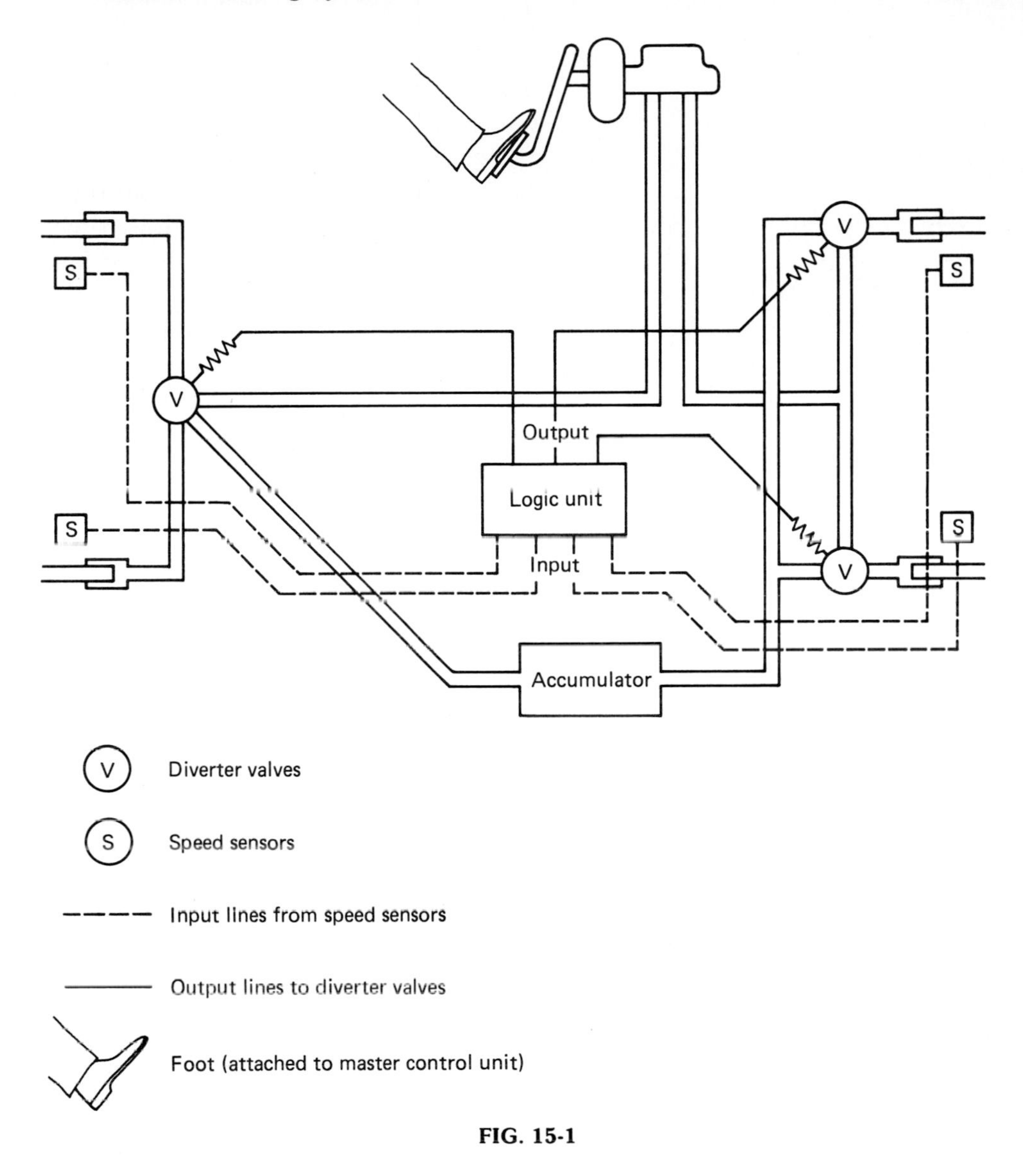

FIG. 15-1

- Wheel-speed sensors
- Control logic unit
- Control valves
- Accumulator reservoir

Typically, the wheel sensors are magnetic induction devices that produce pulsating current signals. The frequency of the pulses is directly proportional to the wheel speed.

The control logic unit is a digital computer. It is usually a stand-alone control device, separate from the engine control computer or other body computers. It receives input signals from the wheel sensors, compares the differences between wheel speeds to built-in reference goals and based on the comparison, sends output signals to control valves. Depending on the system, other auxiliary functions may also be managed by the logic unit.

Solenoid-operated valves control the flow of pressurized brake fluid to the individual brakes. A typical system has one set of valves for each of the front wheels and one set for both rear wheels (operated as a single unit). The valves are located between the master cylinder and the wheels. The valves are located between the master cylinder and the wheels. On command from the computer, a control valve diverts the flow of pressurized fluid from the master cylinder to the brake. A second command allows the flow to continue. The valves may also be commanded to hold the pressure as is (neither diminishing or building up).

An accumulator reservoir of some kind is also required to hold high pressure fluid that has been diverted from the brakes and to help even out hydraulic forces throughout the braking system.

Main differences between systems

One of the main differences between various ABSs is the way they operate in conjunction with braking power assists. In some systems, a power-assist pump is included as a part of the ABS. The pump provides boosted fluid pressure for normal or pulsating braking. In other systems, a more-or-less standard assist mechanism is used (vacuum or pump based). The ABS is incorporated downstream of the assist mechanism and upstream of the brakes. The Pontiac system examined next uses an integral assist pump. The Mercedes Benz system uses a separate, vacuum-powered assist mechanism.

PONTIAC ABS SYSTEM

Basic components

The basic components of the system, as pictured in Fig. 15-2, include:

- Front wheel sensors
- Rear wheel sensors
- Hydraulic unit (master cylinder, integral electric pump, valve block, fluid reservoir, accumulator)
- Electronic brake-control module (EBCM)
- Relays and wiring harness
- Proportioner valve

Electrical function

Fig. 15-3 is a general schematic of the system electrical components. Fig. 15-4 show pressure and fluid level switches in more detail. The following steps trace the operation of the components pictured:

1. When the ignition switch is turned on, power is received at EBCM terminal 2. The EBCM checks itself and all circuits for opens or grounds. If everything is in order, the EBCM powers terminal 8.

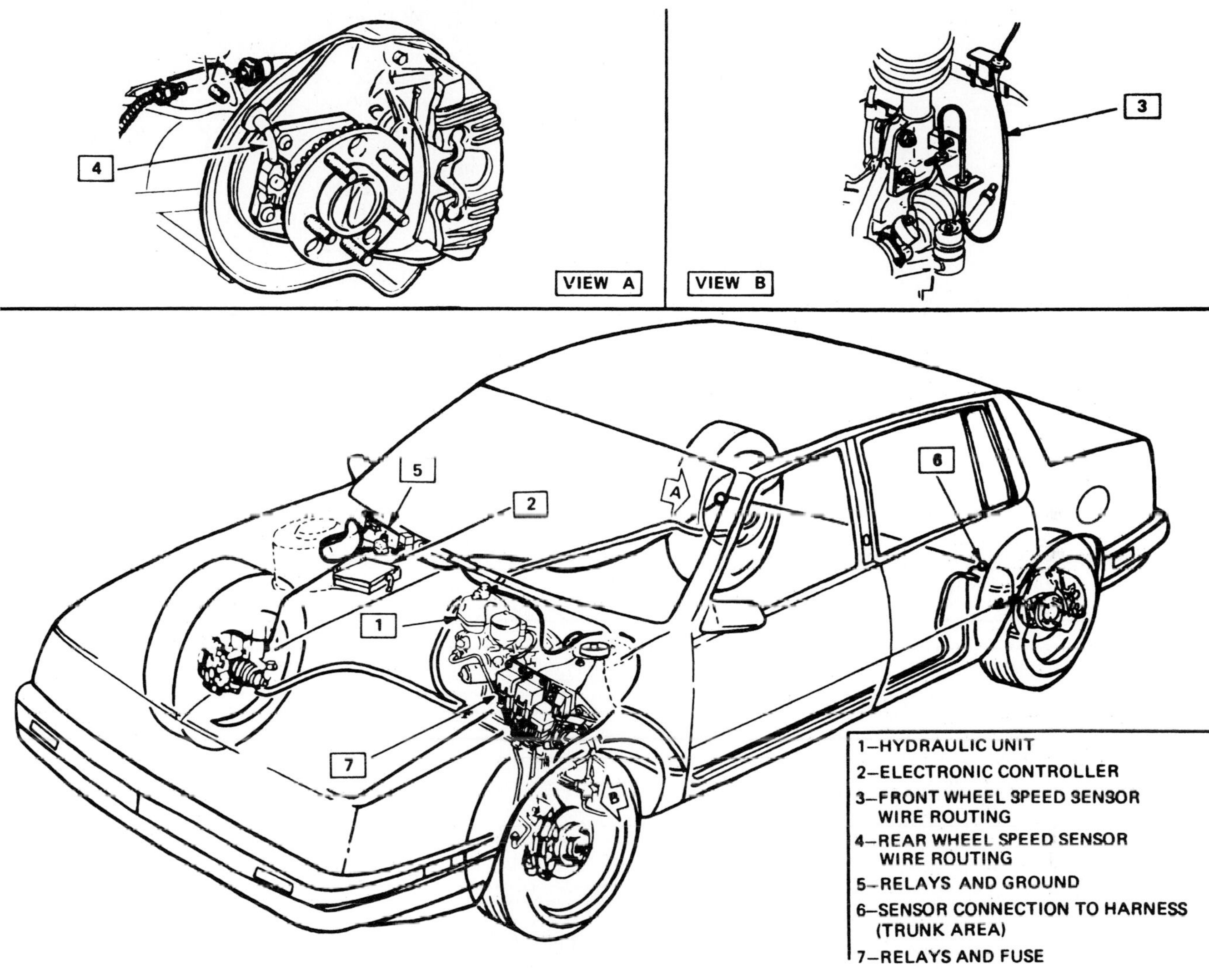

FIG. 15-2 (a) TEVES antilock brake system (ABS) used by GM. (b) Antilock electronic brake system components. (Courtesy Pontiac Division, General Motors Corporation.)

2. Power from terminal 8 energizes the main relay, pulling the switch to the left (as pictured in this drawing). Two things happen as a result: (1) Battery voltage is supplied to EBCM terminals 3 and 20. (2) The ground circuit from the amber antilock indicator bulb is broken, thus causing the bulb to go out. (The antilock brake diode prevents battery voltage from shorting through the indicator circuit, although it allows current to flow toward ground from the other direction). This entire process takes up to 3 seconds.

3. The hydraulic pump relay is on the same circuit as EBCM terminal 2. So, when the ignition switch is turned on (step 1), power is also made available to the pump relay terminal. However, notice that in order for the relay circuit to be complete, the pressure-control switch

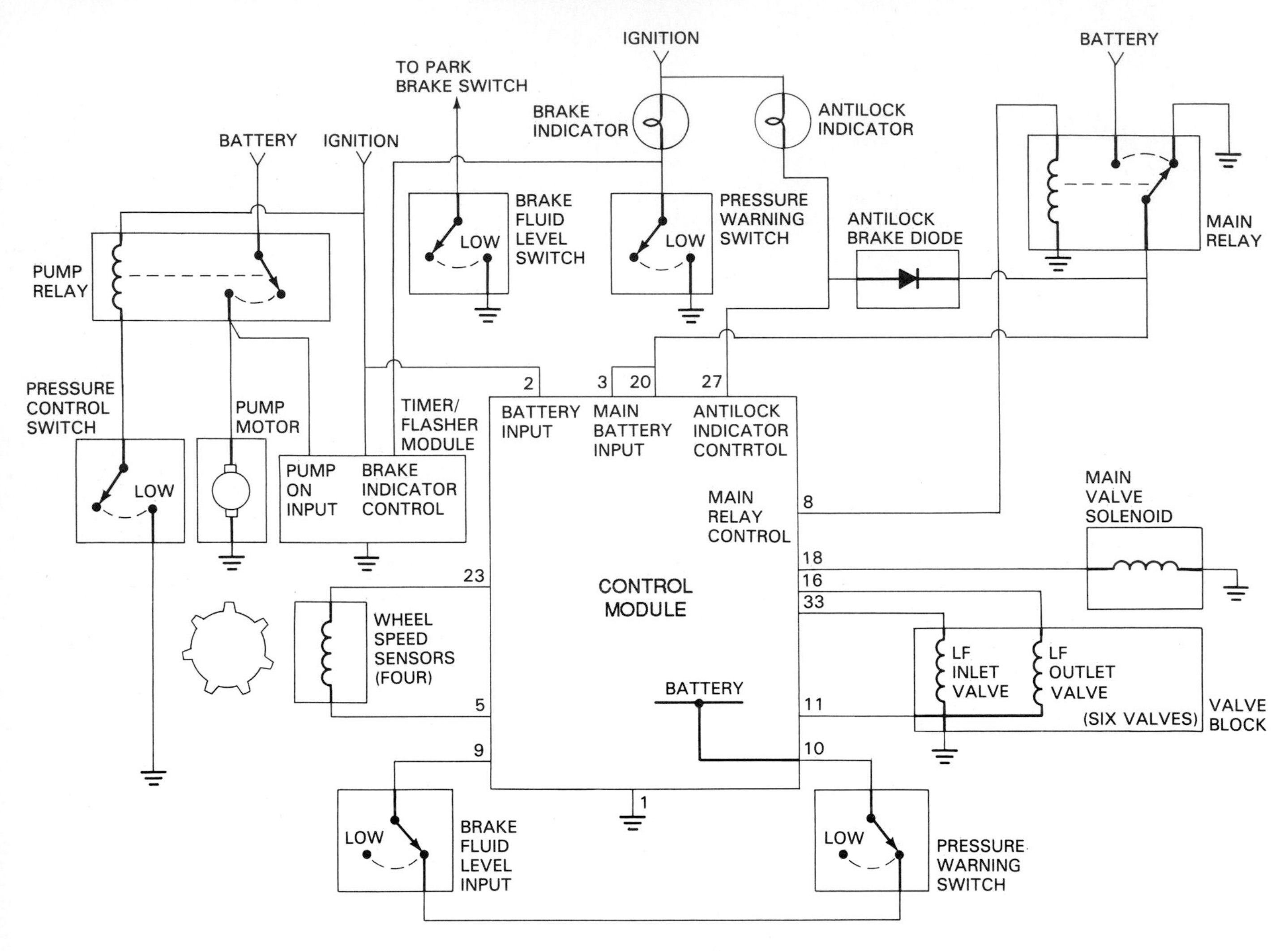

FIG. 15-3 ABS system schematic.

must be closed. Screwed into the pump assembly, this switch closes when the pressure drops below 2200 pounds per square inch (psi) and opens when the pressure exceeds 2600 psi. Within that range, the relay is energized, which causes the pump to operate. As you can see, the hydraulic pump and switch, although part of the system, operates independently of the EBCM.

4. These pressure switches, other switches and indicators are shown more clearly in Fig. 15-4.

a. Power for the red brake indicator light comes from the pump-relay circuit described in step 3. When the pressure falls to 1500 psi, the pressure warning switch closes, causing the brake indicator light to come on. Because it is part of the pump-relay circuit, the brake indicator light operates independently of the EBCM.

b. The pressure-control switch (as described in step 3) and the left low-fluid switch (as pictured in Fig. 15-4) also operate independently of the EBCM. If the pressure-control switch is closed (meaning the pressure is below 2600 psi) the circuit is closed up

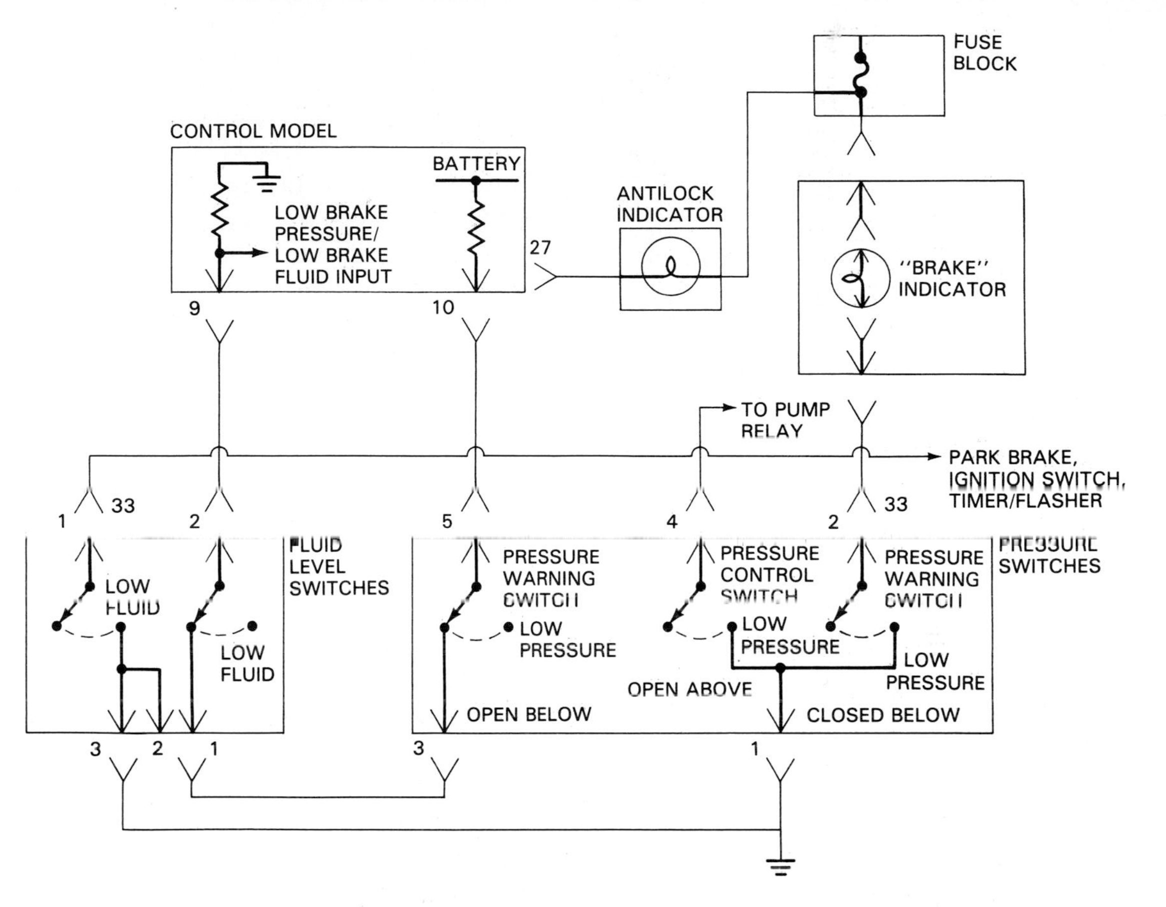

FIG. 15-4 Antilock brake system pressure switches, fluid level switches.

to the left low-fluid switch. If that switch is closed (meaning the fluid level is low), the path to the park brake, ignition switch, and timer-flasher circuit is also closed.

c. The only two switches operating strictly in conjunction with the EBCM are the left pressure-warning switch and the right low-fluid switch (as pictured in Fig. 15-4). Power for these switches comes from EBCM terminal 10 and returns to the EBCM at terminal 9. If either switch opens and breaks the circuit, the EBCM will shut itself down, causing the amber antilock indicator bulb to come on.

Note: Depending on the nature of the fault, the EBCM can put itself temporarily or permanently "to sleep." For example, if radio frequency interference is detected, the EBCM can temporarily inhibit itself and turn on the amber indicator light by grounding its circuit through terminal 27. However, if the problem is severe (low pressure and/or fluid) the system will shut itself down completely through the main relay via terminal 8. If this happens, the antilock will remain on until the fault is corrected and the ignition key recycled.

5. Returning to Fig. 15-3, information about wheel performance comes from the four wheel-speed sensors. There are two input terminals on the ABCM for each wheel sensor, terminals 4 and 22, 5 and 23,

6 and 24, and 7 and 25 (only 5 and 23 are shown). The higher the frequency of the voltage pulses coming from the wheel sensors, the faster the wheels rotate. If one wheel begins to rotate sufficiently slower than the others, it will be presumed to be locking up. At that point an output signal will be sent to the appropriate valve block solenoids.

6. ECBM terminals 11, 15 and 33, 16 and 34, and 17 and 35 (only 16, 33 and 11 are showing) control the valve block solenoids. There are three sets of two valves each, an inlet and outlet valve for each front wheel and one set for both rear wheels. Terminal 11 is a redundant ground; each solenoid is also internally grounded at the valve block.

7. ECBM terminal 18 controls the main solenoid.

GM wheel sensors

Like the wheel sensors used by other manufacturers, GM's sensors operate on the magnetic induction principle. As a toothed sensor ring rotates past a stationary sensor, voltage is induced in the sensor. The frequency of this alternating current is directly proportional to the speed of the wheel.

Figure 15-5 shows a GM front wheel sensor. The magnet and induction coil is located on the steering knuckle. The toothed ring is positioned on the drive axle.

Figure 15-6 shows a GM rear wheel sensor. The toothed ring is bolted to the back of the wheel hub. The sensor is mounted on a bracket that is held stationary by two of the nuts that also hold the

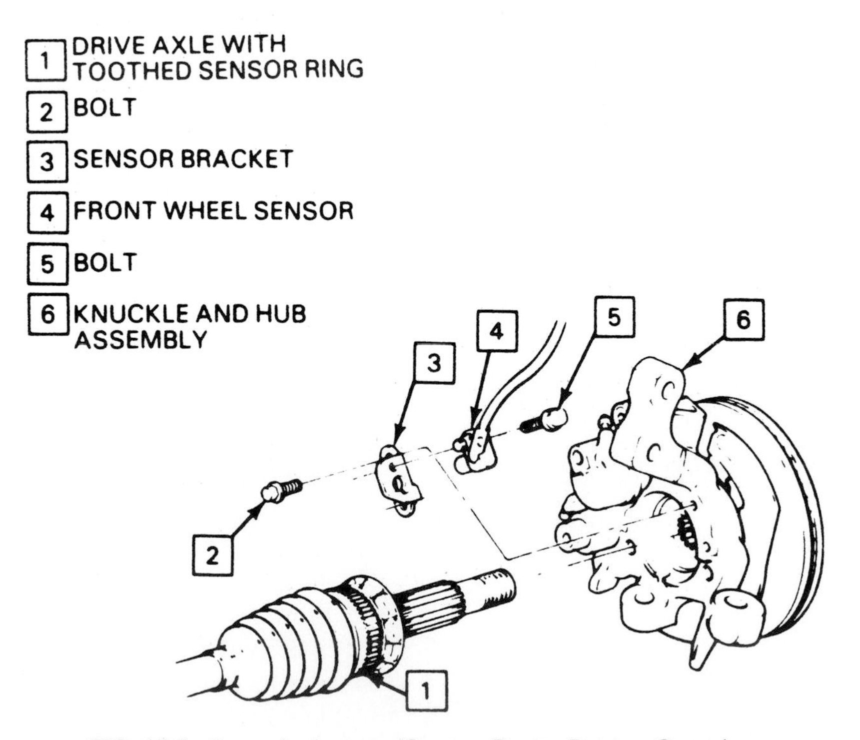

FIG. 15-5 Front wheel sensor. (Courtesy Pontiac Division, General Motors Corporation.)

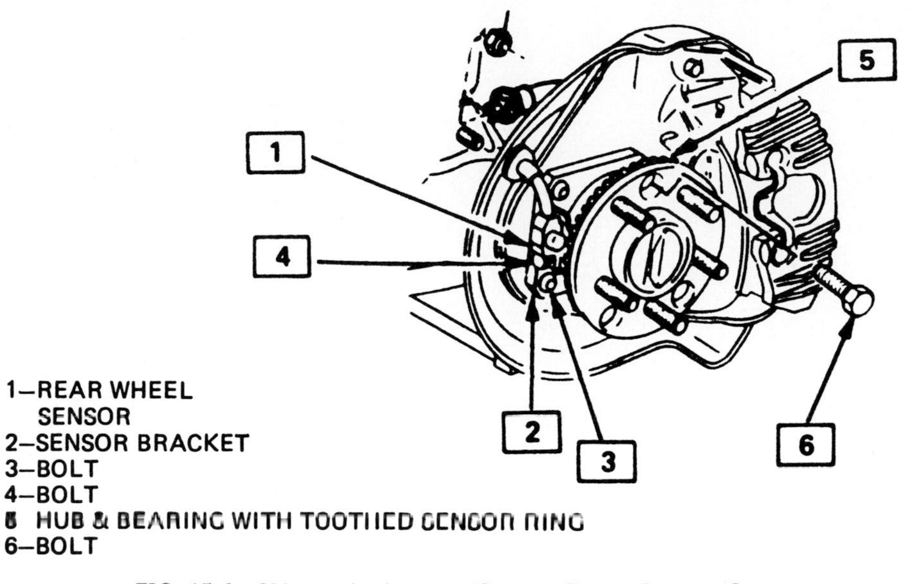

FIG. 15-6 GM rear wheel sensor. (Courtesy Pontiac Division, General Motors Corporation.)

bearing retainer in place. A slot allows the sensor position to be adjusted. The cable harness connects the sensor to the EBCM.

GM hydraulic unit

The hydraulic unit used by several GM products is shown in Fig. 15-7. The unit includes the:

- Valve block
- Master cylinder
- Fluid reservoir
- Accumulator
- Pump and electric motor
- Fluid-level switch
- Pressure switch

As noted before, the pump operates independently of the EBCM to provide power boost to braking operating in normal and emergency operation. In this respect, it is similar to a typical hydro-boost type of system. It operates over a range of 2030 psi to 2610 psi. The pump often runs for short periods to keep the accumulator charged.

The accumulator is a nitrogen-filled chamber. The chamber is separated into two parts by a thick, brake-fluid-resistant, rubber diaphragm. One side contains nitrogen gas, the other, brake fluid. One main purpose of the accumulator is to even out hydraulic pressure through the system when the solenoid valves interrupt the flow. Also, should the pump fail, the accumulator can supply pressure for several applications of the power brakes.

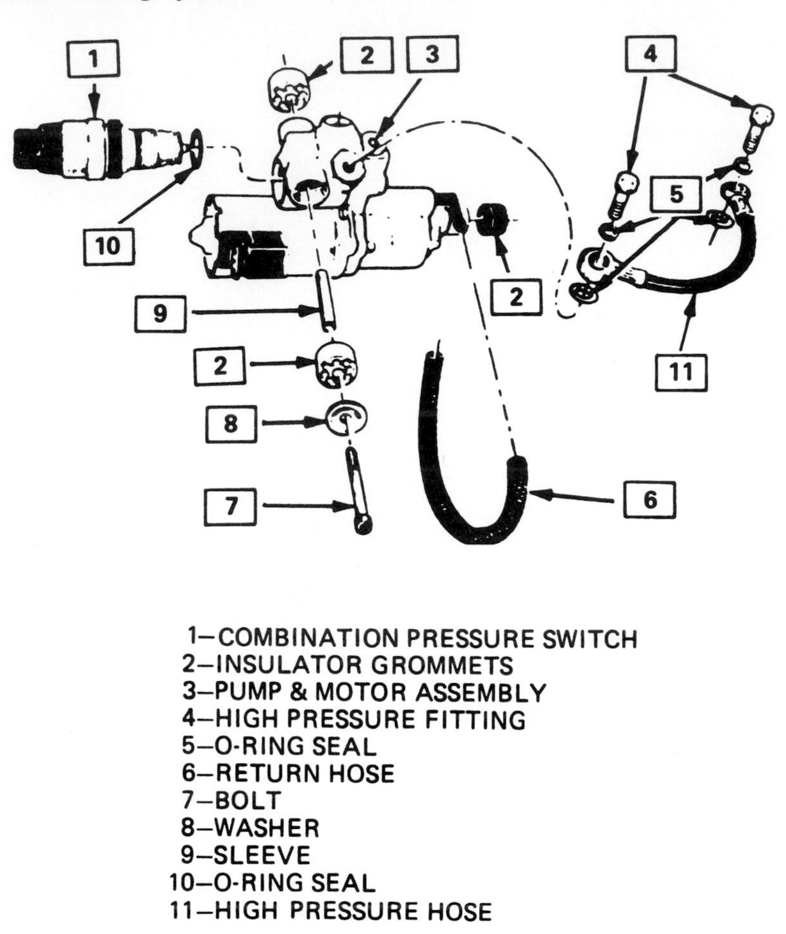

FIG. 15-7 GM hydraulic unit. (Courtesy Pontiac Division, General Motors Corporation.)

MERCEDES BENZ ABS

Mercedes Benz electronic control unit

The Mercedes Benz system is controlled by a single board computer called the *electronic control unit*. The board has printed circuits and components on both sides and is enclosed in a light alloy housing. The functions of the control unit are divided by Mercedes into three categories.

Signal Conditioning. Signal conditioning converts the signals from the wheel sensors into a form usable by the logic section. This filter compensates for distortions in wheel-speed signals that result from production tolerances and movements in the steering knuckle.

Logic Processing. Logic processing examines the filtered wheel signals to determine if the wheel is slipping. Based on the examination, output signals are sent to the solenoid valves in the hydraulic unit. These hydraulic functions can be generated in the brake calipers of the wheel brakes:

- Pressure maintenance
- Pressure reduction
- Pressure buildup

Safety Processing. Safety processing checks for faulty signals within and outside of the electronic control unit. In addition, the safety circuit:

- Intervenes in extreme driving conditions, such as when the vehicle is aquaplaning
- Monitors the battery voltage to see if it is within specified limits
- Performs a testing role called BITE (built-in test equipment)

Whenever a fault is detected, the safety circuit shuts down the ABS and informs the driver by turning an indicator lamp on.

Mercedes hydraulic functions

A simplified schematic of the Mercedes Benz ABS hydraulic system is shown in Fig. 15-8. Three fast-acting solenoid valves are used, one for each of the front brakes and one for both rear brakes. Each solenoid valve can provide the three functions introduced earlier:

- *Pressure buildup.* No current is applied to the solenoid; pressure from the hydraulic assist is routed directly to the wheel. Braking effort increases.
- *Pressure hold.* Half of maximum current is applied to the solenoid. The fluid line from the solenoid valve to the brake is closed. No fluid is allowed to escape to the reservoir, nor does any additional fluid enter from the master cylinder. Braking effort remains the same.
- *Pressure reduction.* Full current is applied to the solenoid. The fluid from the brake is routed through the solenoid to the reservoir. As a result, there is a reduction of pressure in the brake line. Braking effort is reduced.

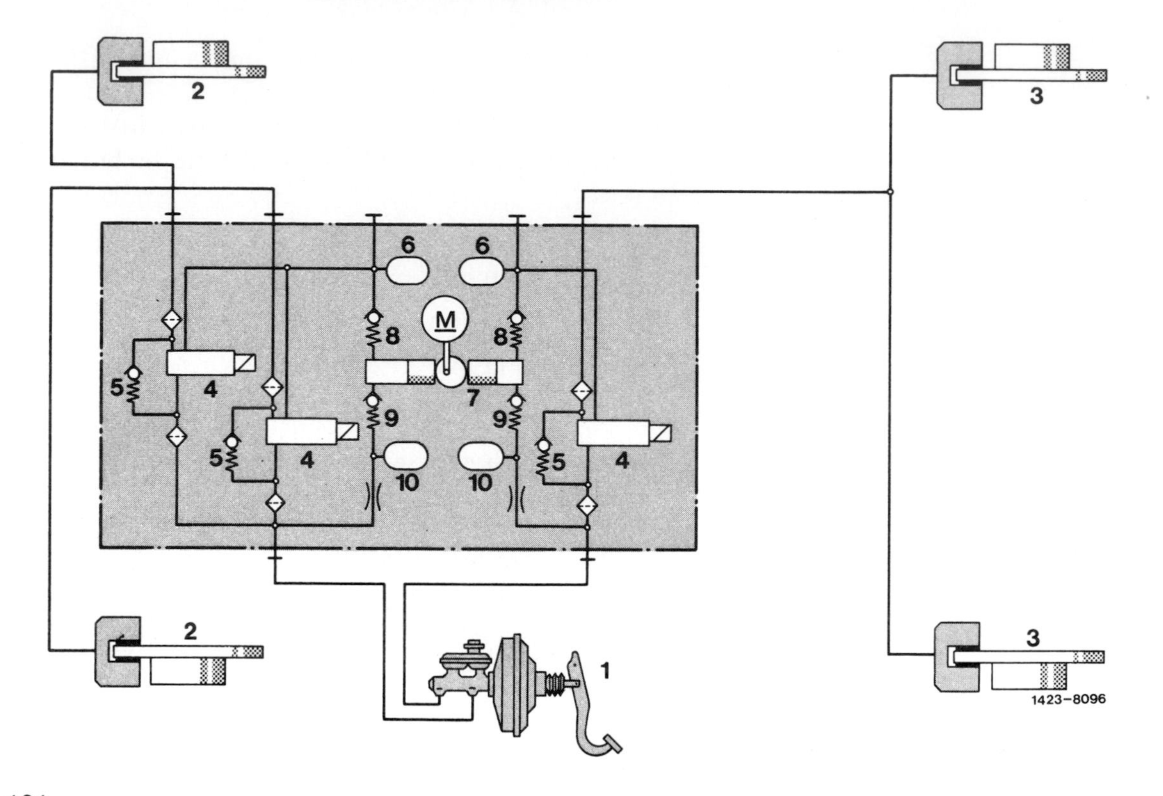

Fig. 121

1 Brake booster with master cylinder
2 Front wheel brake
3 Rear wheel brake
4 Solenoid valve
5 Check valve
6 Pump reservoir
7 Return pump
8 Pump input valve
9 Pump output valve
10 Silencer

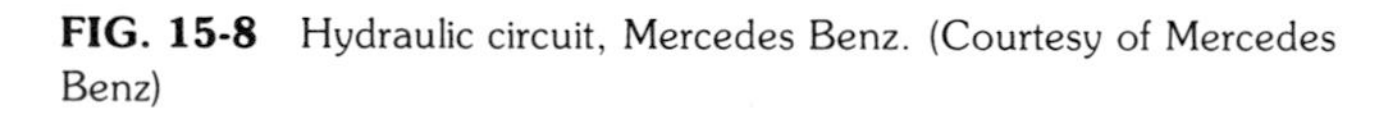

FIG. 15-8 Hydraulic circuit, Mercedes Benz. (Courtesy of Mercedes Benz)

4 Solenoid valve
6 Pump reservoir
7 Return pump
8 Pump input valve
9 Pump output valve
10 Silencer
15 Relay for return pump
16 Relay for solenoid valve

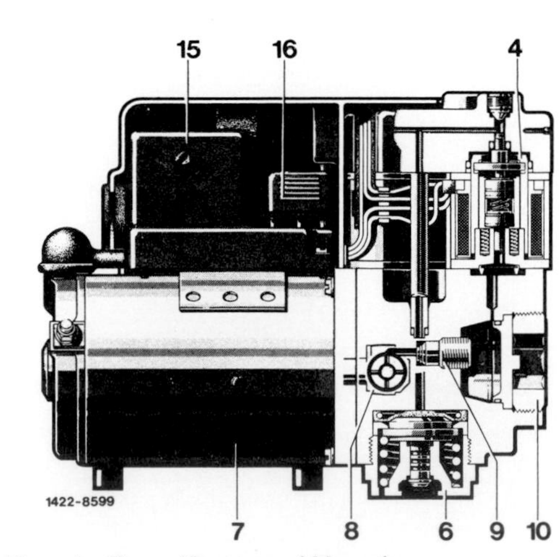

FIG. 15-9 Hydraulic pump, Mercedes Benz. (Courtesy of Mercedes Benz)

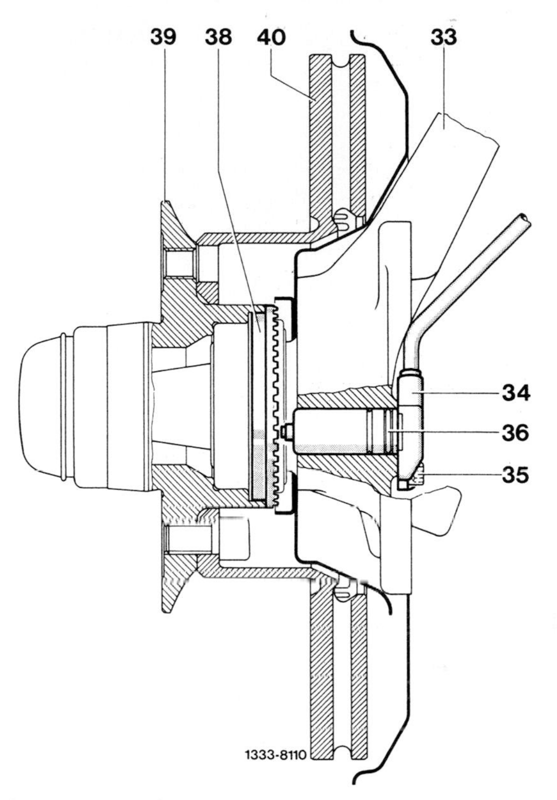

FIG. 15-10 Front wheel sensor, Mercedes Benz. (Courtesy of Mercedes Benz)

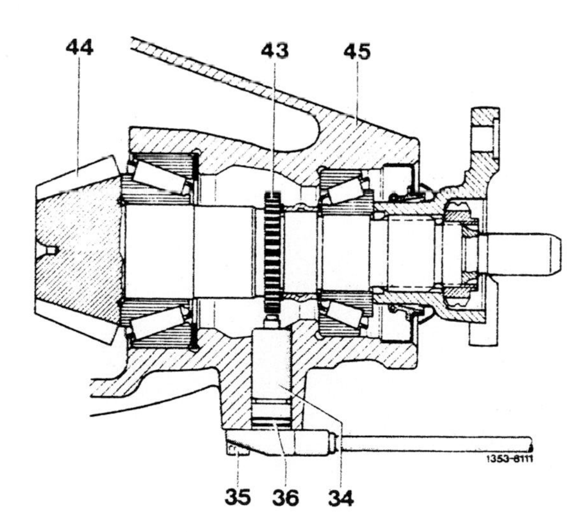

34 Speed sensor
35 Hex. socket screw
36 O-ring
43 Toothed wheel (rotor)
44 Drive pinion
45 Rear axle housing

FIG. 15-11 Rear wheel sensor, Mercedes Benz. (Courtesy of Mercedes Benz)

The return pump (7) is used to maintain brake pressure in the control lines. The input and output valves (8, 9) and the check valves (5) allow flow in only one direction and thus work with the solenoids to determine fluid flow. For example, fluid can flow only from the pump through the output valves (9). Fluid flows into the pump through the input valves (8). A cutaway view of the pump is shown in Fig. 15-9.

Mercedes wheel sensors

Figure 15-10 shows a Mercedes front wheel speed sensor and toothed rotor. Notice that the rotor is machined into the wheel hub.

Only one sensor is used for the rear wheels. It is positioned on the pinion shaft as shown in Fig. 15-11. Each gear ratio is associated with a specific toothed wheel, each with a certain number of teeth.

16

Overview of Testing and Diagnosis

The last chapters in the book cover a variety of related subjects, including:

- A review of meters and circuit testing (Chapter 17)
- An introduction to the diagnostic and test procedures used to troubleshoot and repair automotive computer systems (Chapter 18)
- An introduction to "off-board" diagnostic computers (Chapter 19)

As you study these chapters, the abstract explanations of the previous chapters will start to come into clearer focus. However, before putting this information to practical use, it will be helpful to continue the "philosophizing" we started in Chapter 1. We need to think briefly about different approaches to problem solving and repair.

In Chapter 1, two ways of looking at the world were proposed: the digital view and the analog view. This view can also be extended to the world of automotive repair.

Analog (shade-tree) approach

The analog approach is a fancy name for the time-honored tradition of shade-tree technology. Shade-tree work is based on the mechanic's experience and a certain "feel" for automobiles that comes from that experience. Test equipment and specifications may or may not be used; it depends on the individual practitioner. However, what is certain is the shade-tree's distrust of words. A dyed-in-the-wool-shade-tree mechanic does not like to read and does not like to theorize about car repair. This person simply likes to fix things. Digitizing explanations are left for engineers and philosophers.

A few shade-tree mechanics may be found under literal shade trees, as pictured in Fig. 16-1. Most, however, work in garages, service stations and dealerships. All of us, at one time or another have (or will be) shade-tree operators.

In the past, using an analog, or shade-tree, approach was not such a bad thing. This was especially true when it came to relying on the tricks and skills picked up from years of experience. A good shade-tree mechanic could often get a car back in operation before a novice with a lot of book learning could even diagnose the problem.

Unfortunately, the classic shade-tree, or analog approach is no longer adequate for modern computer-controlled systems. The equipment changes so often that it is difficult to develop a feel for repairs based strictly on working experience. Also, it is difficult (but probably not impossible) to develop analog instincts for the digitized, computerized systems used today.

Digital approach required

As we noted in Chapter 1, in order to fix digital devices you must use digital logic. This does not mean that a good shade-tree operator does not do logical work. What it does mean is that the instinctive, analog logic of the shade-tree practitioner will not work in situations where the mechanic has no direct experience. Nowadays, the logic must be applied beforehand and must be based on a digital understanding of the systems involved and a thorough study of the problem. Figure 16-2 pictures the logical (as opposed to shade-tree) mechanic.

FIG. 16-1 Shade-tree operation.

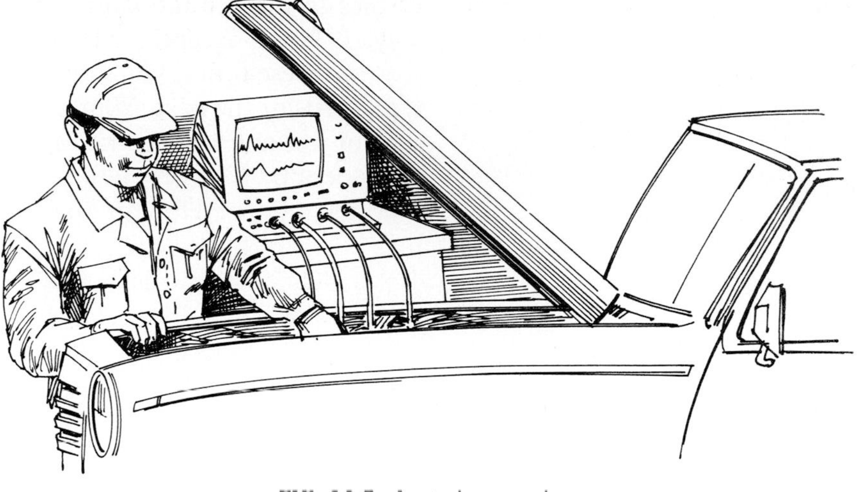

FIG. 16.2 Logical approach.

Is mechanic's logic enough?

Of course, all this sounds great in theory. The real question is, How can you learn enough about these complex electronic systems to solve problems on your own?

The answer, practically speaking, is that you cannot, at least not to start with. Acquiring the expertise needed to develop your own test procedures is simply not feasible for the average technician. It would take too much time.

The solution is to rely on the digitized experience of others. This expert knowledge comes in the form of manuals, diagnostic charts, tables, guides, even computer programs (called expert systems). Your task is to follow manufacturer's procedures, performing every step exactly as it is described. At times, knowing how to follow instructions will be more important than knowing what the instructions mean.

Is expert's digital logic enough?

The clever reader might now ask, "OK, if the factory materials are going to tell me how to diagnose problems anyway, why do I need this book? Why do I need to know anything more about computer systems?"

Quite frankly, some experts might answer that you do not need to know anything more. They would repeat what we have just noted . . . that you cannot be reasonably expected to develop your own procedures. They would go on to state that all problems can be fixed by exactly following the test procedures they have developed.

The authors of this book would agree, but only up to a point. It is certainly important to follow the manufacturers recommendations because most problems are covered by an existing test procedure. However, in the real world, not all problems are covered. As we noted in Chapter 1, the digital view is never 100% complete. Despite our best efforts to digitize the world, it remains an analog, complex place, always offering exceptions to any rule or ready explanation.

At these times, you must work on your own, just like those shade-tree artists of the past. Your only hope then is to know something about the subject and to become, at least for the moment, a modern shade-tree operator. The very best mechanics of the late twentieth century will be those who develop an analog instinct for digital systems.

17

Review of Meters and Circuit Testing

Meaningful testing has always depended on using the correct tools in the proper manner. This is especially true when checking computer-related systems. A hit-or-miss casual approach simply will not work. More problems will be created than solved.

In this chapter we review some common types of test devices and general testing procedures.

THREE BASIC TYPES OF ELECTRICAL TEST METERS

As you are probably already aware, the three basic types of electrical test meters are voltmeters, ammeters, and ohmmeters. In many instances, these meters are combined into one case with switches for selecting the desired function. These units are called *multimeters.* A number of meters use a needle pointer or ''hand'' to portray readings. Other meters use digital readout screens. The next several paragraphs review the operation of dial-type meters.

GENERAL CONSTRUCTION AND OPERATION

All dial-type meters are constructed in a similar manner. The basic movement includes a permanent fixed-position magnet and a movable electromagnet with a needle attached. Current from the test circuit flows through the electromagnet, causing it to be surrounded by lines of magnetic force. These force lines interact with the force lines surrounding the permanent magnet. The resulting attracting and repulsion cause the electromagnet and attached needle to move in proportion to the current flowing through the electromagnet.

To ensure accuracy, the electromagnet is balanced between two jeweled bearings. The electromagnet is also spring loaded so that the attached needle rests (or is "pegged") toward one end of the scale (normally the left).

Current for the electromagnet comes from two flexible leads which are attached to the circuit being tested. Current variations in the test circuit alter the ampere-turns of the electromagnet, thus affecting its strength. The strength of the electromagnet determines the way it responds in the presence of the permanent magnet, which, in turn, determines the position of the needle.

The outside of the meter contains various scales and control knobs. If the meter is a multifunction unit, controls will be provided for selecting either voltmeter, ammeter, or ohmmeter functions. Most meters also provide controls for selecting the test range best suited for the conditions being measured. Turning the selector knob changes the resistance of the input circuit of the electromagnet, thereby altering the response of the magnet to the circuit being tested.

Test leads are usually identified as being positive or negative. Positive leads are colored red and/or marked with a plus (+) sign. Negative leads are colored black and/or marked with a negative (−) sign. Various kinds of ends are found on test leads. Some have sharp, probe tips; others have alligator clips. The particular application for a meter (voltmeter, ammeter, ohmmeter) is determined both by the internal wiring of the meter and by the way it is attached to the circuit being tested.

Voltmeters

A typical voltmeter is pictured in Fig. 17-1. Voltmeters are always attached in parallel to the circuit being tested. In other words, the test circuit is left completely hooked up and the voltmeter leads are connected on either side of the element being tested.

Ammeters

Figure 17-2 shows a typical ammeter. Notice the two shunt branches used to determine the test range, one branch for testing 0 to 6 amperes and the other circuit for checking 0 to 30 amperes. Aside from these internal differences, one of the most significant distinctions between an ammeter and voltmeter is the way in which the test leads are attached to a circuit. Ammeters leads are connected in series with the test circuit so that all the current being measured flows through the meter.

Ohmmeters

Unlike voltmeters and ammeters, ohmmeters are *never* hooked into a "live" circuit. To do so could cause damage both to the meter and the circuit. Instead, an ohmmeter, as pictured in Fig. 17-3, has its own power source, often a regular penlight battery. The battery sends a known current flow through the test leads into the detached ends of the circuit being tested. The resistance of the circuit affects the flow of current back into the meter, which, in turn, determines the behavior

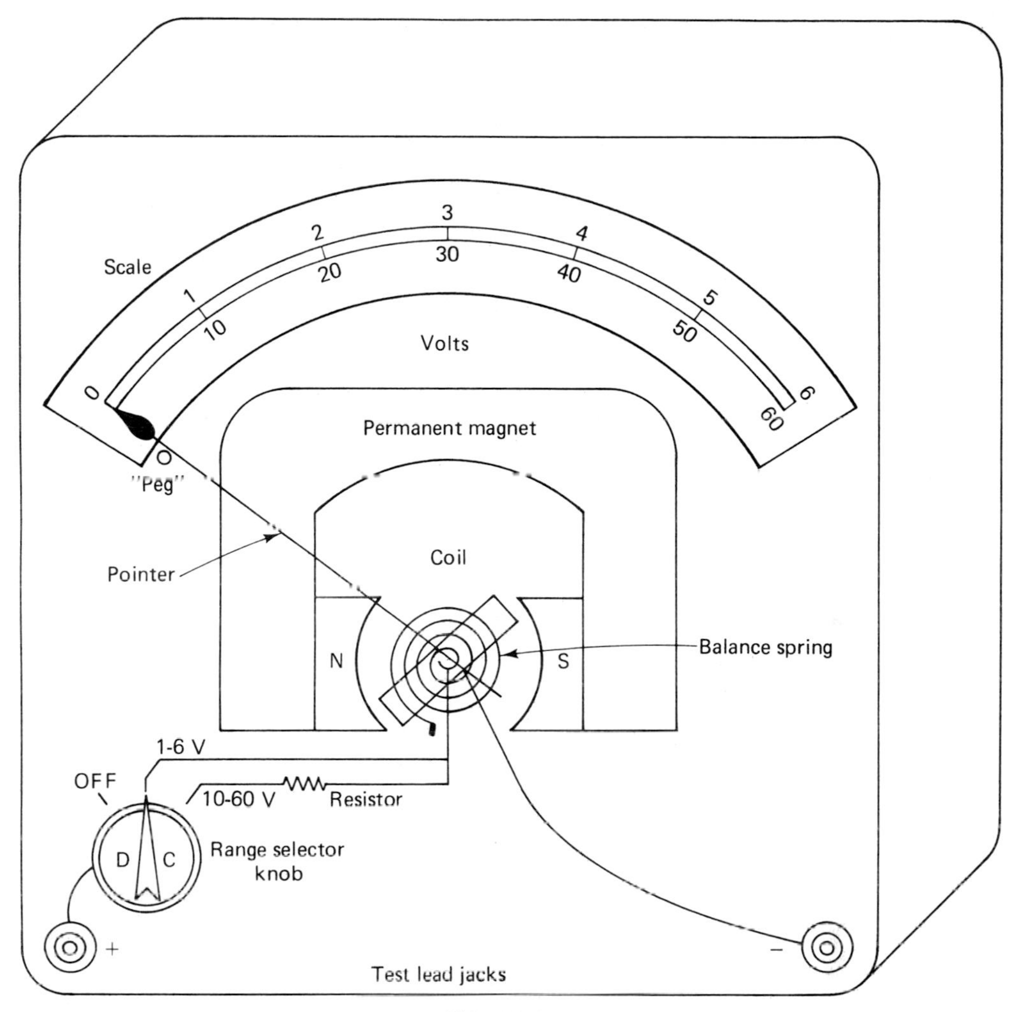

FIG. 17-1 Voltmeter.

of the electromagnet and attached needle. Each time an ohmmeter is used, it must be calibrated. Ohmmeters made by different manufacturers are sometimes used or calibrated in different ways, so be sure to check the operating instructions before employing an unfamiliar unit.

DIGITAL METERS

The most obvious difference between digital and dial-type meters involves the way in which test results are displayed. Dial-type meters are analog devices, representing test results on a divided scale. Digital meters, on the other hand, display specific readings, often accurate to +0.1%. Internally, digital meters are also different from dial-type meters. Instead of using two sets of magnetic fields, digital meters use analog-to-digital converters, similar in some cases to the ones used in on-board computers. Because of these differences, digital meters are often more accurate than dial-type meters and can be used over a wider range. For this reason, critical, computer-related testing sometimes requires a digital volt/ohmmeter with a minimum independence range of 10 megohms (Fig. 17-4).

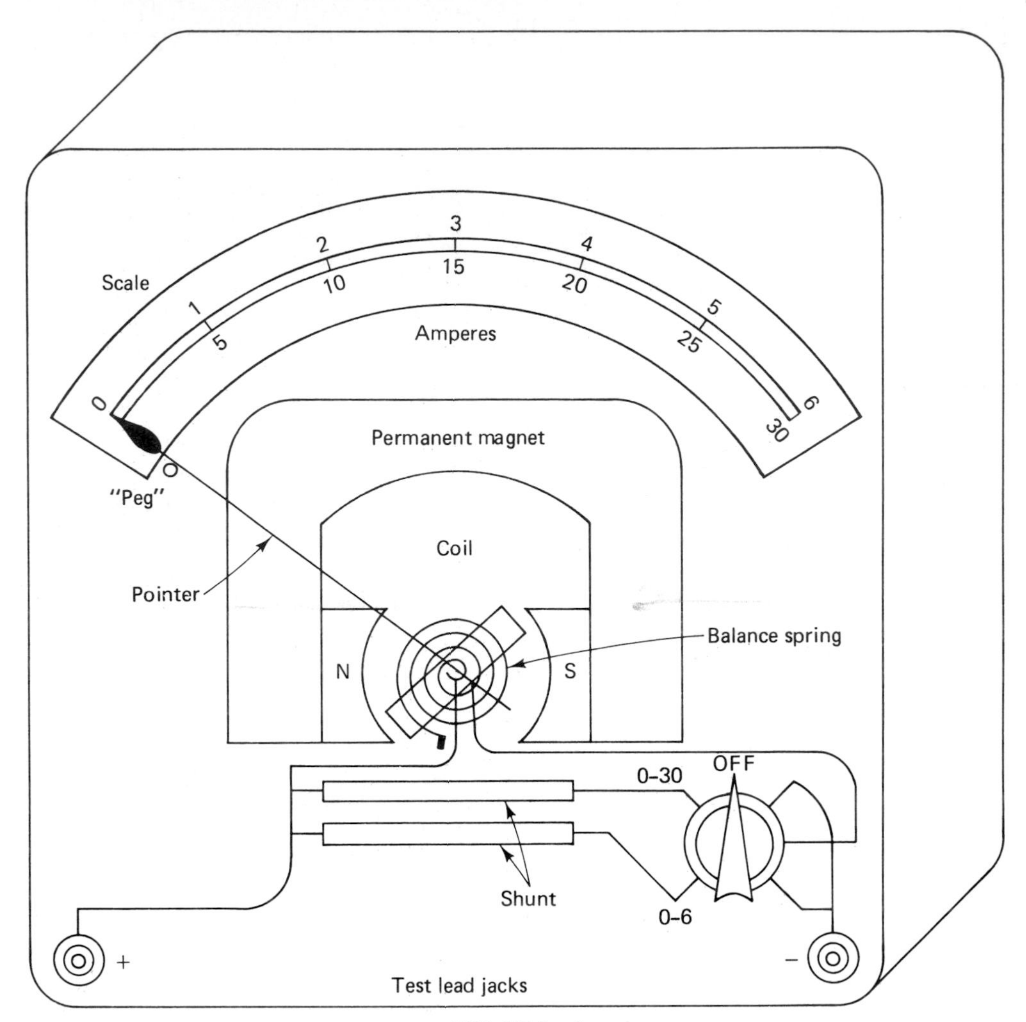

FIG. 17-2 Ammeter.

OTHER TYPES OF TEST DEVICES

Test lights

Test lights, which have been used for years, are among the most common electrical test gear. They are still used for some computer-related diagnosis.

A test light is basically two wires connected to a light bulb. Some mechanics make their own. Unpowered test lights are used to check for power (Fig. 17-5). When connected in parallel or series to a live circuit, the light will usually come on. Powered test lights, which contain a small pen-light battery, are used to check for continuity (Fig. 17-6). The leads are connected to both ends of the circuit or element being checked. If the circuit is complete, the bulb will light up.

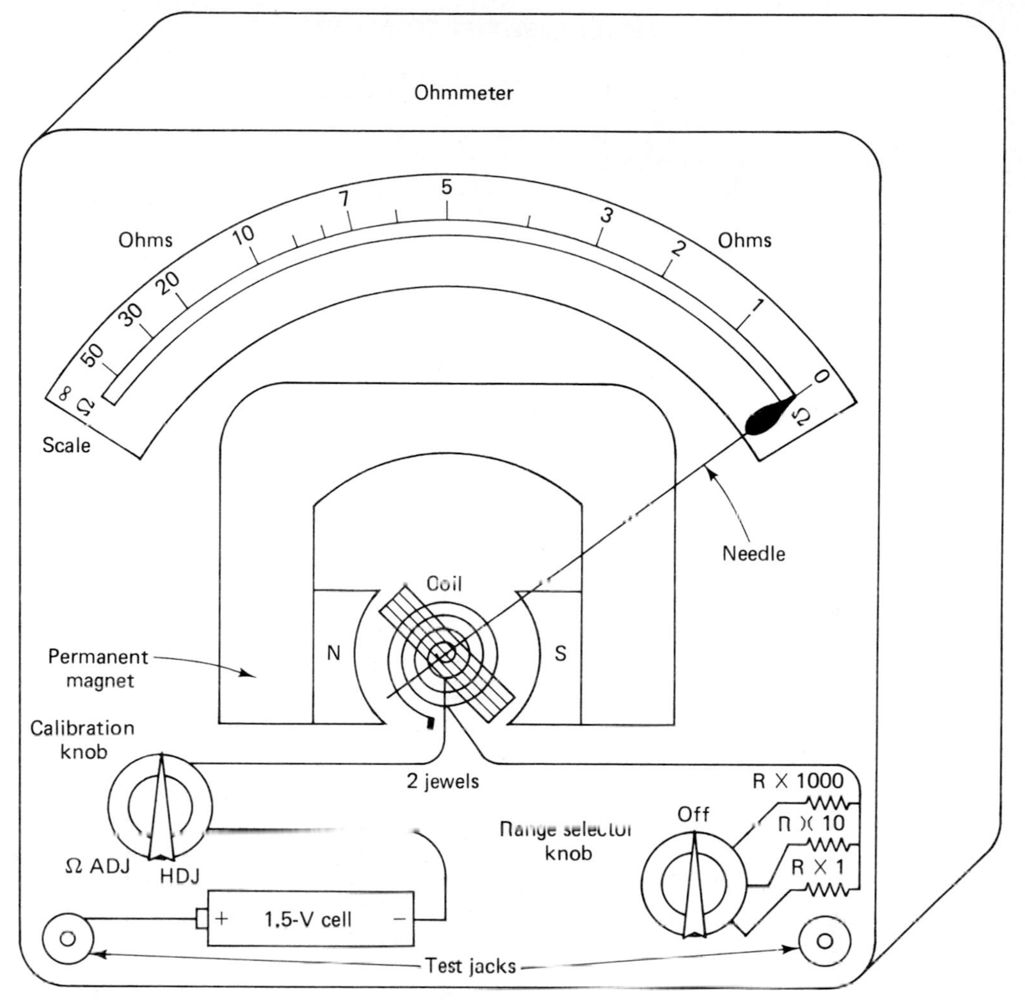

FIG. 17-3 Ohmmeter.

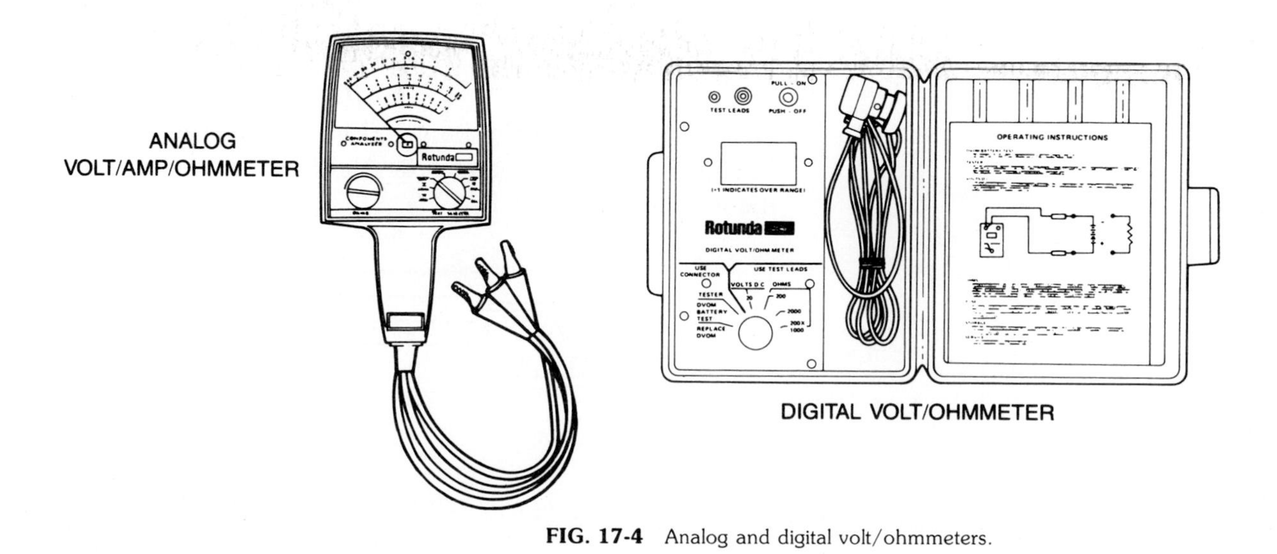

FIG. 17-4 Analog and digital volt/ohmmeters.

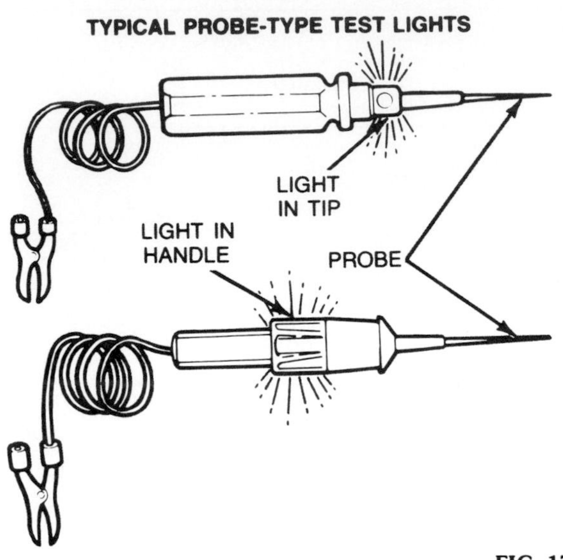

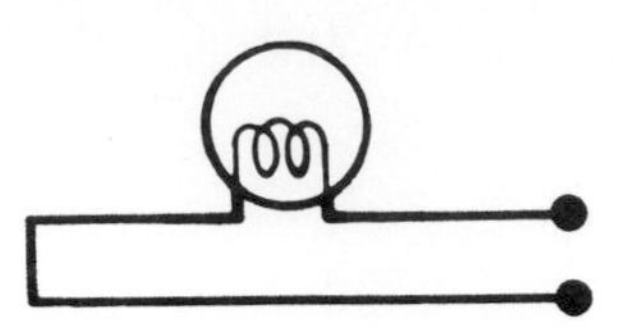

FIG. 17-5 Unpowered test lights.

Test leads and jumper wires

These components (Fig. 17-7 and 17-8) are used to extend the range of a meter's lead or selectively to bypass (or short out) a section of circuit. They are often used in diagnosing computer-controlled devices when circuits must be disconnected and meters or test lights must be connected at various locations.

Dwell tachometer

This familiar test device (Fig. 17-9) is also used in computer-related testing. The tachometer, as in the past, gives engine speed in rpm. However, as we will see in Chapter 13, the dwell scale is no longer used to check ignition dwell. It is employed for other purposes. That is because dwell is not adjustable in computer-controlled ignition systems.

Vacuum pump

Hand-operated vacuum pumps (Fig. 17-10) are often employed in testing air- or vacuum-operated devices, such as the EGR and air management systems.

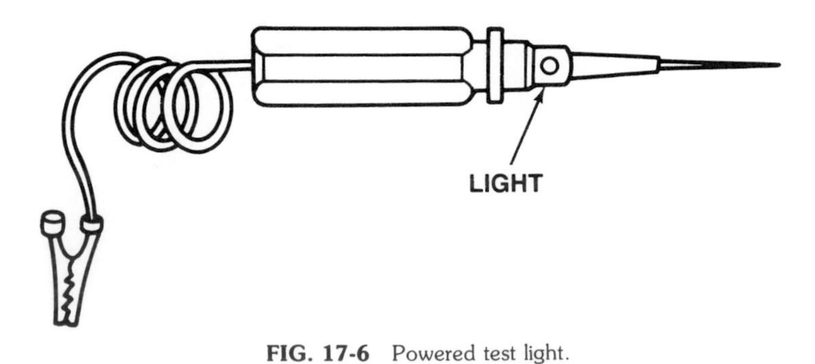

FIG. 17-6 Powered test light.

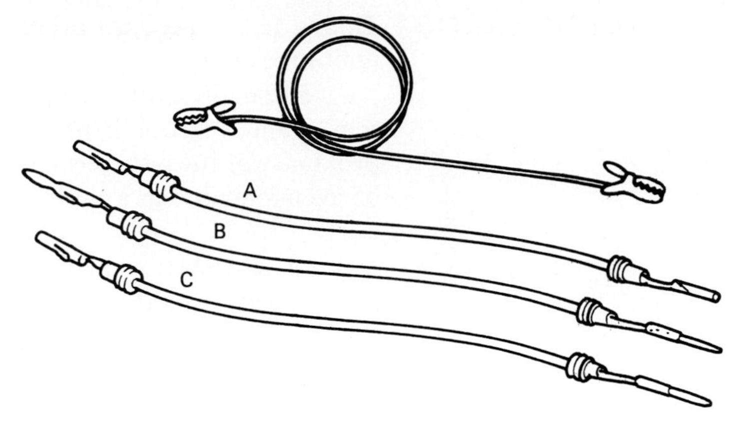

FIG. 17-7 Clip jumper wire and test leads.

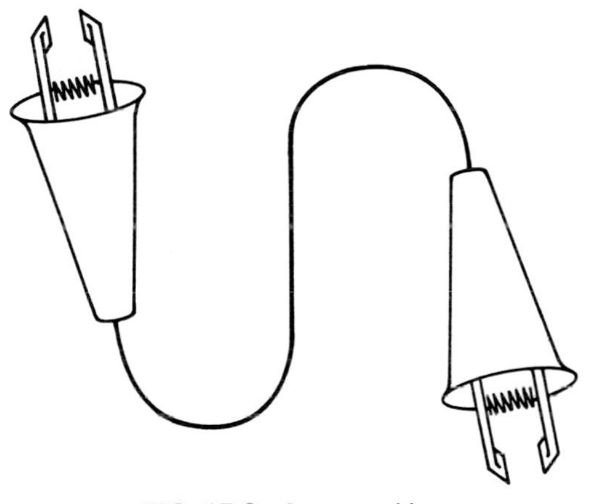

FIG. 17-8 Jumper cable.

FIG. 17-9 Tach/dwell meter.

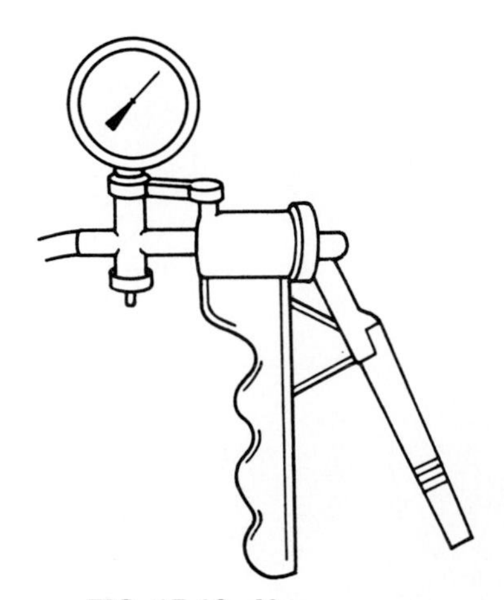

FIG. 17-10 Vacuum pump.

TIMING LIGHTS

Timing lights are used primarily (but not exclusively) to check engine ignition timing.

The main components in a typical timing light include a pickup sensor and a strobe light. The sensor is connected to or placed in the proximity of the number one spark plug wire. Here is a typical sequence of events involving a timing light:

1. Whenever a high-voltage surge passes through the wire, the sensor sends a signal to the strobe, causing it to produce a pulse of light.
2. At the instant the light pulses, the spark plug fires and the crankshaft pulley, the flywheel, and other parts of the power train rotate to certain points. Any of these rotating parts illuminated by the timing light pulses will appear to stop.
3. The relationship between stationary reference marks and marks scribed on the pulley or flywheel is visually checked.
4. If the timing is not within specification, the distributor is loosened and the base is rotated back and forth until the marks line up.
5. When the marks are properly aligned the distributor is retightened.

A powered timing light is shown in Fig. 17-11.

Note: This procedure applies only to vehicles that actually have a distributor. As noted in previous chapters, some newer solid-state engines do not use distributors at all.

Most modern timing lights use an external power source for the strobe light, either the vehicle's battery or a 110-volt ac receptacle. The number 1 spark plug sensor is an inductive clamp. The lines of magnetic force produced by the high-voltage surge passing through the wire induce a small signal-level voltage output in the sensor.

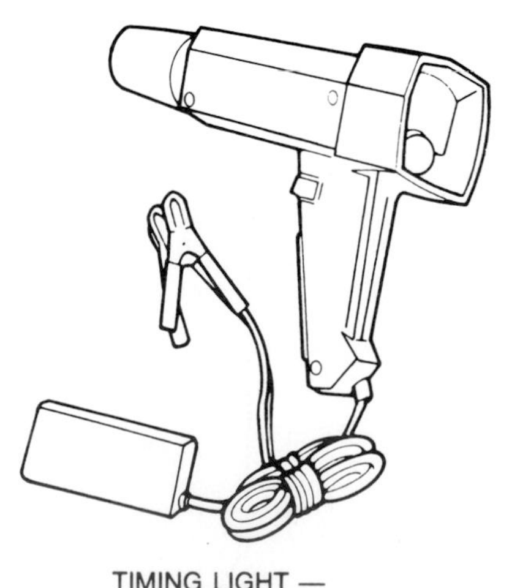

TIMING LIGHT —
ROTUNDA 59-0006

FIG. 17-11 Powered timing light.

The newest timing devices do not use lights at all. Instead, a magnetic probe is inserted into a holder adjacent to the crankshaft pulley. The crankshaft pulse ring used by the vehicle's own sensing system causes magnetically induced (or Hall-effect) signals to be produced by the probe. These signals are sent to an engine analyzer, which displays the timing data as digital values.

GROUND TIP SPARK PLUG

In the preelectronic automotive era, a technician would commonly check a malfunctioning engine by removing the spark plug or coil wire, holding the tip near the engine block, and then checking to see if a spark appeared when the engine was cranked. This was a quick and effective way to tell if the ignition system was working, at least well enough to produce a spark.

However, this procedure is no longer recommended with electronic engine control systems. If the technician tries to force a spark across an impossible gap, an arc may occur inside the distributor, possibly damaging delicate electronic control equipment.

To check the ignition in this time-honored manner, a ground tip spark plug, like the one shown in Fig. 17-12, is used. The ground tip plug is placed on the exposed end of a spark plug or coil wire, and the tip is then clipped to a good, clean engine ground. While the engine is being cranked, the electrodes are checked for arcs.

Note: Do not disable an ignition for testing purposes by grounding a spark plug or coil wire. Refer to the manufacturer's manual for the correct procedure.

HOW TEST DEVICES ARE USED

For the remainder of this chapter, we will look at some common applications of the test devices just introduced. Chapter 18 describes actual computer system testing and diagnosis.

Testing headlights with an unpowered test light

We begin with the simplest type of test device, the unpowered test light. In this exercise, the tester is being used to check for current flow up to the headlights (Fig. 17-13). If the test light fails to come on, you would know something is wrong on the positive side of the headlight circuit.

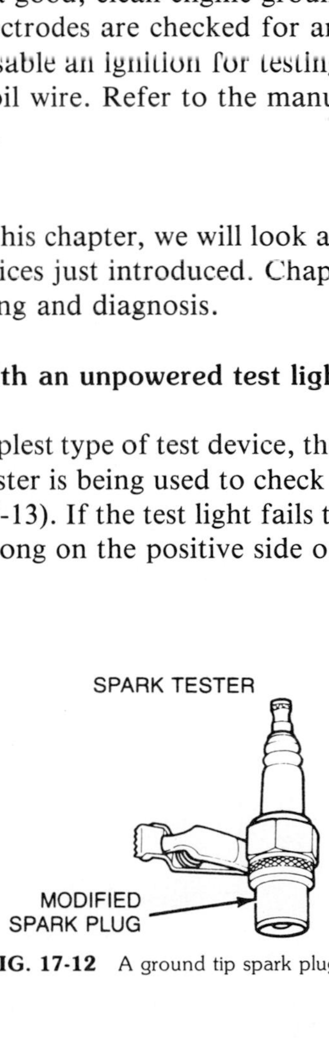

FIG. 17-12 A ground tip spark plug.

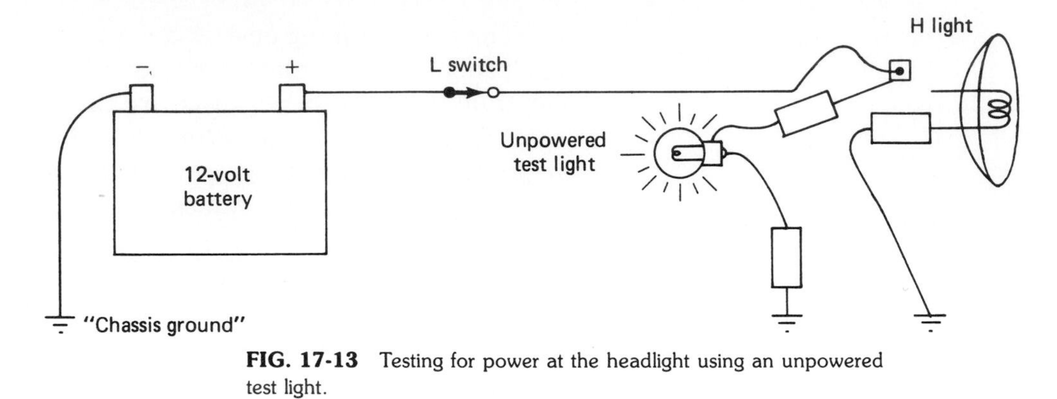

FIG. 17-13 Testing for power at the headlight using an unpowered test light.

Testing headlights with a powered test light

The powered test light in this example is used to determine if the negative side of the headlight circuit is complete (Fig. 17-14). If there are no breaks or regions of excessively high resistance, current will flow and the test light will shine. Such a test can be used to check continuity in almost any circuit.

Testing the ignition coil with an ohmmeter

The ohmmeter in this test is being used to check the resistance of a coil's primary windings (Fig. 17-15). Two types of coils are shown; however, the same procedure can be used to check both. Here is the general procedure for using common types of ohmmeters:

1. Calibrate the meter by placing the prods together and turning the adjusting knob until the needle is at the proper calibration point (usually zero).
2. After calibrating the meter, place the tips of the prods against the two small primary terminals.
3. Unless you know the general resistance to expect, turn the selector knob to the lowest range, usually R × 1 (the resistance reading times 1, in other words, the actual reading on the scale).

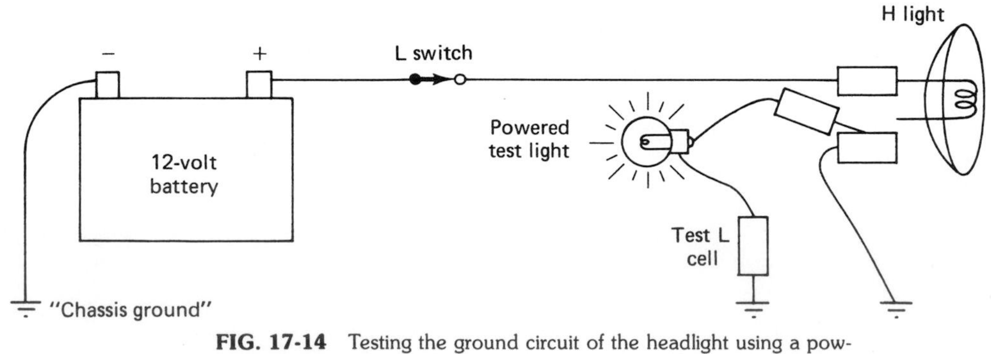

FIG. 17-14 Testing the ground circuit of the headlight using a powered test light.

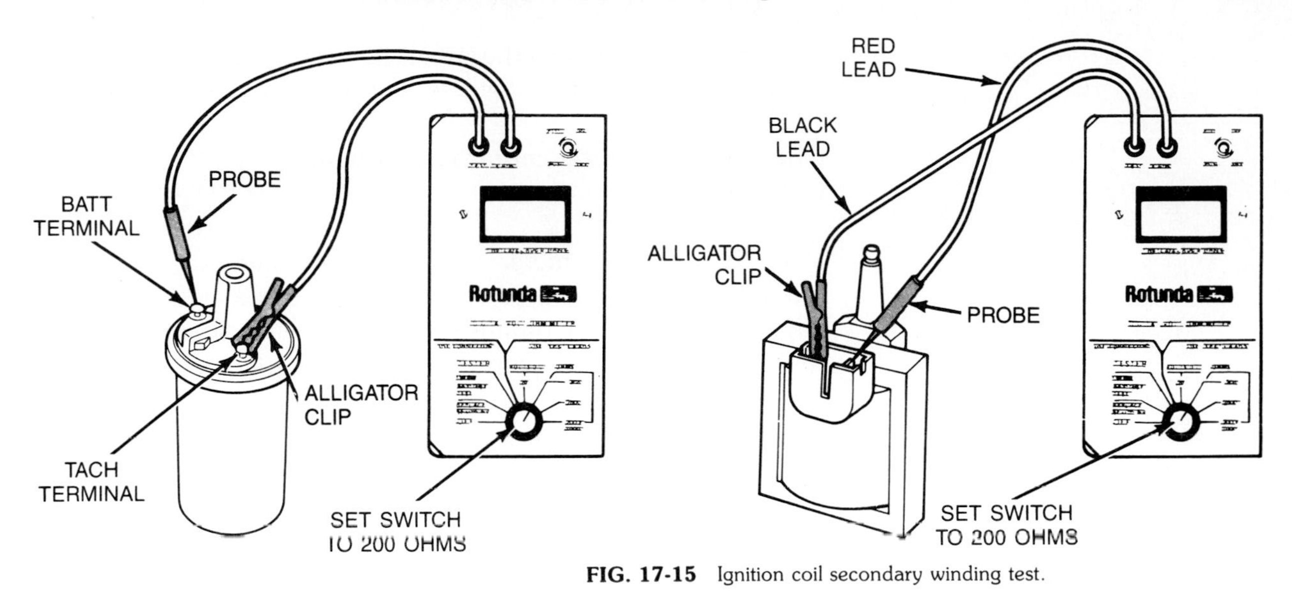

FIG. 17-15 Ignition coil secondary winding test.

4. If the meter hand moves off the scale, adjust the selector until the needle stays within the scale. If that turns out to be R × 10, you would multiply the scale reading times 10. If the adjustment is R × 100, you would multiply the scale reading times 100.

CHECKING THE HEATER MOTOR WITH AN AMMETER

This test, illustrated in Fig. 17-16, shows an ammeter is being used to check the current flow through a heater motor circuit. As noted before, the ammeter must be hooked in series with the circuit being tested so that all the current flows through the meter. In the example shown here, the ammeter leads are connected between the positive post of the battery and the circuit fuse. The ammeter reading is compared to the specification for the circuit.

When performing tests like this, avoid trying to measure more current than the ammeter is capable of handling. Also be aware that some ammeters use clamp-on, inductive pickup sensors, thereby elim-

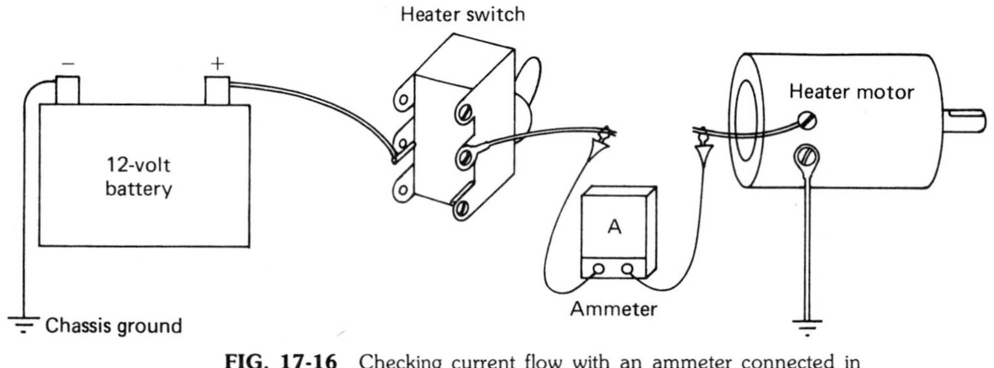

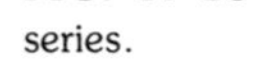

FIG. 17-16 Checking current flow with an ammeter connected in series.

inating the need to open the circuit for testing. Current flow is indicated by the strength of the magnetic field passing through the wire.

Checking the starter motor circuit with a voltmeter

Here we have a voltmeter connected on either side of a starter motor solenoid. It is being used to check the resistance in the circuit. You might think that an ohmmeter should be used. However, since the starter motor is the largest user of battery energy, the cable has a very low resistance, on the order of 0.0008 ohm in some cases. Many ohmmeters do not have scales that read this low. The solution is to check for voltage, since it will increase when ohms diminish and decrease when ohms goes up.

This test reading is known as the *voltage drop*. It is an indication of the pressure needed to overcome the resistance of a given section of circuit. Voltage drop specifications are provided by many manufacturers for various circuits. A general rule of thumb says that each cable or switch in many circuits will have a voltage drop of 0.1 volt. All the voltage drops must add up to the total voltage impressed on the circuit.

The test shown in Fig. 17-17 is conducted in this manner:

1. Select the lowest scale on the voltmeter.
2. Connect the meter leads as shown in the figure.
3. Temporarily disconnect the ignition so that the car will not start.
4. With the engine cranking over, read the voltmeter scale. The reading will be the voltage drop.

Checking continuity with an ohmmeter

Continuity can be checked in many ways. Here an ohmmeter is used to check the continuity of the cables in a wiring harness (Fig. 17-18). To perform the test, first calibrate the meter as noted before. Then, after placing the range knob in the R × 1 position, place one of the tester probes on one of the cable connectors. To determine if the cable is complete, place the other probe at the other end of the cable. A com-

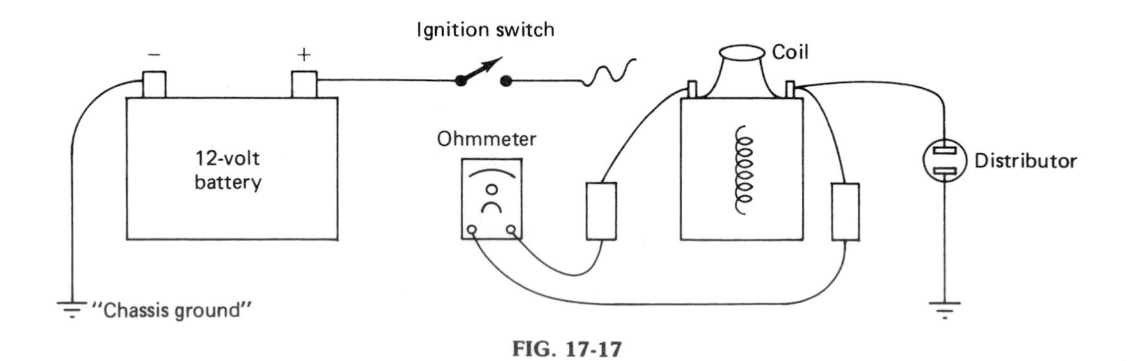

FIG. 17-17

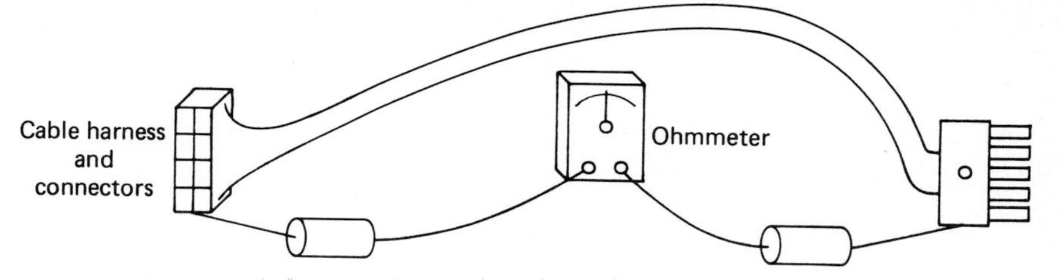

FIG. 17-18 Using a voltmeter to check voltage drop with the engine cranking cover.

plete cable will allow current to flow and will produce a reading on the tester scale. This reading can be compared to the manufacturer's specifications.

At this point, you might be wondering why an ohmmeter is required for continuity testing why, for instance, a powered test light is not used. Simply this: if the resistance in the circuit is high, enough current might not pass for the bulb to shine, even though the resistance is within specifications. This is especially true for test lights using a $1\frac{1}{2}$-volt power cell.

Checking continuity with a nonpowered test light

In this test, a nonpowered test light is used to check for continuity through the terminals in a switch (Fig. 17-19). While one probe remains in contact with ground, the other probe is moved from terminal to terminal. The test light will not shine when connected to an incomplete circuit (assuming that the switch is on). The same general procedure can be used to check continuity in any circuit. Starting with the connection nearest the battery, move the tester probe from connection to connection away from the battery. The first connection where the bulb fails to shine indicates a malfunction.

Checking vacuum leaks with a hand-operated vacuum pump

Not all computer-related testing is done with electrical test gear. It is sometimes necessary to check the operation of vacuum- or air-operated

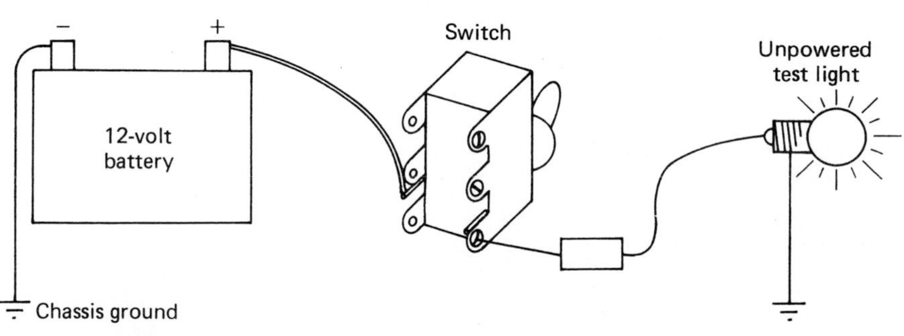

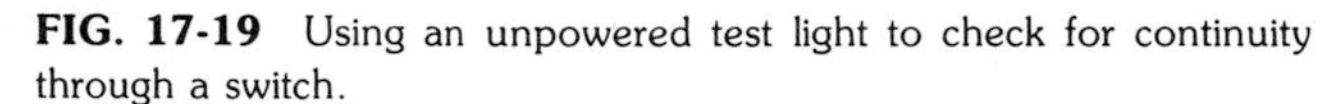

FIG. 17-19 Using an unpowered test light to check for continuity through a switch.

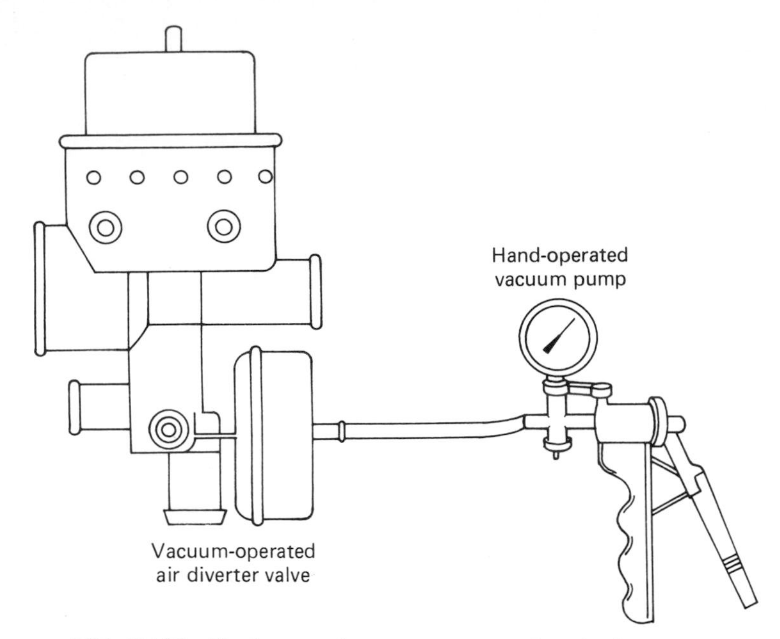

FIG. 17-20 Hand-operated vacuum pump used to check vacuum-operated actuators, vacuum lines, and switches.

output devices. In this example, the pump is used to ''pull'' a vacuum on the air switching unit (Fig. 17-20).

Then the leakage rate is determined by observing the vacuum gauge to see how long it takes for the vacuum to be dissipated. If the rate exceeds the manufacturer's specifications, a problem is indicated.

Vacuum gauges

A vacuum gauge such as the one shown in Fig. 17-21 can be connected anywhere to check vacuum leaks.

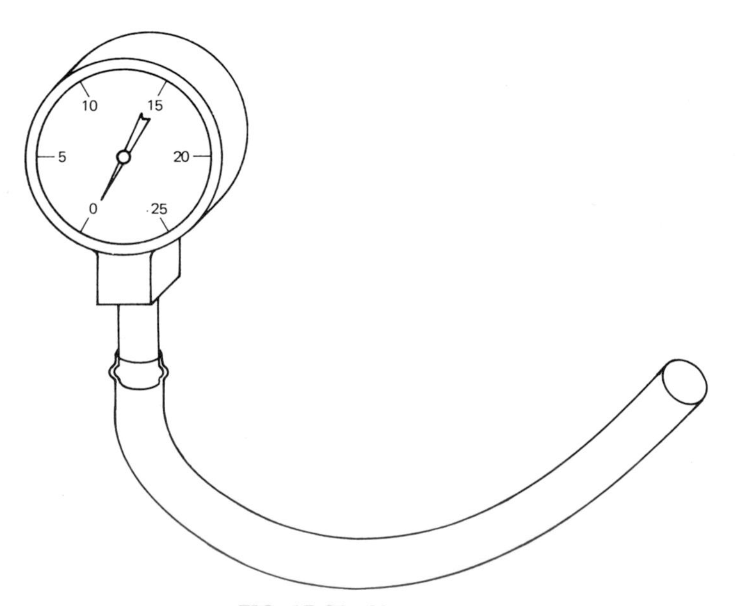

FIG. 17-21 Vacuum gauge.

Note: In addition to these common test devices, mechanics may also need the special test equipment that manufacturers make available to dealers. Such equipment, which ranges from the simple to the complex, is too varied to list here. However, most of the equipment operates in a manner similar to the test devices described in this chapter.

18

Summary of Diagnostic and Test Procedures

Directions for performing formal diagnostic and test procedures are provided by paper-based manuals or by computerized engine analyzers. The actual diagnostic procedures are similar in both cases; the major differences are: (1) Paper-based systems typically require you to do more of the information gathering, (2) computer-based diagnostic systems, in order to avoid confusion, present the diagnostic procedure one step at a time. (In a paper-based system, all you have to do is look to the next pages to see the entire diagnostic procedure—which *could* be confusing.)

This chapter introduces diagnosis and testing from the orientation of paper-based manuals. Aspects of several representative diagnostic systems are examined. The next chapter looks at computerized engine analyzers.

INFORMATION REQUIRED

Before doing any sort of diagnosis you need information about the problem. Usually, this information comes from a number of sources:

1. Customer comments
2. Visual inspection
3. On-board diagnostics
4. Testing

Customer comments

The first place to go for information is to a person who has firsthand knowledge of the problem: the customer. You must actually listen to what the customer says (Fig. 18-1). The report may be garbled and confusing, and perhaps contradictory, but it often holds the key to the problem. Ideally, after listening to the customer's initial comments, you will think for a moment, come up with a preliminary diagnosis, then ask some pertinent questions to help confirm or reject your ideas. Of course, listening to the customer is always a good practice, whether the problem relates to a computer-controlled system or to some other part of the vehicle.

Visual inspection

The next information-gathering activity is the visual inspection. It only makes sense to eliminate obvious sources of difficulty before getting into more complex testing and diagnosis. As before, this recommendation applies to any sort of problem, whether or not it relates to a computerized system.

The following is a checklist of activities that might be performed during the course of a typical visual inspection.

1. Remove the air cleaner and inspect for dirt or foreign material or other contamination in and around the filter element. This applies to injections as well as carburetor-based systems.
2. Examine vacuum hoses for proper routing and connection. Also check for cracked broken or pinched hoses or fittings.
3. Examine each portion of the computer wiring harness. Check for the following at each location.
   a. Proper connections at sensors and solenoids
   b. Loose or disconnected connectors
   c. Broken or disconnected wires

FIG. 18-1 Listening to customer comments. (Courtesy of General Motors Corporation.)

d. Partially seated connectors
e. Broken or frayed wires
f. Evidence of shorting between wires
g. Corrosion

4. Inspect each sensor for obvious physical damage.
5. Reinstall the air cleaner then operate the engine and inspect the exhaust manifold and exhaust gas oxygen sensor for leaks.
6. Repair faults as necessary.

ON-BOARD DIAGNOSTICS

You also use the information provided by the self-diagnostic programming built into many systems. These programs check the operation of sensors (by verifying that the input lies within predefined ranges). Self-diagnostic programs may also check the operation of certain output devices as well as the operation of the on-board computers themselves.

Obtaining diagnostic information

The way you obtain the results of on-board diagnostics depends on the nature of the on-board diagnostic system. Automotive computer systems can handle self-diagnostics in two basic ways. One, which we will call ongoing diagnostics, is to perform the diagnostics repeatedly whenever the vehicle is in operation. The other, which we will call demand diagnostics, is to perform the diagnostics only on demand from an outside control. Examples from Ford and GM illustrate the two basic approaches.

Obtaining demand diagnostics

In a typical Ford demand system, diagnostic programs (which are stored in permanent memory) are initiated only after the technician performs certain actions to the engine. Once started, the diagnostic programs take over the operation of the engine, directing it through preplanned test routines.

If variations from standard results are encountered, service codes are signaled by the pattern of thermistor solenoid pulses. When the EEC tester is used, the pulses flash on the tester's display panel. If the tester is not available, the sound made by the solenoids must be observed.

For instance, the number 1 is represented by this pattern:

- Both solenoids on for $\frac{1}{2}$ second, then both off for $\frac{1}{2}$ second

The number 2 is represented by this pattern:

- Both solenoids on for $\frac{1}{2}$ second, then both off for $\frac{1}{2}$ second
- Both solenoids on for $\frac{1}{2}$ second, then both off for $\frac{1}{2}$ second

Both solenoids remain off for 1 full second before the second digit of a two-digit code is signaled. If more than two codes are present, both solenoids remain off for 5 seconds between codes.

Testing usually takes about 1 minute. After the last service code has been signaled, the vehicle remains in self-test operation for another 15 seconds before returning to normal running. Ford service codes are described in Fig. 18-2.

Ongoing Diagnostics

In GM (and other) ongoing diagnostic systems, problems are signaled to the driver by a flashing CHECK ENGINE light on the dash. A trouble code associated with that problem is also stored in the computer's RAM memory. So long as power is supplied to the computer, the codes remain stored in the memory.

If the problem is intermittent, the CHECK ENGINE light will go out, however, the trouble code will remain in memory as long as power is applied.

Accessing stored trouble codes and other diagnostic information is handled in a number of ways. GM provides two illustrative approaches, one which we will call the basic display, and the other, the advanced display.

Basic Display. The technician grounds a test lead under the dash while the engine is running. The CHECK ENGINE light then flashes on and off. (The exact procedure depends on the engine.) In one example, if the light flashes two times, pauses for a moment, and then flashes three more times, it means that trouble code 23 has been stored. A list of GM trouble codes is shown in Fig. 18-3.

Service code chart

Service code number	Explanation of code
none	No service code output
any	Service code output on one solenoid only
11	EEC system okay
12	Engine RPM is out of specifications
21	Engine coolant temperature sensor fault
22	Manifold absolute pressure sensor fault
23	Throttle position sensor fault
31	EGR Position sensor fails to move open
32	EGR Position sensor fails to go closed
41	Fuel control "lean"
42	Fuel control "rich"
43	Engine temperature reading below 120°F
44	Thermactor air system fault

FIG. 18-2 Ford service codes. (Courtesy of Parts and Service Division, Ford Motor Company.)

TROUBLE CODE IDENTIFICATION

The "Check Engine" light will only be "on" under the conditions listed below while a malfunction exists. If the malfunction clears, the light will go out and a code will set, except for one condition, that is code 12. If the light comes "on" intermittently, but no code is stored, see this symptom under "driver complaint."

The trouble codes indicate problems as follows:

TROUBLE CODE 12	No reference signal to the ECM. This code will only be present while a fault is present. It will not be stored with an intermittent problem.
TROUBLE CODE 13	Oxygen sensor circuit. The engine has to run for about 5 minutes at part throttle before this code will show.
TROUBLE CODE 14	Shorted coolant sensor circuit. The engine has to run two minutes before this code will show.
TROUBLE CODE 15	Open coolant sensor circuit. The engine has to operate for about five minutes before this code will show.
TROUBLE CODE 21	Throttle position sensor or WOT switch (when used) After 10 seconds and below 800 RPM.
TROUBLE CODE 23	Open or grounded Carburetor M/C solenoid.
TROUBLE CODE 32	Barometric pressure sensor (BARO) output low.
TROUBLE CODE 32 & 55 (At Same Time)	Grounded +8V, V REF. or faulty ECM.
TROUBLE CODE 34	Manifold Absolute Pressure (MAP), sensor output high. After 10 seconds and below 800 RPM.
TROUBLE CODE 44	Lean oxygen sensor. The engine has to run for about 5 minutes in closed loop and part throttle at road load before this code will show.
TROUBLE CODE 44 & 55 (At Same Time)	Open tan wire to oxygen sensor or faulty oxygen sensor.
TROUBLE CODE 45	Rich oxygen sensor. The engine has to run for about 5 minutes in closed loop and part throttle at road load before this code will show.
TROUBLE CODE 51	Faulty calibration unit (PROM) or installation.
TROUBLE CODE 52 & 53	Faulty ECM.
TROUBLE CODE 54	Faulty M/C solenoid and/or ECM.
TROUBLE CODE 55	Faulty oxygen sensor, open MAP sensor, open in sensor return wire to ECM, or faulty ECM.

FIG. 18-3 GM trouble codes. (Courtesy of Delco Electronics Division, General Motors Corporation.)

Advanced Display. One example of an advanced display is the dash-mounted CRT used on late-model Buick Riveras. Looking something like the display of a personal computer, the CRT is normally used to display speed, fuel consumption, and other information needed by the operator. Touching the screen at designated locations causes different display "pages" to appear, one page for different categories of information.

Diagnostic information is displayed when the operator touches the climate-control page's OFF and WARM indicators at the same time and holds the contact until a double beep is heard. The display then goes into *service mode.*

Once in the service mode, trouble codes can be viewed for the engine computer (electronic control module) and the body computer (body computer module). In addition, various test types as well as specific tests may be selected. This advanced display system performs the same style of function as the computerized engine analyzers described in the next chapter.

Using the diagnostic information

After trouble codes and other on-board diagnostic information has been obtained, the data are then used in diagnostic procedures like those described in this chapter.

As you might imagine, data obtained from on-board diagnostics are very valuable in servicing computer based systems. However, not all possible causes of problems will be covered by these built-in diagnostic procedures. In some cases it is necessary to obtain independent test results.

TEST RESULTS

Test results are obtained by using test instruments in a manner prescribed by the manufacturers. Usually, the test procedure is included in an overall diagnostic sequence. The results of the test are compared to the manufacturer's specifications. The outcome is then used as a basis for deciding what path to follow in the remaining part of the diagnostic procedure.

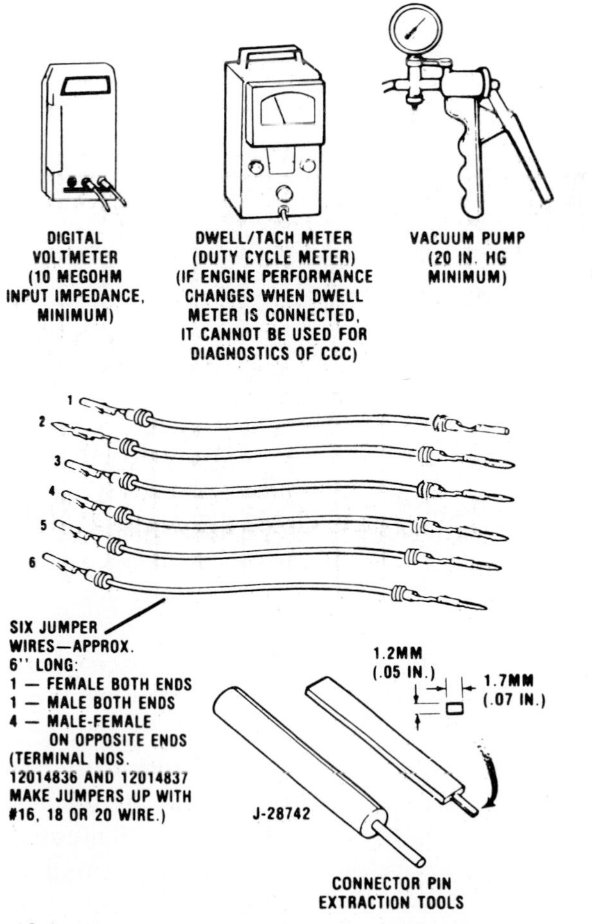

FIG. 18-4 GM test equipment. (Courtesy of Delco Electronics Division, General Motors Company.)

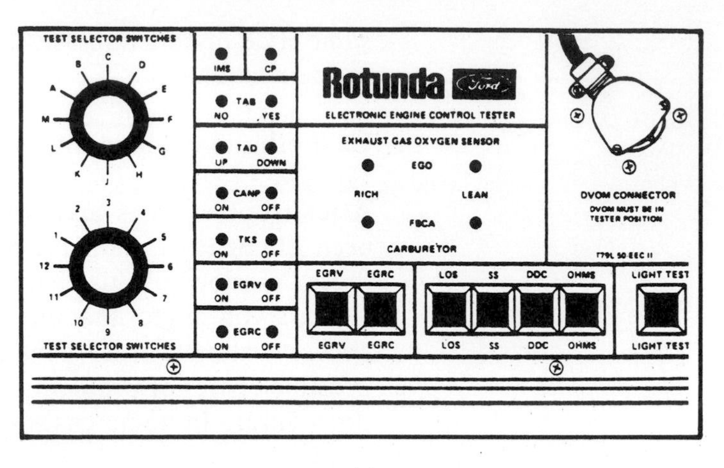

(a)

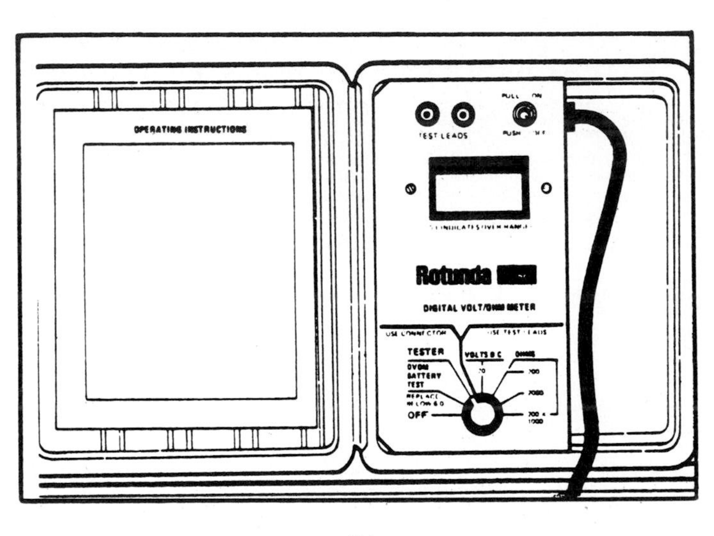

(b)

FIG. 18-5 (a) Rotunda diagnostic tester; (b) digital volt/ohm meter; (c) tachometer; (d) vacuum gauge; (e) vacuum hand pump/tester; (f) special fuel injection tester harness; (g) fuel injection tester harness. (Courtesy of Parts and Service Division, Ford Motor Company.)

The particular test equipment used depends on the manufacturer and on the test being performed. Common test devices used by GM and Ford are shown in Figs. 18-4 and 18-5. (Similar test devices are used by other companies.)

DIAGNOSTIC AND TEST PROCEDURES

First, let us see what is meant by a diagnostic procedure. If you scan the manufacturer's guides, charts, checklists, and so on, you will notice that they contain tests and questions regarding various aspects of the vehicle's operation. Depending on the answers to these questions and the results of the tests, you will move in one direction or another through the diagnostic procedure. The ultimate objective is to pinpoint a particular component that requires repair or replacement. In some cases, the defective component is identified early in the procedure. In other cases, the procedure must be followed all the way to the end. In certain instances, the procedure does not apply at all and the problem will not be pinpointed.

Most diagnostic procedures have two principal features: they are hierarchical and branching (Fig. 18-6). By *hierarchical* we mean that the procedure starts out by examining general conditions, then, by the process of elimination, narrows the possibilities down to one culprit. *Branching* refers to the way you move through the diagnostic charts, tables, or steps. Rather than proceed in a straight line, you are likely to jump back and forth through the material as possible causes of the problem are eliminated.

Because of the hierarchical and branching features, no two diagnostic procedures are likely to follow exactly the same steps. However, up to a point, the overall diagnostic sequence is similar for most computer controlled systems.

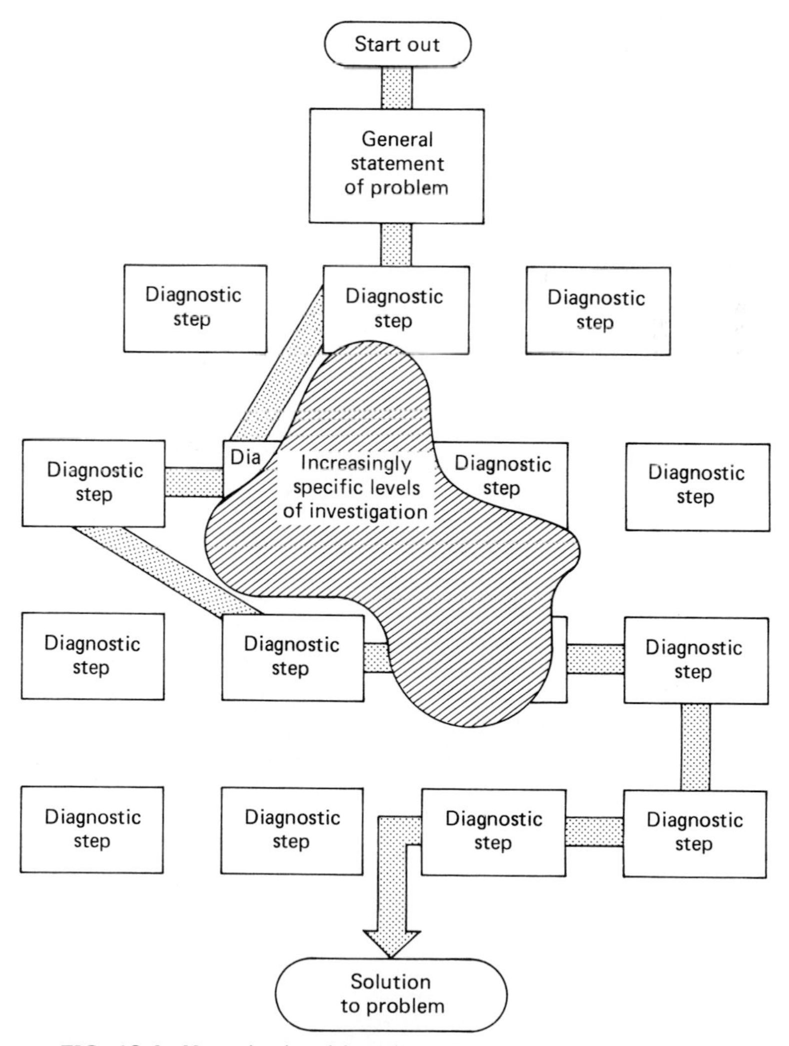

FIG. 18-6 Hierarchical and branching nature of diagnostic procedures.

General procedures

1. Usually, you start out by listening to the customer complaint. A tentative diagnosis of the problem may be made at this time.
2. A visual inspection of the engine compartment is performed next. Particular attention is paid to the components associated with the tentative diagnosis made earlier.
3. Any defects spotted in the visual inspections are repaired.
4. Since many problems are not caused by malfunctions in computer-related systems, all other possibilities are examined and any defects repaired.

Manufacturers' procedures

From now on, the diagnosis concentrates on computer-related systems. The particular procedures followed will depend on the manufacturer's recommendations.

GM Procedure. Usually, GM starts all computer-related diagnosis and testing with a Diagnostic Circuit Check (Fig. 18-7). This set of procedures checks to make sure that:

1. The engine self-diagnostic system is working.
2. If the self-diagnostic system is working, what trouble codes, if any, are present.

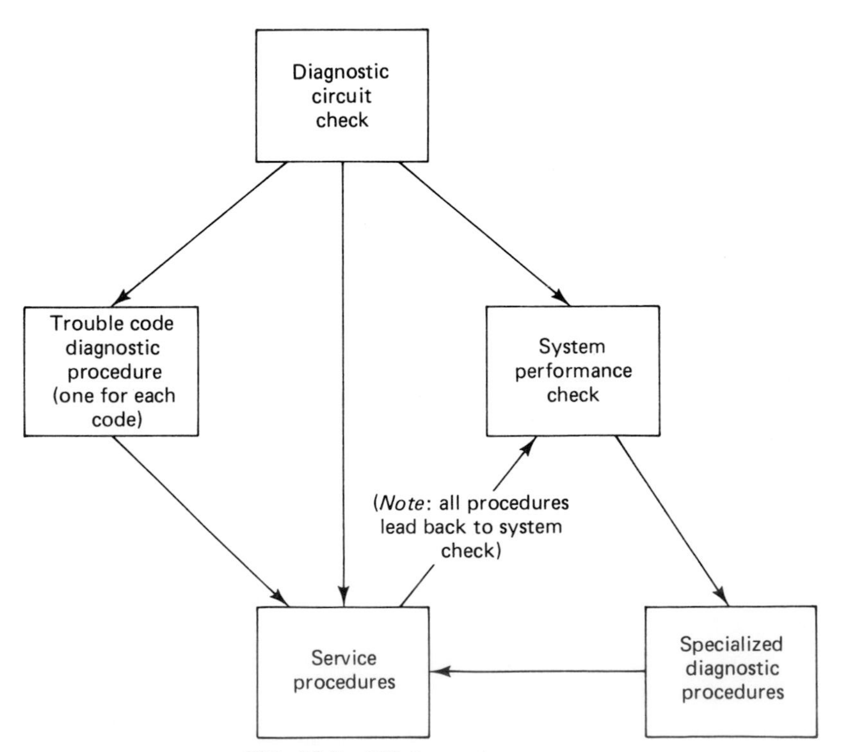

FIG. 18-7 GM diagnostic sequence.

3. If no codes are present, what other general problems may be present.

Depending on the results of the circuit check, the technician may be directed to:

1. Detail diagnostic checks for trouble codes encountered.
2. Repair procedures to replace a defective ECM (computer).
3. Service procedures to fix loose or defective connectors.
4. Another generalized diagnostic sequence, called the System Performance Check.

If directed to the system performance check, the mechanic will work through another series of diagnostic steps that examine additional aspects of the ECM and related components. During the course of this procedure, the mechanic might end up performing repairs, replacing the ECM, or proceeding to additional, specialized diagnostic checks.

At the conclusion of any diagnostic procedures, when all the repairs have been made and the engine seems to be working properly, GM recommends that the System Performance Check be carried out one more time to make sure that the components are indeed working properly.

Ford EEC III Diagnosis. After a visual inspection, Ford testing generally proceeds to one of three diagnostic procedures: (1) a Self-Test for engines that will run, (2) a No-Start Test for carburetor-equipped engines that will not run, or (3) a Diagnostic Chart for fuel injector-equipped engines that will not run (Fig. 18-8).

1. *Self-Test.* This procedure may employ the Rotunda T791-50-EECII Diagnostic Tester. The tester is hooked up to the engine after the vehicle has been operated long enough for the radiator hose to become hot and pressurized. Then, with the engine running at idle speed, a hand-operated vacuum pump is attached to the barometric sensor vent outlet. The sensor vacuum is pumped down to 20 inches Hg and held there for 5 seconds. This reading, which is below any possible normal barometric pressure, causes the test cycle to begin (with or without the tester hooked up).

Responding to programs running in the vehicle's on-board microcomputer, the engine goes through a series of controlled operations. Information going to and from the sensors and output devices are compared with preset values stored in the computer. The results of this comparison, whether ''OK'' or out-of-spec, are signaled to the technician by thermactor solenoid pulses. Particular pulse patterns relate to certain service codes. After the test cycle has been completed, the technician checks the service codes against a diagnosis guide in the tester's operating manual.

2. *No-Start Test.* These diagnostic procedures check for computer-related conditions that might cause the no-start condition. The No-Start Test will lead the technician to the appropriate repair operation

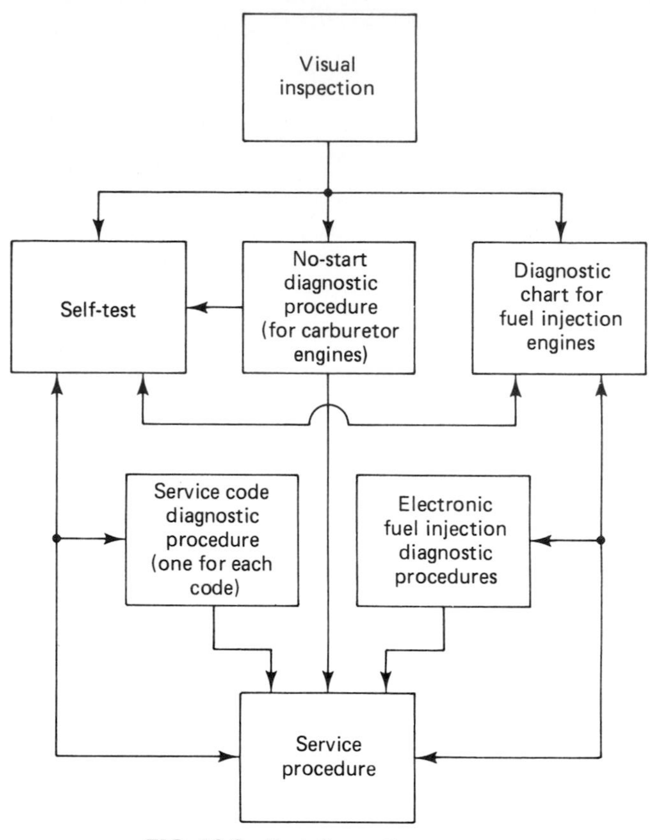

FIG. 18-8 Ford diagnostic sequence.

and (if the operation is successful) back to the self-test procedure for final confirmation of the work performed.

3. *Diagnostic Chart.* This procedure checks for a spark from the ignition coil and for fuel coming from the injectors. Depending on the results, the mechanic is directed to the No-Start Test or to an additional set of procedures called the Electronic Fuel Injection Diagnosis.

USING A DIAGNOSTIC REPAIR MANUAL

All these procedures are described in detail in the manufacturer's diagnosis and testing manuals. Of course, as you may have already discovered, it is not always easy to use one of the manufacturer's manuals. You may understand individual sentences and paragraphs, but the whole thing still may not quite fit together. You may have trouble deciding where to get started or where to turn for the procedure that applies to your particular problem.

Multiple levels of information

One secret for understanding the manufacturer's materials is to realize that most diagnostic and repair manuals deal with several levels of information at the same time (Fig. 18-9).

Most manuals include at least some reference or background information. In other words, the manual explains the components covered in the diagnostic procedures. Sometimes the explanation will be

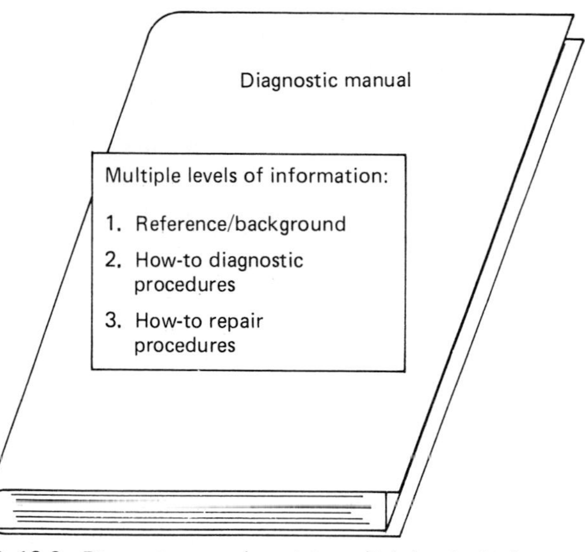

FIG. 18.9 Diagnostic manuals contain multiple levels of information.

in-depth and may include detail specifications for various types of engines. However, usually, that sort of detailed information is covered in manuals devoted strictly to repairs.

Reference or background information does not necessarily tell you how to do anything—simply how things work. It is the next level that tells you how to perform tasks—in this case, how to perform diagnostic procedures.

The how-to directions in diagnostic manuals may take several forms. Some how-to instructions are simple numbered steps. Other instructions are organized into table form, where each test or question, the possible results, and the mechanic's actions are described in separate columns. Other how-to instructions appear as charts, resembling program flowcharts.

Occasionally, diagnostic manuals may include a third level of information. This level constitutes how-to instructions for conducting actual repair procedures. However, most repairs are not explained in detail. It is assumed that the mechanic either knows how to perform the task, or that it is covered in some other document.

The important point to remember is that the different levels of information require different approaches by the reader. When you read reference or background information, you are trying to see how components work and how they relate to one another. You may also be looking for particular facts about a component whose basic operation you already understand. So, you probably will not read every word, only those passages that relate to your particular needs. The organization of reference material usually lends itself to this sort of information-gathering process.

How-to guides, on the other hand, are meant to be read word for word and followed exactly in the order presented. You cannot skip around through the material; you must do exactly what the directions tell you to do.

The trouble with many diagnostic and testing manuals is that they mix reference and how-to information without giving the reader a clear idea about the organization of the different parts. As a result, the wrong approach may be taken to the manual. In particular, how-to directions may not be followed exactly. So, given the fact that manuals may be confusing, it is up to you to figure out what is what and to decide on the correct approach to the different parts.

Representative manuals

To help you in this task, we examine briefly parts of several representative manuals.

GM BCM/PFI ELECTRONICS MANUAL

This manual is designed to be used in conjunction with the Buick Rivera on-board diagnostic system described earlier. About two-thirds of the manual deals with fundamentals and one-third with system diagnostic procedures. Some of the topics included are:

- Computer basics (parts, memory, clocks, communication)
- The body computer and its functions
- Ported fuel injection
- Component assemblies

GM POCKET DIAGNOSTIC CHARTS

Over the past few years, GM has provided several small pocket diagnostic charts. These guides contain summary information for a number of procedures. The table of contents shown in Fig. 18-10 is typical.

FORD "SERVICE HIGHLIGHTS"

The Ford service manual is more reference-oriented than the GM manual. As noted in the table of contents pictured in Fig. 18-11, four of the six main sections deal with system description and operation features. Another section (the last) provides how-to instructions for performing certain repair operations. Only one section (the second) deals specifically with diagnostic information.

The diagnostic section itself is broken into six main parts. The first part pictures and briefly describes the test equipment required. The next part shows how to hook up the test equipment. The third part tells how to perform the visual inspection. The last three parts describe, in order: the Self-Test Procedure, the No-Start Test, and the Electronic Fuel Injection Diagnosis. It is interesting to note that each diagnostic procedure is treated in a different way in the manual.

The Self-Test Procedure is actually more of a reference guide. Mechanics are directed to the Test Operator's Manual for specific diagnostic procedures. However, mechanics who do not have access to that manual can extract enough information to do a certain level of diagnosis.

The No-Start procedure is organized into what is sometimes called a "step-go-to-step" table (Fig. 18-12). The table is divided into three

FIG. 18-10 Table of contents from "Pocket Brains." (Courtesy of Delco Electronics Division, General Motors Corporation.)

columns, headed STEP, RESULT, and ACTION. The operations described in the STEP column are numbered and each describes a particular diagnostic test or question. The next column lists the possible RESULTS of each step. The third, ACTION column, directs the mechanic to another numbered STEP, to a particular repair operation, or to the Tester Operator's Manual. Most actions eventually branch back to step 1, to confirm that any repairs made have been successful.

The Electronic Fuel Injection Diagnosis contains commands and directions arranged in a paragraph and list format. Major groups of

TABLE OF CONTENTS

FIG. 18-11 Table of contents from Ford diagnostic manual. (Courtesy of Parts and Service Division, Ford Motor Company.)

related activities are grouped under common headings. Most of the directions are branching; that is, you are directed to go from one step or heading to another depending on the results of your testing.

This last diagnostic guide may appear to be the easiest because it is not organized into an unfamiliar format. However, it is actually more complicated because it contains more descriptive text and because the connections between the various "branches" are not illustrated graphically. However, like any diagnostic guide, the best approach is to start at the beginning and to follow the directions wherever they lead you.

SPECIAL MANUFACTURER NOTES

The following paragraphs list special features of the GM and Ford systems mentioned previously in this chapter. For complete descriptions of these and other systems, refer to the manufacturers' manuals.

Step	Result	Action
1 Attempt to start engine	Engine starts	Go to self test for system check
	Engine does not start	Go to 2
2 Check vehicle battery (VBAT) voltage · Set tester switches to "A" and "1" · Record battery voltage	10.5 volts or more	Voltage ok; go to 3
	Less than 10.5 volts	Voltage too low; stop testing – go to tester operator's manual

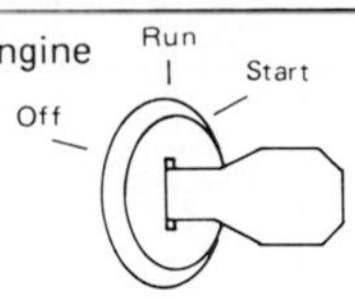

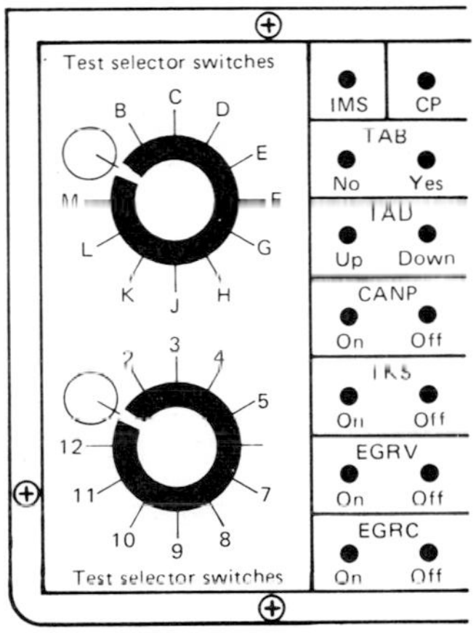

(a)

Step	Result	Action
3 Check EGR valve position (EVP) sensor voltage: · Set tester switches to "A" and "9" · Record voltage	1.95 volts	Voltage ok; go to 4
	Not 1.95 volts	Voltage out of limits; go to tester operator's manual

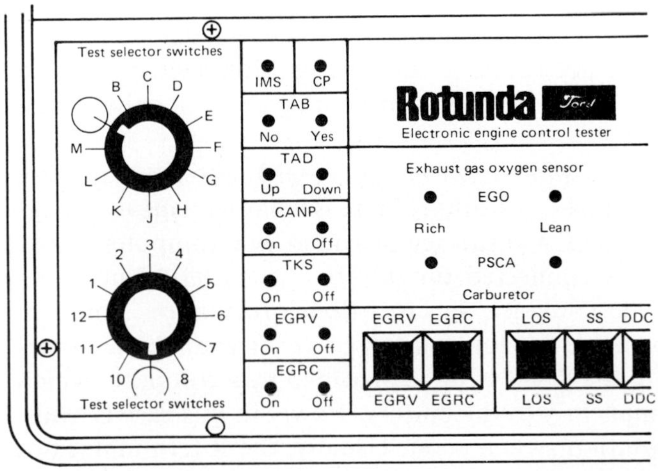

(b)

FIG. 18-12 Sample "step-go-to-step" diagnostic guides. (Courtesy of Parts and Service Division, Ford Motor Company.)

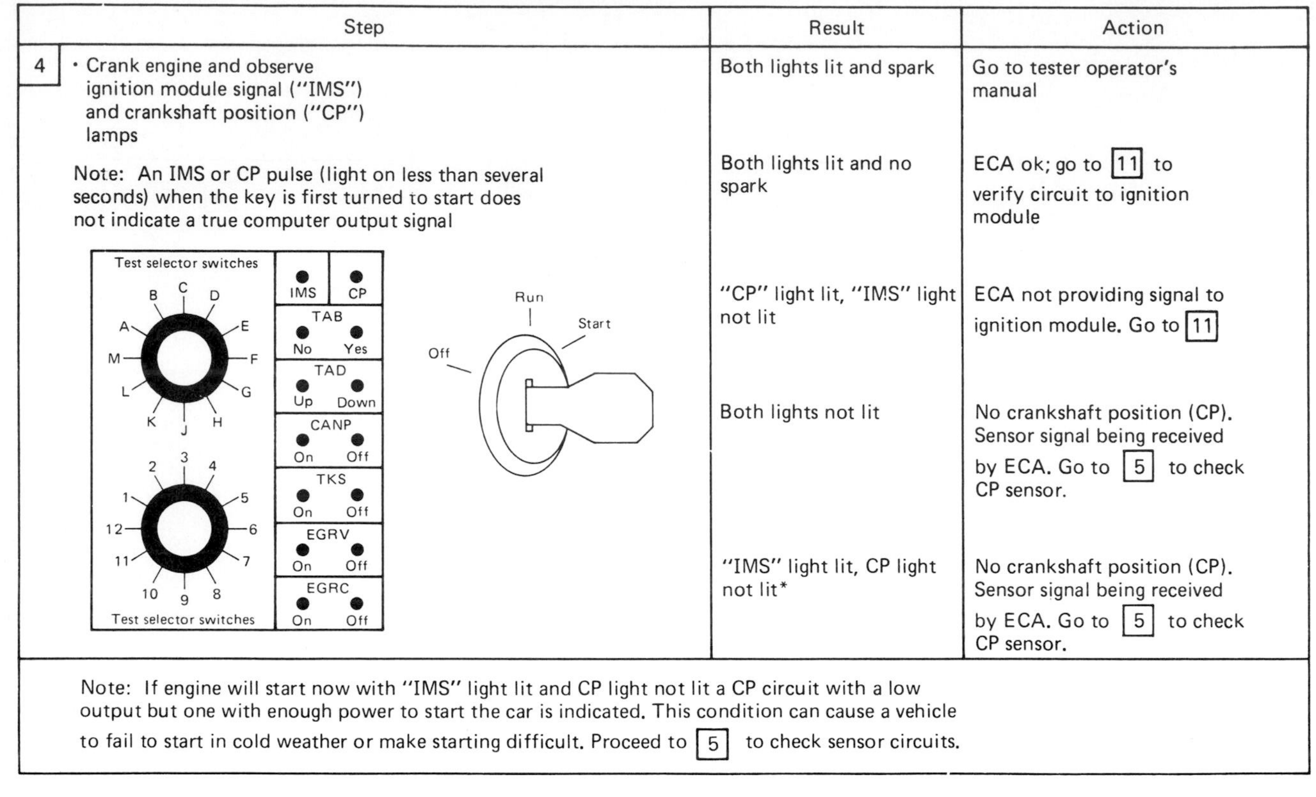

	Step	Result	Action
4	• Crank engine and observe ignition module signal ("IMS") and crankshaft position ("CP") lamps Note: An IMS or CP pulse (light on less than several seconds) when the key is first turned to start does not indicate a true computer output signal	Both lights lit and spark	Go to tester operator's manual
		Both lights lit and no spark	ECA ok; go to 11 to verify circuit to ignition module
		"CP" light lit, "IMS" light not lit	ECA not providing signal to ignition module. Go to 11
		Both lights not lit	No crankshaft position (CP). Sensor signal being received by ECA. Go to 5 to check CP sensor.
		"IMS" light lit, CP light not lit*	No crankshaft position (CP). Sensor signal being received by ECA. Go to 5 to check CP sensor.
	Note: If engine will start now with "IMS" light lit and CP light not lit a CP circuit with a low output but one with enough power to start the car is indicated. This condition can cause a vehicle to fail to start in cold weather or make starting difficult. Proceed to 5 to check sensor circuits.		

(c)

FIG. 18-12 Continued

GM

1. *Checking the trouble code display.* There are several built-in checks of the trouble light bulb and circuit. The light should always come on when the ignition switch is turned to the "on" position with the engine not running. In addition, trouble code 12 (one flash, a pause, and two flashes) should be displayed whenever the test lead is grounded, the engine is not running, and the ignition switch is in the "on" position. (Trouble code 12 means something else when the engine is running.)

2. *Long-term versus short-term memory.* GM uses both of these terms discussing RAM memory storage. Here is what they mean: *Long-term memory* refers to the RAM storage provided by a computer whose "R" terminal has been connected directly to the battery. Supplying constant voltage to the computer ensures that information stored in RAM will remain available (as long as the power is available). *Short-term memory* refers to the RAM storage provided by a computer which is not connected permanently to the battery. Power in this case is available only when the ignition switch is on. Usually, the R terminal is not connected at the factory. This prevents battery drain in vehicles that might not be operated for long periods. The R terminal must be connected by the dealer. Also, after any testing of the repair, the R terminal

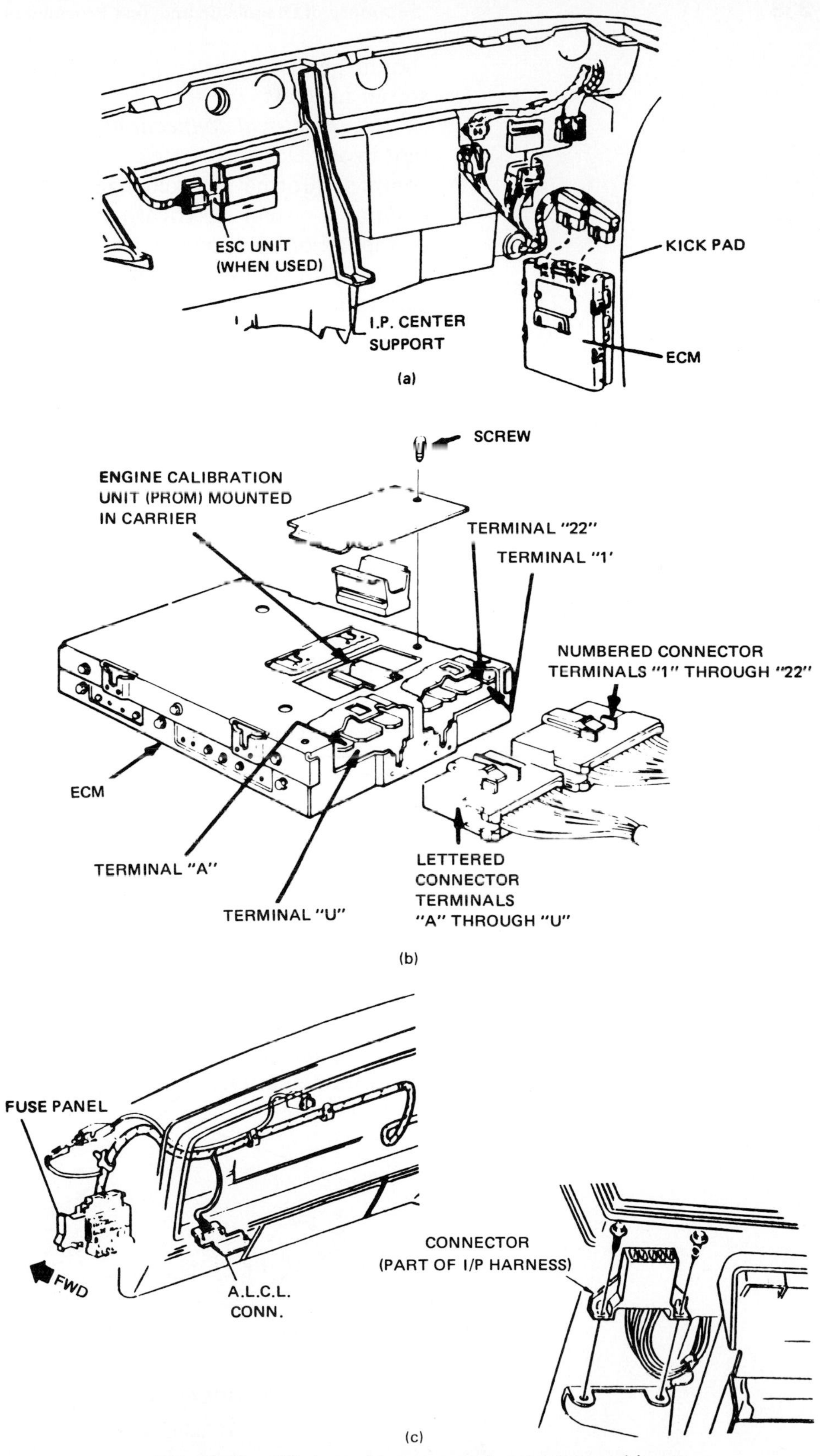

FIG. 18-13 GM electrical harnesses and connections and locations of major components: (a) ECM mounting location, typical; (b) ECM terminal locations; (c) ALDL connector mounting, typical. (Courtesy of Delco Electronics Division, General Motors Corporation.)

should be disconnected briefly. This will clear out any trouble codes stored in RAM.

3. *Electrical connections.* Good electrical connections are important in any electrical system, especially an electronic system involved in complex information exchanges. Figure 18-13 pictures some of the GM computer system electrical harnesses and connectors and the location of the major components.

4. *PROM Installation.* The PROM memory unit, which contains calibration values for particular vehicles, is one of the few computer components that may be serviced at the local level. Usually, if the computer is replaced, the PROM chip is removed and reused on the replacement computer. Because it is so important to computer operation, the

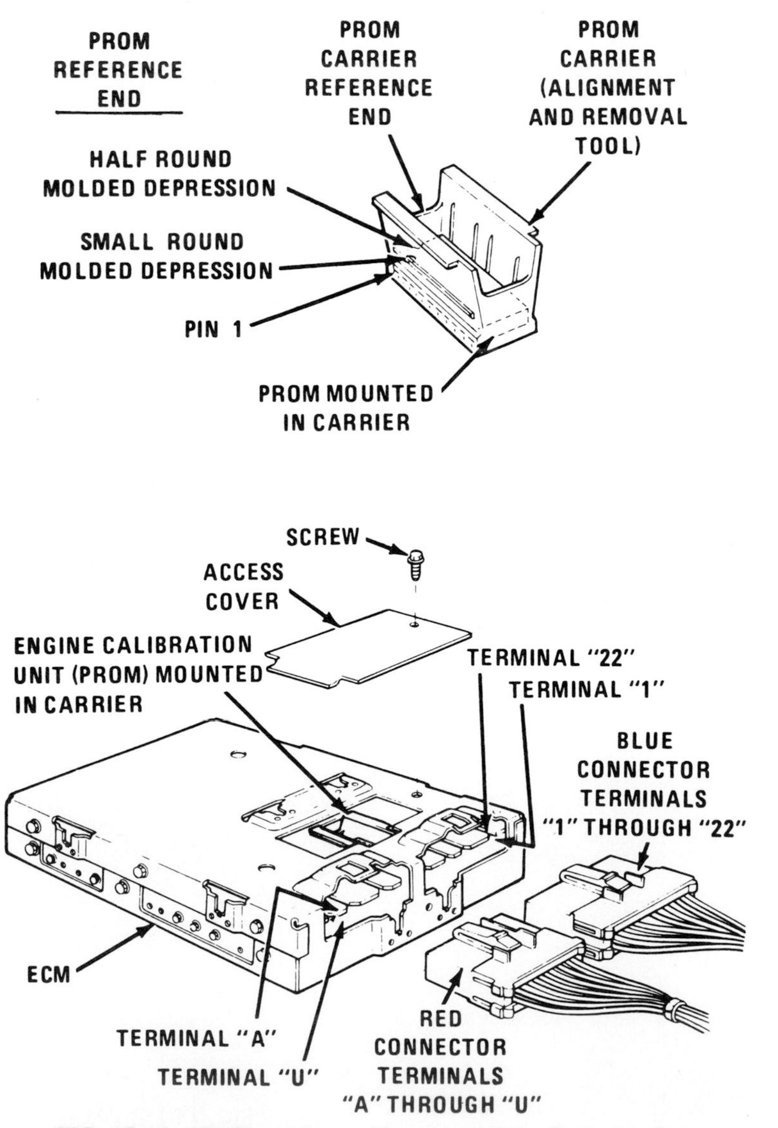

FIG. 18-14 PROM installation. (Courtesy of Delco Electronics Division, General Motors Corporation.)

mechanic must be very careful to install the PROM chip correctly and securely. Figure 18-14 pictures some aspects of PROM installation.

5. *Delayed trouble codes.* Most trouble codes associated with the oxygen sensor will not show up immediately after the engine is cranked. It takes about 5 minutes before the sensor warms up. It also takes several minutes for codes associated with the coolant sensor to appear. A list of GM trouble codes are shown in Fig. 18-3.

Ford

1. *Component locations.* Figure 18-15 shows the locations of major EEC components. As in the GM system, it is very important to make sure that all cables are properly and securely connected. According to Ford, very critical connectors should be protected with a grease moisture barrier.

2. *Connecting the equipment.* Figure 18-16 shows the procedure for connecting the tester to the vehicle.

SAFETY PRECAUTIONS

The same safety precautions that you follow when performing any service operation should be observed when conducting the diagnostic procedures referred to in the preceding test. At the minimum, you should:

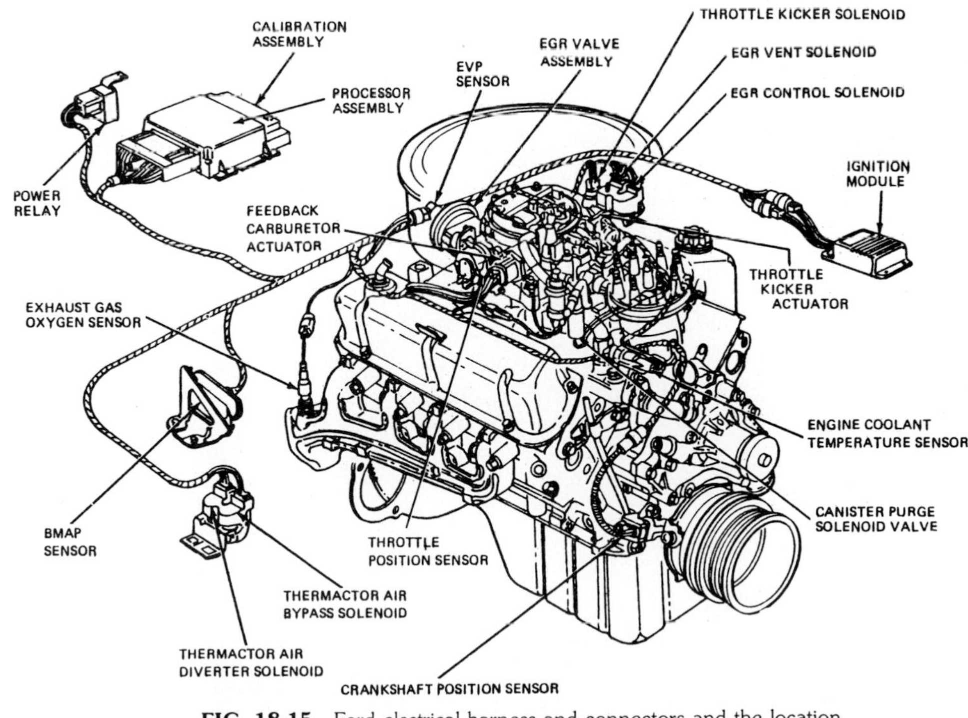

FIG. 18-15 Ford electrical harness and connectors and the location of the major components: EEC system installation. (Courtesy of Parts and Service Division, Ford Motor Company.)

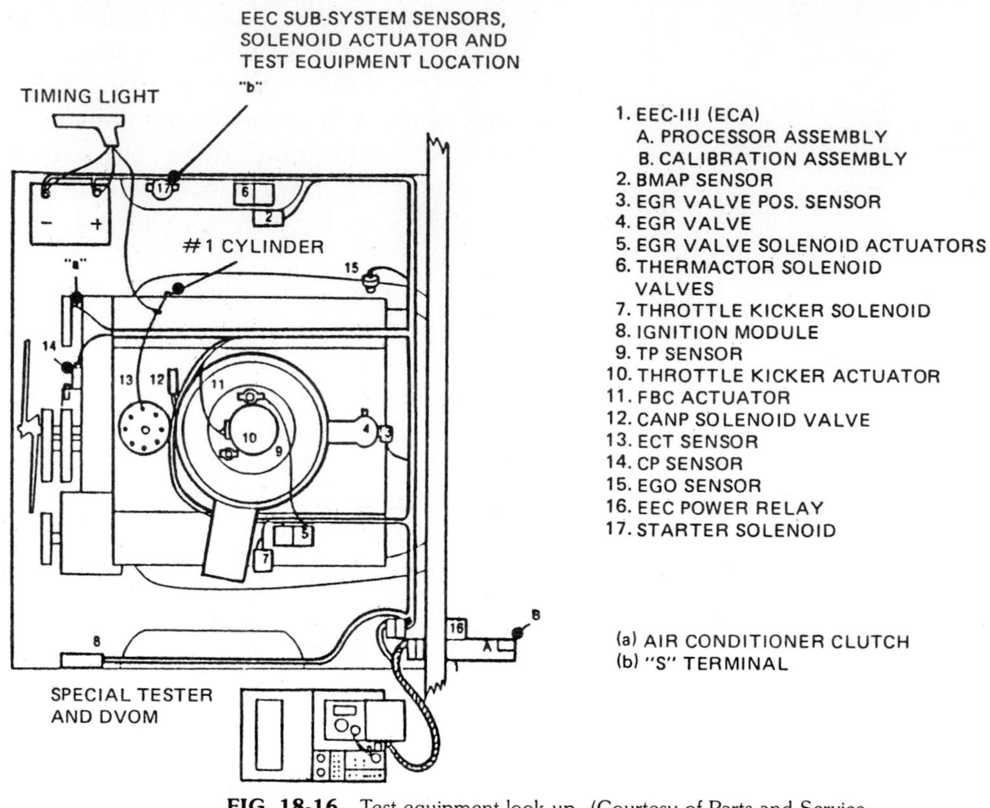

FIG. 18-16 Test equipment look-up. (Courtesy of Parts and Service Division, Ford Motor Company.)

1. Block the vehicle's front wheels.
2. Apply the parking brake.
3. Put the transmission selector in PARK.
4. Unless otherwise instructed, turn the accessories off and close the door.
5. Make sure that the work area is adequately ventilated.

19

Major Diagnostic Systems

INTRODUCTION

As noted in Chapter 18, diagnostic procedures are usually presented in the form of printed manuals or charts. Typically, they call for the use of hand-held test devices like those described in Chapter 17.

Although such procedures are superior to primitive shade-tree techniques, there can still be problems. As we have already noted, the procedures are sometimes complex. Also, in a noisy, dimly lit shop, it is not always easy to read the greasy, well-worn shop manual in which the procedures are contained. Finding the right manual and specs may not be possible. Plus, even given the right manual, specs, and environment, the available test equipment may not be sufficient to diagnose the problem. Sometimes, help is required.

As a result, many automotive manufacturers and electronic firms have produced major hardware and software diagnostic systems. Usually managed by separate diagnostic computers, these systems automate certain aspects of the diagnostic process. Most systems maintain a database of test specifications for a number of vehicles. Some systems provide a communications link to a larger information system located hundreds or thousands of miles away.

This chapter reviews the evolution of major diagnostic systems and examines examples of systems in current use.

DEVELOPMENT OF ENGINE ANALYZERS

The term *engine analyzer* has typically been used to describe a large, roll-around console containing a number of test devices. Today's computerized diagnostic systems evolved from engine analyzers, which are still in use.

An engine analyzer might have an oscilloscope, voltmeter, ammeter, ohmmeter, exhaust gas analyzer, timing light, and so on. An elaborate harness and set of connectors is provided to attach the various test devices to the vehicle. Figure 19-1 depicts an early model analyzer.

Aside from possible marketing considerations, the main reason for grouping test equipment in one console is to provide a central location for all the information needed to diagnose a problem. Rather than having to search through the shop for individual pieces of test equipment, all the devices are in a single location, ready to be hooked up and powered up at one time.

In keeping with the idea of an analyzer as an information-collection system, most are designed around a set of test procedures. The main steps in the test are sometimes displayed directly on the equipment. Test details and an explanation of the results mean are contained in manuals. Vehicle specifications are sometimes provided on a card index system that is updated yearly with new cards supplied by the test equipment company.

Often included in the cost of the analyzer is a certain amount of training to ensure that the operator is comfortable with the system.

DIAGNOSTIC COMPUTERS

However, even though it serves as an information focal point, the classic engine analyzer still misses out on a major source of information: the data from the sensors used to supply the test vehicle's on-board computer. Modern diagnostic systems collect this information, using it and other resources to perform such functions as:

- *Data reporting:* Listing information currently coming from the test vehicle's sensors.
- *Data analysis/diagnosis:* Providing computer generated, troubleshooting directions based on symptoms entered by the operator and on data received from the vehicle being tested.
- *Vehicle specifications:* Displaying specifications for various vehicles.

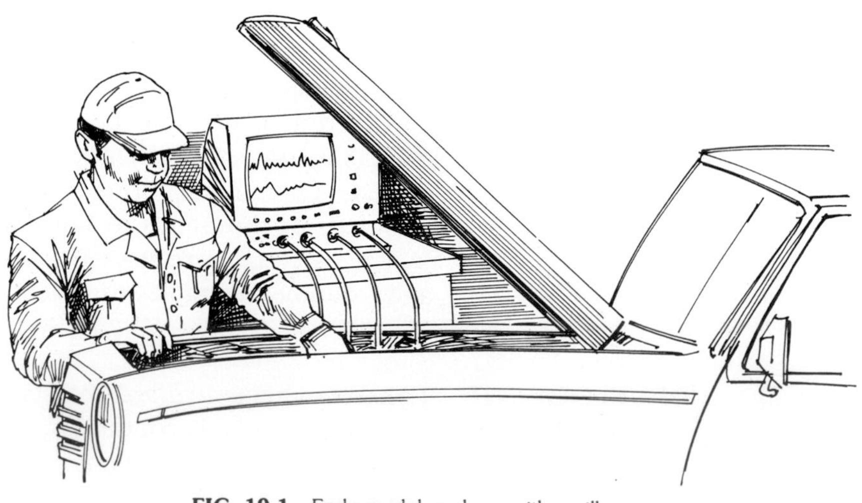

FIG. 19-1 Early-model analyzer with oscilloscope.

- *Automated shop manual:* Displaying repair or other information for a specified vehicle.
- *Communications:* Providing a link between the shop diagnostic computer and a larger computer in a remote location. The larger computer can provide updated information as it becomes available as well as additional diagnostic capabilities.
- *Printouts:* Providing paper copies of information displayed by the computer. This is useful for customer relations.

The next pages describe current examples of diagnostic computers.

THIRD-PARTY ANALYZER

The modular computer analyzer (MCA) from Sun Electric is a good example of a third-party, computer-assisted analyzer (i.e., an analyzer not manufactured by a car company). An MCA is shown in Fig. 19-2.

The computer itself is an IBM Personal Computer (PC) compatible system. This means that it can run any of the software (programs) developed for the IBM PC. When not used for testing, the MCA can be used for word processing, spreadsheets, accounting, and so on.

The system has two diskette drives and a 19-inch, high-resolution color monitor. Problem indicators are highlighted in red. Characters and graphics are oversized for better visibility.

A number of features are claimed for the MCA:

- Since different on-board computers transfer data at different rates, provisions are made for matching the import interface to the transfer rate of the system being tested.
- The operator is given the option of a complete systems analysis or a rapid test, which only takes a few minutes.
- A high-speed printer is used to obtain printouts of test results and other data.
- Specification disks are available that contain information on all U.S. engines produced in the last 20 years.
- The system can be equipped with a modem. This device connects a computer with the phone line, allowing a computer in one location to communicate with another computer somewhere else. Sun uses this link to update the MCA with new data automatically as it becomes available.

Sun also produces a device for testing the new distributorless ignition systems (like the one described in Chapter 7 of this book). It is called the D.I.Link and is used in conjunction with Sun's Interrogator Analyzer. See Fig. 19-3 for the D.I.Link and Fig. 19-4 for the Interrogator.

AUTO MANUFACTURER'S SYSTEM

A major automotive manufacturer has developed a testing and maintenance system with hardware, software, and human components located in the shop and at central locations. The part of the system most often encountered by the average technician is the engine analyzer, or—as it

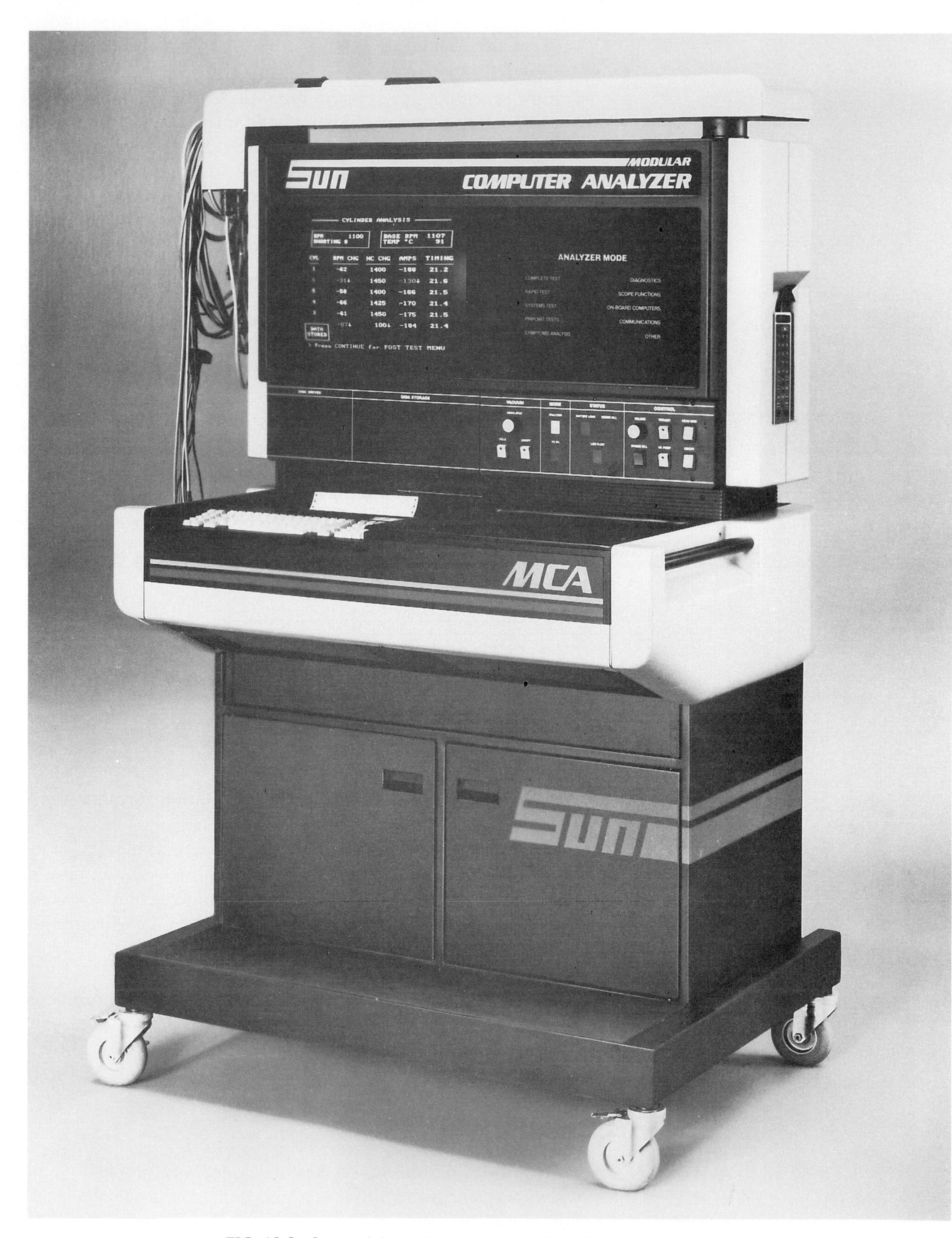

FIG. 19-2 Sun modular analyzer (Courtesy of Sun Electric Corporation).

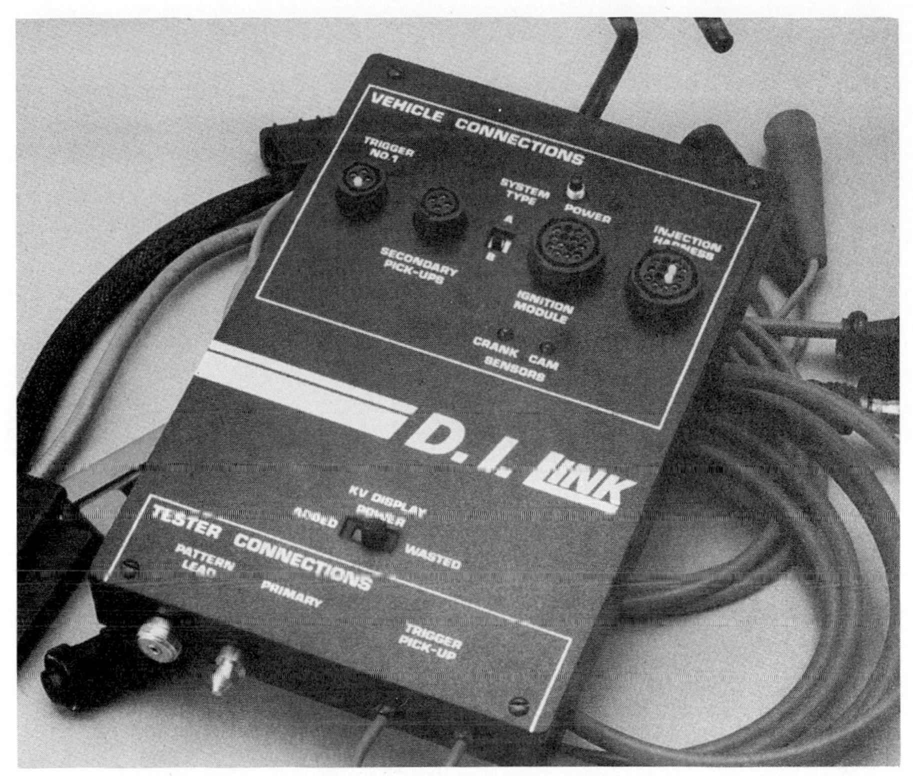

FIG. 19-3 Sun D. I. Link (Courtesy of Sun Electric Corporation).

is called by the manufacturer—the technician's terminal. The technician's terminal is similar, in some respects, to the Sun MCA device. It is used for diagnosing problems, reading data from the vehicle's onboard computer(s), obtaining engine specifications, and communicating with a remote computer. This discussion concentrates on the technician's terminal.

COMPONENTS

The technician's terminal is based on an IBM personal AT computer that has been specially modified for automotive applications and packaged in an industrial grade housing.

The main components of the system include:

- System unit
- Color display monitor
- Touch sensor input frame
- Printer
- Keyboard
- Power-control unit
- Cable test connector
- Accessories
- Cables

Figure 19-5 is the author's rendering of the Technician's Terminal.

The following discussion and illustrations describe the components in more detail:

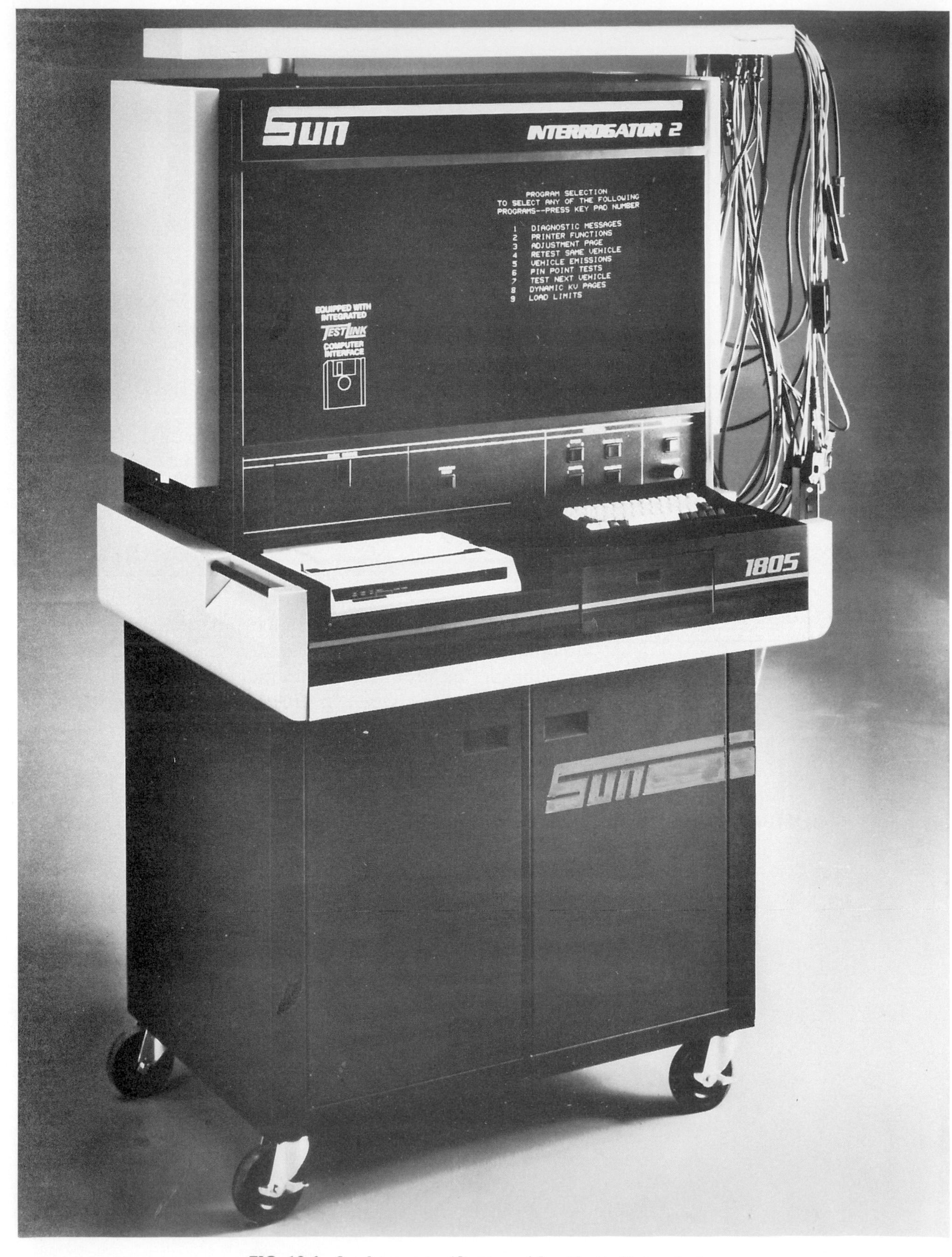

FIG. 19-4 Sun Interrogator (Courtesy of Sun Electric Corporation).

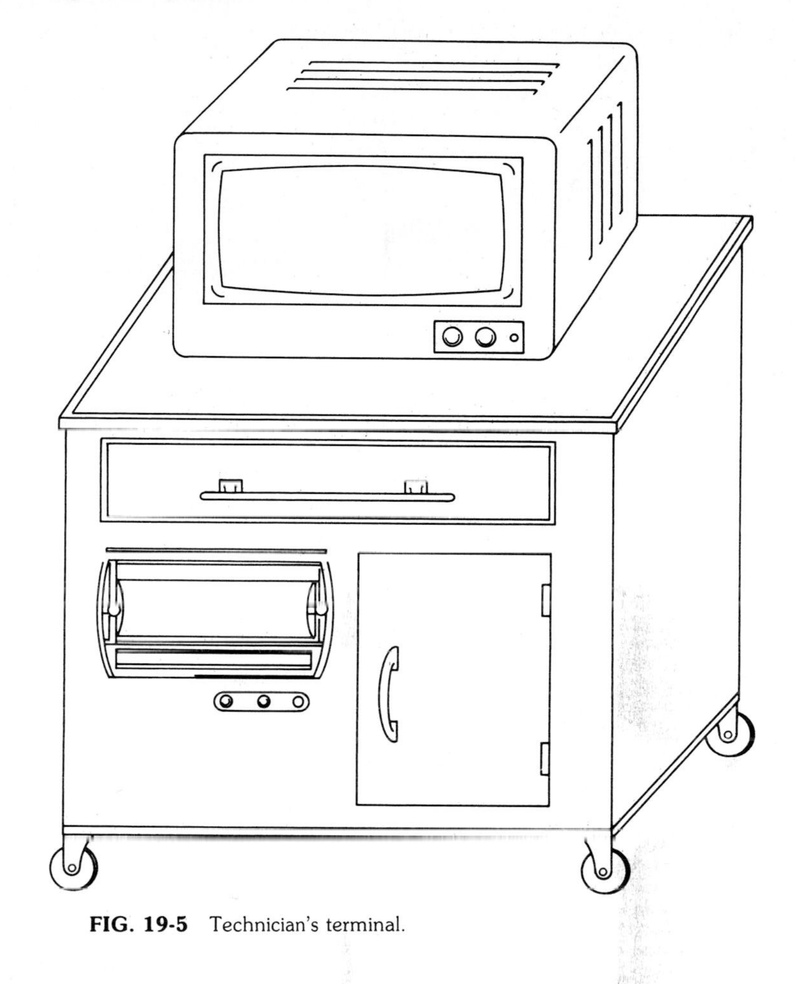

FIG. 19-5 Technician's terminal.

System unit

The system unit is the heart of the system, housing the CPU (central processing unit), the disk drive, memory, and communication control. It is located inside the terminal cabinet.

Color display monitor

The system uses color text and graphics to make the displays easier to read.

Touch-screen grid

The screen is surrounded by an infrared frame, which emits an invisible grid of infrared light beams. When an operator touches the screen, the grid is interrupted at the point of contact. The touch system is used instead of a keyboard for responding to questions and prompts asked by the computer. For example, a question displayed on the screen can be answered by touching either a YES or NO box. Or, selections can be indicated on a menu (list of choices) by touching the box associated with each item on the menu. The author's rendering of the touch-screen grid is shown in Fig. 19-6.

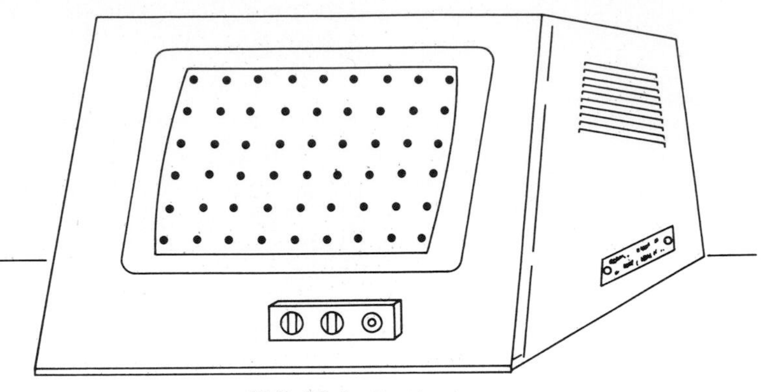

FIG. 19-6 Touch-screen grid

Printer

A thermal *printer* is located on the lower front of the console. It is used to print screen data and is operated by touching a PRINT box on the screen.

Keyboard

The top drawer of the console contains an IBM PC/AT *keyboard.* During normal operation, the technician does not use the keyboard, relying instead on the touch screen to communicate with the system. The keyboard is used primarily by the manufacturer's representative to load new software, or to diagnose the system.

Power control

The *power-control* unit is the central power supply point for the entire system.

Cable test connector

The *cable test connector* is used each day to test the cables that are connected to test-vehicles and to test the related cards contained in the system unit. The system calibrates itself based on variances recorded during this self-test process.

The cable test connector is located on the side of the unit and is supplied with a protective dust cover. When using the test connector, care must be taken to avoid bending the fragile pins on the ends of the cables.

ACCESSORIES

A number of test accessories are available for the technician's terminal. Some are:

- *Vehicle probe interface cables:* Used to test particular engine electrical components. During the diagnostic procedure, the probe required to test a given component is indicated on the screen by the diagnostic function.

- *Breakout boxes:* One is used for circuit testing at the connectors of the on-board engine computer. Another breakout box is available for testing various kinds of fuel injector circuits. Jumpers, inserted into holes in the break out box adapter, are used to select the kind of circuit being examined. The diagnostic function also directs the operator in the use of this equipment.
- *Vehicle service monitor:* A portable recording device attached to the engine's on-board computer. It stores data from the computer as the engine is operated over a period of time. The recorded data is uploaded to the technician's terminal for display and analysis. The service monitor is especially useful for diagnosing intermittent problems that might not show up during regular testing.

CABLES Three main sets of cables are used:

- *Battery connection cable:* Connected to the battery of the vehicle being tested. This cable allows the terminal to test the vehicle's voltage and current.
- *Assembly line data link:* Connected to the serial connector of the vehicle's on-board computer.
- *Probe cable:* Used with the appropriate vehicle probe interface cable to test individual sensors and wiring harnesses. The probe cable has a detachable surge suppressor that protects the terminal from voltage overloads from the vehicle. The surge suppressor must be attached to the probe cable and the probe to complete the probe interface connection.

MAJOR FUNCTIONS

Starting up

Starting up the terminal begins by plugging the system into a 110-volt outlet and turning on the monitor. The system then conducts a number of self-tests and, if necessary, prompts the operator to attach the assembly line data link and probe cables. A prompt also appears if any messages have been downloaded to the terminal from the central computer. After the startup procedure has been completed, the main menu appears.

Function list

The main menu displays a list of functions that are available to the terminal operator. A function is initiated when the operator presses the screen box for the desired function. Additional menus then appear for the tasks and procedures within the function. Following is a brief description of each of the functions.

Diagnostics. The system poses solutions to problems based on information received from the test vehicle and symptoms entered by the operator. In order to get the information it needs, the system guides the operator through various test steps. At the time of this writing, procedures are provided for:

- Diagnostic circuit testing
- Engine-performance testing
- Vehicle-wiring-circuit testing
- Symptom analysis

Service Bulletin. This allows the operator to obtain a listing of a specific service bulletin published by the manufacturer.

Information. This provides operational information.

Terminal Test. This is used to initiate a self-test of the Terminal.

Utilities. This provides a screen display of information being transmitted from the test vehicle's on-board computers through the assembly line data link. This function is also used to upload and display data contained in the portable recorder.

Vehicle Specs. This displays specifications for the manufacturer's vehicles.

Communications. Allows the operator to set up a modem-based, telephone communications link between the technician's terminal and a central computer. The communications link is used primarily to obtain updated service information from the central computer. Depending on the update mode selected, the information will be sent back immediately or will be sent at night when the terminal is not being used for testing.

Index

C

D

E

N

O

P

Q

R

S

T